PALEOECOLOGY OF TERRESTRIAL PLANTS

V. A. Krasilov

PALEOECOLOGY OF TERRESTRIAL PLANTS

BASIC PRINCIPLES AND TECHNIQUES

Translated from Russian by Hilary Hardin

A HALSTED PRESS BOOK

JOHN WILEY & SONS, New York · Toronto

ISRAEL PROGRAM FOR SCIENTIFIC TRANSLATIONS, Jerusalem

Sole distributors for the Western Hemisphere
HALSTED PRESS, a division of
JOHN WILEY & SONS, INC., NEW YORK

Library of Congress Cataloging in Publication Data

Krasilov, Valentin Abramovich.
Paleoecology of terrestrial plants.

Translation of *Paleoekologiia nazemnykh rastenii.*
1. Paleobotany. 2. Paleoecology. I. Title.
QE905.K7213 561'.1 75-12799
ISBN 0-470-50690-3

Distributors for the U.K., Europe, Africa and the Middle East
JOHN WILEY & SONS, LTD., CHICHESTER

Distributors for Japan, Southeast Asia and India
TOPPAN COMPANY, LTD., TOKYO AND SINGAPORE

Distributed in the rest of the world by
KETER PUBLISHING HOUSE JERUSALEM LTD.
ISBN 0 7065 1488 2
IPST cat.no. 22131 5

This book is a translation from Russian of
PALEOEKOLOGIYA NAZEMNYKH RASTENII
(OSNOVNYE PRINTSIPY I METODY)
Vladivostok, 1972

Printed and bound by Keterpress Enterprises, Jerusalem
Printed in Israel

PREFACE

In former times, only thinkers of the caliber of Leonardo da Vinci could discern in fossils the remains of one-time live organisms. This is now everyone's privilege, but we do not always take advantage of it. For the most part fossils are still classified and used in stratigraphy as objects that were never alive. Obviously, if we adopt such an attitude, much of the information contained in them will go to waste. The ability to visualize the living organism outside the framework of the fossil, with all the questions it poses, this is the essence of paleoecology. When we ask ourselves under what conditions a rock with fossils came to be formed, when we try to fix the limits of interregional correlation on the basis of particular guide forms, reconstruct lost links between continents, or envision the present-day population of the Earth and ourselves as the result of natural selection in ancient ecosystems, we turn to paleoecology. It is clear, therefore, that paleoecology is a fascinating science, and yet it is amazing how few specialists there are in this field. The reason is probably that paleoecology is too closely bound up with paleontological systematics. The paleoecologist who does not concern himself with systematics is, as Ager (1963) says, taking the cream from hard-earned milk. Wittington (1964) speaks still more scathingly about the parasitic tendencies of paleoecologists who make use of the determinations of others. But the ethical aspect is not, of course, the main one. To be able to understand function, we must make a thorough study of form. Therefore, the paleoecologist cannot be without an intimate knowledge of the morphology of extinct organisms, and this knowledge can be acquired only by engaging in systematics. If he does not know the weak points in the system and the arbitrariness of many classification decisions, he runs the risk of embarking on a path of false comparisons or of making mistakes in his choice of modern model for his reconstructions. In other words, he must be a systematist, but systematics is such a labor-consuming occupation that it leaves little time for anything else, and the result is that paleoecology becomes a job on the side. As a rule, the paleozoologist-systematist concentrates on one group of fauna, so that to study ancient communities consisting of various groups he has to work in collaboration with other specialists. In this respect, the paleobotanist is at an advantage, for he generally works with all the groups making up a particular flora. It is thus easier for a paleobotanist to go over to a synecological investigation than it is for a paleozoologist. On the other hand, there are some handbooks on paleoecology written by paleozoologists but not one by a paleobotanist. True, much of the apparatus and many of the methods used are the same. The books by Efremov (1950), Gekker (1957), and Ager (1963) are certainly useful also to the paleoecologist dealing with plants. However, a whole range of specific problems naturally lie outside the paleozoologist's field of vision. To some extent this gap can be filled in from the following sources: 1) works on the history of vegetation and paleobotany textbooks (Krishtofovich, 1955, 1958; Scott, 1926; Seward, 1933; Gothan and Weyland, 1954; Mägdefrau, 1956; etc.); 2) summaries on palynology giving information on the types of burial of plants and on the reconstruction of plant communities and their evolution (Sladkov, 1967; Erdtman, 1954; Faegri and Iversen, 1964); the focus of attention here is Quaternary vegetation, peat bogs being the favorite object of study; 3) works on paleoclimatology (Brooks, 1949; Schwarzbach, 1963; Sinitsyn, 1967), where plants make their grand début as climate indicators; 4) publications on paleogeography and facial analysis (Markovskii, 1966; Menner, 1964; Nalivkin, 1956; Rukhin, 1959; Twenhofel, 1926).

But with all this, if plant paleoecology is to develop further, we must systematize, summarize, and reconsider, from the standpoint of recent findings in the field of ecology, the whole mass of paleoecological observations that has accumulated during more than 150 years of paleobotanical research and is scattered in numerous articles and monographs. This book is the first attempt at such a summary. There are no doubt important omissions, and the author will welcome criticism and comments, which will aid him in his revision of the material.

I wish to thank M. N. Gramm, E. V. Krasnov, S. V. Meinen, A. I. Tolmachev, V. A. Vakhrameev, D. P. Vorob'ev, N. N. Vorontsov, and E. D. Zaklinskaya for their suggestions during the reading of the manuscript and my wife for her help in preparing the book for the press.

CONTENTS

Part I

INTRODUCTION TO PALEOECOLOGY

And on the tenth day we touched the land of the Lotus-eaters, men who make food of flowers.

The Odyssey

Although this book is intended for the reader who is acquainted with the fundamentals of ecology, it is not out of place to remind him of the meaning of some key concepts, such as ecosystem, niche, succession, climax, etc., which reflect three basic aspects of the system – structure, function, and history in their application to ecology.

1. FORM, FUNCTION, DEVELOPMENT

The place of ecology in the biological sciences is determined by the level of organization of the material and by the form of relation that are studied. Odum distinguishes eight levels of organization (protoplasm, cell, tissue, organ, organ system, organism, population, ecosystem, biosphere: Odum, 1959), which basically agrees with the classification of other authors. The sphere of activity of ecology embraces various supraorganismic levels, and among these the population-cenotic and landscape-biocenotic, or biogeocenotic are considered to be the principal ones (Vasilevich and Ipatov, 1969). Of all the types of relation existing in such systems – spatial, temporal, trophic, energetic, teleonomic (Kesteven and Ingpen, 1966) – ecology is primarily concerned with those that govern the metabolism and the transformation and accumulation of energy. In essence, ecology is the science of systems (Dice, 1955), and Hedgpeth (1957) rightly points out that the general theory of open physical systems was developed by Bertalanffy mainly on the basis of ecological generalizations.

An ecosystem (ecological system) is often defined as an interacting complex made up of an organism (or organisms) and its environment, the term "environment" being used in the biological sense (habitat) and denoting part of the system, rather than the source of incoming signals as it is understood to mean in the systems theory. We should therefore lean toward another formulation, provided by Tansley (1936): an ecosystem is composed of an "organism complex and physical conditions in the widest sense," or a still more general definition: a system in which at least one of the components is classified as living (Dale, 1970).

Obviously, such systems encompass practically the whole of biology, and the domain of ecology does not have clear-cut boundaries. Chapman was probably right when he wrote that ecology is more a point of view than an independent discipline (see Klaauw, 1936). In fact, ecology studies the same objects as do physiology, population genetics, chorology, biogeography, phylogenetics, and so on. But the ecologist considers these objects as parts of a system, from the standpoint of the mutual adaptations that lend stability to this system. A teleological approach to phenomena under investigation is thus typical for ecology.

Although the ultimate purpose of an ecological study is to discover the conditions under which the system as a whole existed and developed, in certain cases a particular component of the system becomes the focus of attention. This component may be an individual or a population (autecology), a biocenosis (biocenology), a biocenosis and the major abiotic factors (biogeocenology), a biocenosis and all the abiotic factors that are to be taken into account (synecology), and, finally, a specific process (I would call this the field of operational ecology: for example, pollination ecology, migration ecology, etc.).

The content of the different divisions of ecology has been interpreted variously (Dokhman, 1963; Klaauw, 1936; Steubing, 1967; and other publications). Whereas Schröter (Flahaut and Schröter, 1910) construed synecology as the ecology of plant formations and divided it into physiognomic, geographic, and historical synecology, Gams considered it as part of biocenology, and Du Rietz (1921) regarded it as a branch of biosociology that included in addition the systematics of communities, their morphology, symphysiology, syngenetics (phylogeny of communities), synchorology, and synchronology (the theory of successions and paleosociology).

Due to the interplay of the components of an ecosystem, each one can become a source of information on another. It is hardly possible to take all the physical characteristics of an ecosystem into account, and a study of the living component – the biocenosis – is often the only practical means of gaining the most complete picture of an ecosystem.

The form of an ecosystem is defined first and foremost by its spatial limits. Odum (1959) described the ecosystem as "any area of nature that includes living organisms and unliving substances interacting to produce an exchange of materials between the living and nonliving parts." In this context the ecosystem is taken to be the main subdivision of the biosphere. In physical terms the biosphere is the **spatial unit** within which life is possible (the upper boundary being the ozone screen and the lower boundary lying at a depth of 3–4 km: Vernadskii, 1967).

The spatial structure of the biosphere is characterized by a complex combination of continuum and discreteness. It is sometimes contended that the concept of discreteness was born of a methodology of nonrandom, typical samples, whereas random sampling speaks for continuum. Typical samples do in fact overstate discreteness, but they would be altogether impossible if discreteness did not exist. On the one hand, continuum is relative, since it is possible to select areas that are sufficiently homogeneous and clearly defined for them to be considered discrete units, and on the other, discreteness is also relative, because there are no areas that are absolutely homogeneous in terms of a particular sign, and all the spatial elements are linked together by more or less broad transitional zones. The largest units of the biosphere are the oceans and continents (Vasilevich and Ipatov, 1969), and within continents biomes, defined by a type of climax vegetation (Odum, 1959). Biomes are subdivided into smaller units, down to elementary parcels.

The hierarchy of the spatial units reflects the "anatomical" structure of the biosphere, while the interaction of the main components: living, biogenic, and bio-inert substance, this is its "physiology." In the functional aspect, the biosphere (like any ecosystem) can be regarded as an aggregate of interacting factors that have threshold values for a given ecosystem (values at which the ecosystem preserves its structure or at which it exists as such). If we express this concept of the biosphere in geometric terms, using segments of lines that correspond to the threshold values of each factor for life, we obtain a **polymeric ecological hyperspace** which encloses the habitat environment of each species. The **environment** of a species (or population) constitutes the totality of the limits in terms of all the limiting factors, or the part of the hyperspace of the biosphere occupied by the given species. When a species becomes extinct, the place it took up in the hyperspace may become free. In contrast to the environment, the **ecological niche** of a population is its functional status (its "profession": Odum, 1959). The niche is considered an attribute of the population, and in this sense there cannot be two species in one niche, just as there cannot be "free" niches (Valentine, 1969). On the other hand, the niche changes in the course of ontogenesis. Woody plants that pass from one stratum to another in the process of growth, endoparasites that switch hosts, and sessile forms of marine benthos with free-swimming larvae illustrate this.

According to the professional index, ecological niches and the life forms corresponding to them are divided into groups that refer to vertical trophic levels between which the exchange of materials and energy takes place. Since the biomass of each level depends on the productivity of the level preceding it, the whole vertical structure has the form of a pyramid, with the autotrophic organisms, green plants, at the base. The elementary unit of such a vertical structure is called a **consortium**.

Each level of the vertical structure consists of ecological niches, which are not identical (the coexistence of such niches would contradict the principle of exclusion) but more or less overlap each other. The horizontal structure of a level is therefore determined mainly by competitive relations. When species have to share an ecological hyperspace, some prove capable of restricting the spread of others. This phenomenon is called dominance (Waughton and Wolf, 1970). Thus, each trophic level has its dominant species.

The vertical and horizontal relations taken together define the functional structure of the community or biocenosis. Comparing this structure with the physiological structure of the organism, Clements (1916) and his followers treated the community as a quasi-organism. Their views came under criticism from adherents of Gleason's individualistic concept, who presented a multitude of examples showing that species are independent of a particular community. The notion of the independent distribution of species was used as a basis for doubting the validity of the community as a system. It is clear, however, that a community possesses properties that the organisms composing it do not. Clement's point of view still seems to be acceptable if we visualize the community as the totality of interrelated ecological niches and interacting life forms. Here we come up against the problem of the main operative unit. In Dale's opinion (Dale, 1970), the units may be of three types: taxonomic (species), structural (life forms), and functional (herbivores, carnivores, etc.). Here Dale reduces the concept "life form" essentially to a form of growth. A broader interpretation of life forms embraces Dale's functional units, so that only two types of unit remain: structural and taxonomic. A definite assortment of life forms is compulsory for a given community if it is to be energetically productive. Where species are concerned, there is a greater or lesser degree of interchangeability. Some communities include species that are bound together by a long process of coevolution, while others are chance groupings of species with a different evolutionary history. Hence, the stability of the cenotic links and the individualism of the species depend on the history of the community.

Supporters of the quasi-organism-community concept spoke of the ontogenesis of communities passing through juvenile, adult, and senile stages. Tansley (1935) wrote about the natural selection of ecological systems, in the course of which less stable systems are replaced by more stable ones. The too-direct analogies drawn between communities and organisms provoked justified criticism. Yet, on the other hand, it is well known that in a certain locality a particular type of community is maximally stable and able to regenerate itself if subjected to destructive influences. Thus, we may say that a community strives to achieve a steady state, and this enables us to adopt the organismic approach (which does not, of course, mean that we identify the community with the organism) and to analyze the history of the community from the standpoint of the second law of thermodynamics, Cannon's principle of homeostasis and the principle of equifinality.

A community that is undergoing renewal passes through a series of stages called succession series and reaches a state of dynamic equilibrium, or climax. The definitive state (climax community) depends little on the initial state (pioneer group). Dynamic equilibrium is naturally maintained only under certain conditions, and therefore we speak of a climatic climax, an edaphic climax, and so on. In the course of succession the ecosystem increases in complexity. To say that it has become stabilized probably means that it has attained the level of complexity that is optimal for the given ecological situation. Climax communities are considered as ecologically closed, because the ecological space is here filled to maximum capacity, and the inclusion of new components would destroy the dynamic equilibrium.

The straining of the ecosystem toward a state of equilibrium is probably governed by the second law of thermodynamics, but this is not absolutely clear. Some scientists think that since living systems become more complex as they develop, the second law is inapplicable to them. It is also held that the second law of classical thermodynamics is only a specific case of a more general law, which for open systems is formulated as the law of minimization of the growth of entropy (Rapoport, 1969). Sometimes the explanation is sought in the incomparability of the entropies of Clausius, Planck, and Shannon, who seem to be designating different characteristics of different systems. However, they are all isomorphous enough for us to be able to speak of entropy in general as an indicator of the state of a system.

The second law of thermodynamics was derived for closed systems, so that in extending it to other types of systems, we must expect to have to find a more generalized formulation. Common to all the nonequilibrium systems known to us is the striving for equilibrium. To illustrate more clearly why the classical formulation of the second law is a random case, let us imagine the closed system to be an open one, in which the sum of incoming signals is equal to nil. In such a system equilibrium is achieved when its negentropy (the reciprocal of the entropy) goes to zero. But if the sum of incoming signals is above zero (the open system in the literal sense), then with zero negentropy equilibrium will not be achieved. It follows that there must be conformity between the number of incoming signals and the negentropy, and we arrive at the following formulation: **any system tends toward a state of equilibrium at which its negentropy is adequate to the sum of incoming signals.** This ties up well with the long-established empirical rule that the complexity of an organism is directly related to the complexity of its habitat conditions.

2. PALEOECOLOGY

On the face of it, the paleoecologist is a misfortunate ecologist, one whose business is with organisms which he did not know when they were alive and with ecosystems which no longer exist. The situation is indeed a difficult one, but it does have its advantages. The ecologist is able to make a direct study of the form and function of an ecosystem, but where its history is concerned, he is obliged to hypothesize (except in the case of very brief periods). The paleoecologist can only guess at form and function, but he is in a better position to observe long stretches of history. Thus, in paleoecology there are two decisive factors: reconstruction and time. We may state that **paleoecology is the science of the reconstruction of ecosystems of the geological past and of their development in time.** Paleoecology is a broader field than ecology (see also Imbrie and Newell, 1964). Almost all the methods employed in ecology are theoretically applicable to paleoecology, but paleoecology also has methods proper to it alone.

The paleoecologist has at his disposal a certain volume of rock whose components have preserved only the spatial relation. Pushing aside this spatial relation, he has to visualize the interaction which went on in the past, in other words, he ends up with a reconstructed (or retrospective: Kosygin and Solov'ev, 1969) system. I call such reconstructed ancient systems **paleoecosystems** (in contrast to Imbrie and Newell, who consider the paleoecosystem as a facies). It is these that are the subject of paleoecological research.

Information on ancient ecosystems can be gleaned from various characteristics of a **geological facies**, which, according to the "classical" interpretation, denotes a segregation of a stratified series that is as homogeneous as possible in terms of its lithogenetic features (by lithogenetic we are also to understand here the paleontological signs, since ancient organisms were part of the medium of lithogenesis). It is often said that there are landscapes where no sedimentation takes place, and, as a result, we do not have facies corresponding to them (Efremov's infrafacies, 1950). But, in fact, every facies contains some information (it is included in the rock composition, the geochemical characteristics, the granulometric structure, the textural features, and in the organic remains) on two types of continental landscape at the same time: the denudation and the accumulation types. The totality of facies, therefore, reflects the whole system of bygone landscapes and, in the final analysis, the entire ancient biosphere. It must not be thought that no traces at all are left of any particular landscape – this would contradict the systemic notion of the biosphere. The lacunae in the record are due to important losses of information, which occur both when a facial complex is formed (primary information loss) and when individual facies are destroyed (secondary information loss).

The amount of information and, consequently, the possible minuteness of a reconstruction depend on whether the spatial relation between the organic remains in a facies is primary or secondary. If the original distribution of the organisms is preserved (when an entire biotope, let's say, an area of bottom, is buried), the task is far easier. If this is not the case, we can only conjecture the relations during the organisms' lifetime by going on the basis of two assumptions: 1) a bed usually contains remains of organisms that lived close to one another (this paleobotanical reconstructing device is now also employed in the paleoecology of land vertebrates: Efremov, 1950; Shotwell, 1964; 2) recurrence of the same combinations in different beds is evidence of joint habitation.

Reconstruction is based on a comprehensive study of facies, which enables the information gaps to be filled in. Despite their inevitable flaws, paleoecosystems form a relative continuum, which corresponds to the **paleobiosphere**. The boundaries of the paleobiosphere's units, beginning from the major systems of ancient continents and oceans, can be determined only by paleoecological investigations. Here paleoecology touches on paleogeography and paleochorology. Of course, it is possible to identify paleobiogeographic regions and provinces by operating just with lists of taxa and allowing for a degree of endemism. However, the incompleteness of the geological record makes any assessment of the endemism of extinct animals extremely arbitrary, and we run the risk of taking gaps in the record for fundamental chorological differences. To avoid such errors, or at least to reduce the likelihood of their occurring, we must not only take note of the absence of a taxon but also try to find out why it is not present here. The answer to this question can obviously be provided only by paleoecology. While it has been established that there are no stigmariae in the Carboniferous sediments of Siberia, this fact in itself allows us to make only a purely formal discrimination, the meaning of which is unclear. But if we know that stigmariae are rhizomes of Paleozoic mangroves, which apparently did not grow in a temperate climate, the discrimination becomes not only more reliable but also more intelligible (Meien, 1970). Such an approach makes it possible to relate paleochorological features to climate, relief, the distribution of land and sea, and, in the end, to arrive at not a formal but a natural zoning that reflects the patterns of differentiation of the paleobiosphere and its hierarchical structure.

Paleoecosystems form not just a spatial but also a temporal continuum, which also combines with discreteness. The temporal discreteness of a paleoecosystem may be primary, associated with nonuniform development, or secondary, brought about by interruptions; actually, these two factors often act jointly, since a change of the conditions responsible for an interruption can also cause an ecosystem to undergo reorganization.

Becking wrote that a living system is an "event" rather than a "thing" which can be photographed today and tomorrow and identical pictures will be obtained. In describing it, we arbitrarily reduce the event to the level of a thing (Becking, 1946). Perpetual changes, fluctuations, are a necessary condition for the existence of an ecosystem. Fluctuations of the second-order components make little difference to the overall structure of an ecosystem, whereas, for example, a marked change in the population of the dominant group of species may have a critical effect on it. The succession of facies along the vertical informs us of changes of paleoecosystems. We never find completely identical thanatocenoses in two consecutive strata of a geological cross section. This **paleosuccession** (Krasilov, 1969b) in some cases attests to minor deviations from the statistically normal state that do not destroy the dynamic equilibrium, while in other cases it reflects a successional sequence, and in still other instances a sequence of climaxes. However, paleosuccessions are not adequate to the actual changes of ancient ecosystems. Taphonomic factors act as powerful resonators that amplify some changes and ignore others. Only by means of a taphonomic analysis is it possible to determine the boundaries of a paleoecosystem in terms of time.

Introducing the time factor into paleoecology, we find a close connection with stratigraphy. Existing side by side with more or less ephemeral ecosystems were long-persisting ecosystems, a fact demonstrated by the recurrence of analogous combinations of fossil organisms in a stratigraphic section. The higher the status of an ecosystem in the hierarchical structure of the biosphere, the more stable it is. We may apparently use this consideration as a basis for constructing a hierarchical system of stratigraphic units (stratons).

The smallest of these correspond to moments of stabilization of elementary biocenoses, whereas stratons of universal significance reflect the equilibrium state of the biosphere as a whole.

A good deal has been achieved in stratigraphy with the use of the traditional methods of tracing index forms or comparing lists of fossil organisms. But the stratigraphic schemes worked out with these methods are not free of the usual shortcomings of a formal classification. Due to the lack of correspondence between a paleosuccession and the actual time sequence of paleoecosystems, the boundary line drawn according to a clear-cut change of fossil assemblages in a number of cases corresponds only to second-order changes of the conditions of burial. It is only when we have determined the cause of a paleosuccession that we can get at the meaning of stratigraphic discontinuity, which may be related to local, regional, or global factors and accordingly have a partial, regional, or universal significance. If we approach the matter thus, we are getting nearer to solving a basic problem in stratigraphy: analysis of chronological continuum and identification of the natural stages in the evolution of the biosphere.

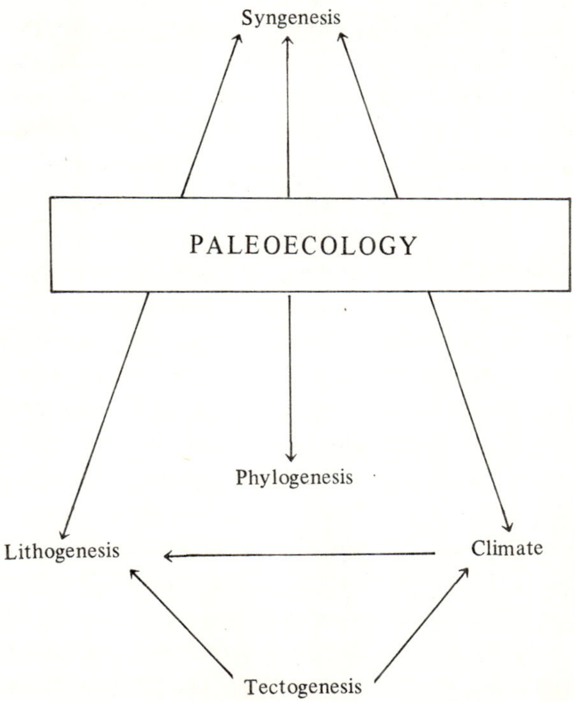

Figure 1. Sphere of activity of paleoecology

Hence, paleoecology plays the same role in stratigraphy as it does in paleogeography (Figure 1). It is essential for building up a natural classification (that is, one based on causal links). All classifications have two aspects: a utilitarian – the breaking up of sets into subsets for convenience of use or storage, and a theoretical – a study of the patterns of differentiation. Between these there is a close bond, but the practical task can be solved in the first approximation by means of a formal estimate of similarity, whereas the theoretical task necessitates revealing the causes of the similarity; or, in other words, it calls for a study of the structure of the system, as determinable from the interaction of its components.

Harper (1967) wrote that the theory of natural selection is an ecological theory developed by the most eminent of ecologists. In fact, Darwin's theory treats the evolution of organisms as the result of interaction between the different components of an ecosystem. However, Darwin inherited the uniformitarian views of Lyell where the physical factors of evolution are concerned. What the evolution process meant, in his opinion, is that organisms attained an ever-increasing degree of harmony with their life conditions. Some scientists believe that a biochemical transformation of energy preceded the appearance of the first organisms, in other words, that ecosystems existed before organisms. The very first primitive ecosystems apparently possessed homeostatic properties. A change in physical parameters in the course of evolution of the outer envelopes of the Earth destroyed the dynamic equilibrium and led to a succession of ecosystems, each more complex than the preceding (admittedly, we still do not have an adequate description of this process, which was neither one-directional nor rectilinear). Thus, **an ecosystem may be considered as an elementary evolution unit, phylogenesis as part of syngenesis, and the origin of the species as the result of a changed ecological niche**. Such an approach to the problem of evolution will be strictly ecological. However, the science of evolution has a multitude of ramifications. We may speak of the origin of flowering plants, basing ourselves on comparative morphology or by manipulating telomes, without concerning ourselves with the paleoecological situation, but if we do this, we will not gain a full idea of the evolutionary event.

And so, founded on the ecology of recent organisms, systematics, comparative morphology, and facial analysis, paleoecology in turn serves as the foundation for paleogeography, stratigraphy, and phylogeny (Lawrence, 1971).

Taphonomy can hardly be called a science on its own, but is rather a part of paleoecology. It seems that the most flourishing branch of paleoecology is paleoautecology, the founder of which is considered to be Kovalevskii. Counted among the pioneers of paleocenology are Forbes, Walter, Vogand, Abel, Dollo, Weigelt, and other scientists (see Imbrie and Newell, 1964).

It is justified to separate the ecology of plants and that of animals in view of their essentially different role in the biochemical cycle of matter and those features of the population structure that are related to uninterrupted (or stagewise) ontogenesis of plants, the combination of different modes of reproduction that is characteristic for them, etc. In paleoecology we have still more reasons to make such a separation: 1) the different distribution of habitats with respect to sedimentation basins (especially in terrestrial plants as compared with marine animals) has the result that the cenotic structure is not always reflected with the same degree of fullness in beds; 2) the insignificant degree of completeness of the plant organism and the detachment, in the process of life activity, of organs that are often specially adapted for transport by wind and water determine the mode of transition to the fossil state (in animals the detachment of organs, during molting for instance, has a limited significance); 3) remains of animals are preserved thanks to the presence of firm skeletal tissue. Soft tissues and organisms without a skeleton are preserved relatively rarely. With plants the principle of fossilization is different: the degree to which they are preserved depends, at least in many cases, not on the durability and chemical stability of the tissue but on the chemical interaction of the remains with mud solutions. This is why in plants there are no such disparities in the likelihood of burial as exist between animals with and without a skeleton; 4) a great difference from land animals is also the huge amount of plant material that gets buried, plus its rock-forming value.

The reconstruction of life forms and communities of ancient plants is based on all information on a bed received during facies and taphonomic investigations. Then the reconstruction itself becomes the source of additional information on the environmental conditions. The paleoecologist is thus constantly solving a direct and inverse task, guiding himself by the principle of uniformity.

3. THE PRINCIPLE OF UNIFORMITY

Leonardo da Vinci, Steno, Lomonosov, Hutton, von Hoff, Kant, and other thinkers saw in the present the key to understanding the geological past (see Vysotskii, 1961; Guntau, 1963). Charles Lyell transformed this concept into a basic working method of scientific geology. He spoke of the "uniformity of nature," which served as a basis for defining the "uniformitarianism" that those unfavorably disposed toward him ascribed to his teachings. Lyell proposed that all motive forces are identical and that past and present geological processes show a uniform activity. His views constituted in a large measure a reaction to the catastrophism of Cuvier and Alexander Brongniart (in his youth Lyell was their supporter). Influenced by Cotta (1846) and Darwin, Lyell's works became infused with the notion of evolution. At the same time, Thomson (Baron Kelvin) roundly criticized the theory of a uniform rate of geological processes and the perpetuity of geological time, which contradicted the second law of thermodynamics. Grossly erring in his assessment of the age of the Earth and upholding the biblical myth on the creation of life, Kelvin did, however, contribute toward consolidating the concept of directional development in geology (Hubbert, 1967). That put an end to naive uniformitarianism.

It gradually became clear that the epoch in which we live cannot serve as a standard for all times. If we speak of "typical" or "statistically normal" geological conditions, we see that our epoch markedly deviates from them, at least as far as climate is concerned ("Die Gegenwart ist um so weniger geeignet, für alle Kräfte und Vorgänge der Vergangenheit Beispiele zu liefern, als sie keine normale Warmzeit, sondern eine glaziale Ausnahmzeit darstellt": Kummerow, 1932, p. 565). This key, therefore, does not fit every lock. Furthermore, in history circumstances never repeat themselves exactly. This rule of nonidentity of conditions ("environmental uniqueness": Nairn, 1965) is the logical conclusion to be drawn from the law of the irreversibility of evolution.

So what about the principle of uniformity? There have been repeated attempts to give the dictum "the present is the key to the past" a nonuniformitarian meaning. The constancy of physical and chemical laws has been spoken of, but such an attitude makes the principle of uniformity too abstract and to all intents and purposes useless (even accepting that the sphere of activity of the principal laws is in fact known to us). It has been said that an analogous result (or end product) is evidence of similar "experimental" conditions (Nairn, 1965). But here, evidently, no account was taken of equifinal systems, in which the final state depends little on the initial state or on the conditions of development. All the same, the principle of uniformity as a methodological device does preserve some significance, since a study of the geological past begins with the choice of a present-day model. In view of the nonidentity of past and present conditions, we are obliged to introduce certain correctives into the model. From the different variants available for the interpretation of a given phenomenon, we select the one that requires the minimal deviation from the actual model at hand. Here the principle of uniformity merges with the logical principle of the simplest decision (Albritton, 1964). The choice of the simplest decision (which does not necessarily turn out to be the right one) is dictated to us by the strategy of the sensible gambler in the "game" with nature.

Apparently, in terrestrial paleoecosystems all the major components have changed in one way or another during their existence (their appearance is approximately dated to the Silurian period). We can definitely say that the plants, animals, and climate have themselves undergone change. It is also assumed that there have been changes in the atmospheric composition, the geological substrate, as well as in such factors of general significance as the size of the Earth, the magnetic field, the structure of the continental and oceanic crust, the position of the continents, etc. Here we shall give a very brief description (since we shall return to them in the following chapters) of the principal factors and presumed motive forces of the evolution of ecosystems. Let us remember that no theory of the evolution of Earth as a cosmic body has yet been developed, though there is no lack of hypotheses ("denn je weniger man weiss, desto leichter entsteht eine Hypothese": Schwarzbach, 1968, p. 251).

a) **Thermal regime.** Proceeding from the laws of thermodynamics which he formulated, Clausius postulated the heat death of the Universe. Kelvin, as we have said, regarded the development of Earth as a process of dissipation. These ideas influenced minds at the end of the 19th century and contributed to the popularity of the contraction hypothesis. Cooling and the contraction that is associated with it were considered the source of the stresses in the Earth's crust. It was thought that heat losses are compensated only by solar energy, a good deal of which goes for the movement of the atmosphere and hydrosphere. The following circumstances led to the collapse of this particular house of cards: the new cosmogenic hypotheses, the discovery of radioactivity as a major source of heat, and, finally, the data obtained by geologists attesting to massive glaciation in the Precambrian (and subsequently in the Silurian and Late Paleozoic), that was probably comparable in scale to the Quaternary glaciation or even more tremendous (Lungersgauzen, 1963; Rudwick, 1963).

b) **Differentiation of the mantle.** Many scientists consider the differentiation of the mantle to be the driving force in the evolution of the outer envelopes of Earth (Berg, 1966; Vinogradov, 1962; Kropotkin, 1961a, b; Lyustikh, 1956, 1961; Magnitskii, 1958, 1961; Peive, 1961; Sobolev, 1963; and others; these are just a few publications, in which the reader will find numerous references). Degasification of the mantle and the migration of lithophile elements toward the outer envelopes change the chemical composition of the crust (and consequently the geological substrate of biocenoses), hydrosphere, and atmosphere. Another result of the differentiation is that the Earth's crust undergoes a structural alteration. According to some calculations, the mantle supplies the crust annually with $1-2$ km^3 sialic substance, i. e. the current volume of sial, 5×10^9 km^3, must have been formed in approximately

3×10^9 years (Levin, 1966). The growth of continents is believed by some to be an irreversible process of transformation of primary basaltic crust into sialic crust (Berg, 1966; Gill, 1965), whereas other scientists contend that the degradation of sial occurs simultaneously on account of rupture, collapse, and basification (Belousov, 1955; Nekhoroshev, 1963; Sheinman, 1959). Finally, the migration of elements upon differentiation of the mantle may in itself be a cause of deforming stresses (the concentration of radioactive elements in certain zones leads to warming up and expansion; the movement of siderophile elements toward the core is thought to be a possible cause of contraction: de Sitter, 1964).

The differentiation of matter and the increase of sialic masses may cause the ocean basins to deepen and thus generally enhance the polarization of the megastructure of the Earth's crust (Khain, 1968). The theory that the deep ocean depressions are relatively young is upheld (though from different standpoints) by Berg (1966), Dietz (1962), Rukhin (1959), Heezen (1959), Jordan (1966), and others.

c) **Convection currents.** It is the heterogeneous structure of the lithosphere that is believed to be the cause of convection in the subcrustal matter. The boundary between crust and mantle is fixed by the Mohorovičić discontinuity, which passes on average at a depth of 10–12 km from the ocean surface. The oceanic basaltic crust (with a seismic wave propagation rate of 6.7 km/sec) has an average thickness of 5 km; above it in the oceans lies a layer about 1.7 km thick that represents in physical terms a mixture of basaltic and sedimentary material. Under the continents the Mohorovičić discontinuity lies at a depth of about 35 km. The continental crust with a seismic wave propagation rate of 5–6.5 km/sec at a depth of 15–35 km is similar to the basaltic crust in its physical properties. It has been established that the upper mantle is also heterogeneous.

The current trend in geophysics, which has been dubbed the "new global tectonics" (Isacks et al., 1968), grew up on the basis of hypotheses of convection cells (Holmes, 1931), ocean floor spreading, and drift. Mid-oceanic ridges (this name is arbitrary, because the ridges continue under the continents) are considered as zones in which the lithosphere is stretched and split up into segments – lithosphere plates. Breaking off at the sides of the median ridges, the plates build up at the rear on account of substance supplied by the mantle. In oceanic troughs girdling continents and island arcs, the lithosphere plates are submerged and absorbed by the asthenosphere (Figure 2). The complete rotation cycle of the plates is calculated to extend over 250×10^6 years (Dewey and Horsfield, 1970). Continental drift is thought to result from movement of the rigid plates caused by convection currents. The tectonic structure of many regions has already been successfully interpreted

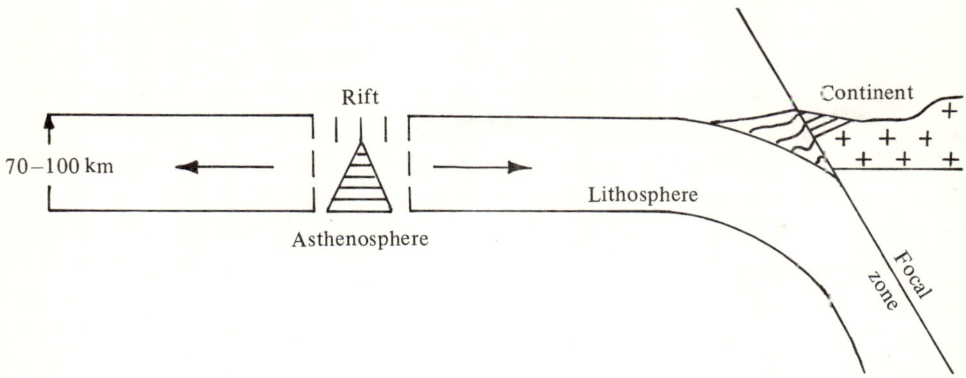

Figure 2. Diagram showing the development of lithosphere plates

from the viewpoint of lithosphere plate dynamics. The boundaries of these plates are determined according to the distribution of seismic zones.

The system of convection cells may have been relatively simple in the early stages of evolution. At that time the sialic blocks constituted a single mass or, more likely, two supercontinents of approximately equal area: Laurasia and Gondwana. The fragmentation and drifting apart of the primary supercontinents were the result of the gradual complication of the convection cell system.

d) **Expansion of the Earth.** Most supporters of the "new global tectonics" consider that the Earth's radius has not changed and that the growth of the lithosphere plates in the zones of mid-oceanic ridges has been fully compensated by their absorption in troughs. However, Meservey (1969) points out that the assumed migration of the continents is impossible on the present-sized Earth for topological reasons. The expansion of the Earth is related to a lowering of the "gravitational constant." Proceeding from the theory of relativity, Dirac advanced the hypothesis of a change in the value of the gravitational constant. This gave rise to far-reaching conclusions on the expansion of the Earth or on pulsations in connection with directional change (Jordan, 1966) or in relation to pulsations of the gravitational field. The cause of the pulsations may be a change in the position of the solar system relative to the center of the galaxy with a period of about 280×10^6 years (Steiner, 1967); Lungersgauzen (1963) thought the same phenomenon to be a possible cause of glaciation. There are as yet no strict geological proofs of expansion. Extension along rift zones and contraction in geosynclines as a rule develop in parallel (Khain, 1969). It is difficult for the time being to make an accurate quantitative assessment of them.

e) **Rotation forces.** One of the possible consequences of the evolution of the gravitational field is a change in the speed of the Earth's rotation (Kropotkin and Trapeznikov, 1963; Nairn, 1965). Examination of growth rings of ancient corals seems to bear out the fact that rotation has slowed down (Runcorn, 1967). Gravitational-inertial forces are considered to be a source of horizontal stresses in the crust, in which many scientists see an alternative to the hypothetical mechanism of convection currents or a complement to it (see Peive, 1961). It has been noted that the planetary system of fractures, which includes the rift zones and mid-oceanic ridges, has a definite orientation, related to the position of the rotation axis (Voronov and Nezametdinova, 1970). Consequently, a change in the position of the rotation axis will produce a corresponding change in the direction of the force field (Dolitskin and Kiiko, 1963). Astronomical methods have established a slight shifting of the poles (in the 20th century at a rate of about 111 mm a year: Mikhailov and Shcheglov, 1969). The trajectory of the poles, reconstructed according to paleomagnetic data, probably reflects the evolution of the system Earth–Moon–Sun. Here we enter wholly into the world of hypotheses, one of which assumes an episodic rapprochement of the Moon and Earth exceeding the Roche limit, with catastrophic consequences: an increase in the height of the tides to 6 km, stoppage of the Earth's rotation and the liberation of a massive amount of rotation energy, together with a warming up and even evaporation of the World Ocean. Other investigators believe that the effect of this rapprochement was not so catastrophic, although a study of stromatolites gives some basis to speak of an increase in tidal amplitude in the Late Precambrian (Alfvén and Arrhenius, 1969).

f) **Cosmic rays.** The intensity of cosmic rays is subject to seasonal and solar-diurnal fluctuations. In addition, this intensity is higher at the poles than at the equator, so that a displacement of the poles would have an effect on this factor too. Marked changes in cosmic radiation intensity may be associated with solar flares and disturbances of the magnetic field. Schindewolf (1950) and some other scientists assume that reversals of the magnetic field went along with irradiation that was catastrophic for many organisms. In fact, mass mortalities do coincide with reversals, but radiation cannot be the cause of them because not only terrestrial but also marine organisms die off, and the latter are protected from irradiation by the water column.

g) **Development of the organic world as a geological factor.** With the growth of biomass, not only did the bulk of organogenic rocks increase but also the overall influence of the biogenic transformation of substances on the chemical composition of the outer geospheres and on the sedimentation process was enhanced. Entire geological formations (for example, the Precambrian iron ore formations) "died out" and were replaced by new ones.

With the development of land vegetation, the rate of erosion decreased on the whole, the ratio between physical and chemical erosion was reduced, and, probably, stable water streams appeared for the first time (Seward, 1933; Meien, 1969). But the most important result of the development of life both for the organisms themselves and for their surroundings was the appearance of free oxygen in the atmosphere. The original atmosphere probably consisted of methane with an admixture of nitrogen and hydrogen (several models exist). An insignificant amount of oxygen was produced upon photochemical reactions (Vinogradov, 1962), but permanently existing free oxygen is a product of photosynthesis. The first autotrophic plants emerged at least 2000×10^6 years ago (according to determinations of the absolute age of very ancient rocks with stromatolites: Cloud, 1965). According to the Berkner-Marshall theory (Berkner and Marshall, 1965), the oxygen content reached the Pasteur point (1% of the present content) at the beginning of the Cambrian, and this caused a considerable increase in the abundance and diversity of organisms: the number of animal classes doubled at the boundary between the Proterozoic and Cambrian, and the number of genera rose by 227.5% (Müller, 1964). At the beginning of the Ordovician the oxygen content rose to 10% (the Berkner-Marshall point according to Brönnimann), an ozone screen was formed as a barrier against ultraviolet rays, and it became possible for the land to be inhabited. During the Carboniferous the content of free oxygen was more or less the same as it is now. Other authors hold that the Pasteur point was passed back in the Precambrian. Wangeroff proposes a figure above 10% for the Cambrian, this, in his opinion, being necessary for the formation of firm skeletal tissue (see Brönnimann, 1966). Some geologists associate a broad distribution of red sediments with a high oxygen content in the atmosphere. In that case we should have to assume that the present-day level was reached, or even surpassed, as early as in the Devonian. Hence, oxygen was apparently not a limiting factor for terrestrial plants. It has been observed that in a time of catastrophic mortalities the organisms that suffer most are those with a high metabolic rate. In view of this it is thought that mortalities are related to a diminished oxygen content (McAlester, 1970). However, a drop in temperature must also have an analogous effect.

The carbon dioxide content in the atmosphere is considered by many to be a vital factor of evolution (the founder of the CO_2 theory was Arrhenius, 1896; it is discussed in Brooks, 1949; Gol'bert et al., 1968; Rukhin, 1959; Dorman, 1968; Mathieu, 1952; and other publications). The CO_2 content in the atmosphere and in the ocean is in a state of dynamic equilibrium and can probably serve as an example of an open equifinal system. Under stable conditions an additional influx of CO_2 into the atmosphere (about 10^8 tons annually are received as a result of mantle degasification) cannot alter the equilibrium. However, the rate at which carbonates are deposited and the rate at which carbon is fixed by plants (the main consumer of carbon dioxide is phytoplankton: see Tappan, 1968) depends on the temperature, so that an overall modification of the conditions may alter the CO_2 content, in which case about 10^6 years are required for equilibrium to be restored (Dorman, 1968).

Each of the factors mentioned above and their various combinations may produce changes in climate, relief, distribution of land and sea, and geological substrate, and lead to a radical reorganization of terrestrial ecosystems. A common theory should explain both the progression of ecosystems from simple to more and more complex and the periodicity of this process. The progressive differentiation of the mantle, the growing complexity of the convection cell and lithosphere segmentation system, the increasing heterogeneity of the megastructure of the Earth's crust and of relief contrasts, plus the steady march of development of the organic world, all these apparently lead to a greater complexity of the overall ecological situation and, accordingly, to an enhanced information capacity of ecosystems, to a rise in their negentropy. The periodicity probably mirrors both the internal rhythm of development and the external influences. The cosmic phenomena of which we spoke above, superposed on the endogenous factors, may serve as the cause of the periodicity of geotectonic processes that have a multi-faceted influence on syngenesis. The principle of the nonidentity of conditions explains the emergence of unusual life forms and of "forbidden" combinations of species in communities of the geological past. At the same time, in explaining deviations from a present-day model, we must apparently reckon with the possibility that the reaction of organisms to analogous conditions has changed.

4. HISTORY OF PLANT PALEOECOLOGY

The scope and aims of this book, and also the limited possibilities of its author, preclude a minute survey of the history of paleoecology, which according to tradition should begin with Xenophon and Herodotus, who made the first guesses about life in the past on the basis of fossil finds, and should go right up to the present day. Instead, we shall dwell on just some of the periods which are, in my opinion, the most interesting ones.

Almost simultaneously with the birth of paleobotanical systematics (considered to date at 1820) or even somewhat earlier, but of course before the

Darwinian revolution, fossil plants assumed a role in the natural sciences of indicators of paleoecological conditions. In 1811 George Cuvier and Alexandre Brongniart established in a section of Tertiary deposits in the Paris Basin an alternation of marine and continental sediments which brought them to the idea of catastrophes. They identified sediments of freshwater basins from remains of land vertebrates and plants. Charles Lyell, the father of scientific geology, used this method widely. Thus, in 1826 he wrote that coal-bearing sediments in many places show signs of freshwater origin and the plant remains preserved in them give proof of the existence of land during the period of their formation. Since the tracing of ancient shorelines is the primary task of not only the paleogeographist but also almost all geologists, whatever their field, fossil plants have rendered invaluable service to geology even within the framework in which it was used by Cuvier and Lyell. Soon, however, it became clear that their usefulness lay not merely in their enabling us to distinguish paleoland from paleosea. They have a direct bearing on the problem of the origin of coal, in which the German geologist Gerhard in 1870 identified metamorphosed peat bog. In 1804 Schlottheim determined that leaf impressions from Carboniferous sediments belong to extinct plants. His conclusions were soon confirmed by Sternberg, Lindley, Hutton, and Adolphe Brongniart. We have already mentioned that plant ecology was nurtured on the soil of phytogeography. In just the same way, changes in the groupings of fossil plants in time and space stimulated the development of paleoecology. The lateral and vertical variability of ancient vegetation is so intimately connected with changes of climate that research in both fields progressed in parallel. Paleoecology received a vital injection following the discovery of Tertiary arctic floras containing a subtropical element and clearly not corresponding to the present-day climate in these latitudes. Credit for discovering and describing the arctic floras is due to Oswald Heer, who may rightly be considered the founder of the paleoecology of terrestrial plants. Heer's many-volume work "Flora Fossilis Arctica" was largely responsible for introducing the idea of the evolution of climates into geology (indeed, possibly the general notion of development). But were not the tropical plants transported by currents from the south? (Let us note that with respect to the *Nipa* fruits in the London Clays doubts of this kind are expressed also today.) And does the accumulation of "imprints" give an adequate picture of the ancient vegetation? In answering these questions, Heer found some general principles of paleoecological analysis: 1) a certain combination of fossils is often repeated in two or more beds; the same numerical ratios also recur. Therefore, a bed possesses a structure. Moreover, an approximate structural ratio may be found between a fossil assemblage and a particular type of recent vegetation (if not the Arctic or Central Europe, then let's say Florida);

2) plant assemblages from different rocks (for example, from the marine and freshwater mollasses of Switzerland) differ from each other; these differences reflect both the differentiation of the ancient vegetation and the conditions of burial; 3) a study of the structure of a bed is based on an analysis of the population and spectrum of the life forms (for instance, according to Heer, phanerophytes account for 6/7 of the Tertiary flora of Switzerland); 4) information losses on burial may be made up by introducing corrective coefficients (920 species from the mollasse deposits constitute one third of the Miocene flora of Switzerland) and interpolations for other elements of the ecosystem: remains of ungulates attest to forest-free spaces, while the high iron content in the Atanekerdluk deposits is proof of a forest type of vegetation (at least, so Heer thought).

The same principles lie at the basis of the paleoecological reconstructions of Göppert, Weber, Gardner, Ettingshausen, and others. The influence of catastrophism is apparent in their works. The formation of thanatocenoses is treated as the outcome of a sudden, one-act burial ("durch plötzliches Anschwellen der Wasser": Weber, 1852; "durch eine plötzlich eintretende Katastrophe": Göppert, 1855). Whence the intensified efforts to pinpoint the season (even the month) of burial.

In the mid-19th century Steenstrup (1842) gave the first scheme of the succession of forest types in Northern Europe during the Holocene. Grisebach (1844) interpreted it as reflecting postglacial warming. Having described the first fossil pollen, Göppert (1836) also made a contribution to Quaternary climatology. He set out a general scheme of the evolution of the Tertiary climates of Europe (Göppert, 1855: the subtropics with palms, Proteaceae, and breadfruit tree in the Eocene; deciduous and conifer species with subtropical forms in the Miocene; the elimination of subtropical plants and the spread of the Platanaceae in the Pliocene).

The first summary of paleophytogeographic studies of the 19th century is that by Engler (1879–1882). Drawing on the experience of his predecessors – de Candolle, Hooker, Bentham, Unger, and others, he propounded a number of "leitende Ideen," which have not lost their significance up to the present. Engler asserted that we can understand the patterns of recent plant distribution only if we know their geological history. Regions in which the vegetation has not been destroyed by recent geological catastrophes show a greater floristic wealth and a higher genus/species ratio than younger formations. On the other hand, with the colonization of new territories, where the vegetation has been wiped out by geological catastrophes, new plant varieties are able to establish themselves. The center of their settlement generally does not coincide with the area in which they attained their maximum development and diversity. Especially important for geographic zoning are genera that are dominant in some areas and rare in others. There are no clear-cut boundaries between floristic regions, and the degree of their differentiation changes from epoch to epoch. Engler arrived at a three-dimensional classification of vegetation and noted that with a cooling of the climate, mountain races of plants sometimes move down into the lowlands. This idea gained impetus in paleobotany.

The relationship between a paleoclimatological evaluation and the **hypsometric level** of habitat of an ancient community was first established by Nathorst (1888). Studying Pliocene thanatocenoses on the Japanese islands (the Mogi and Taka shima taphofloras), that had been formed under conditions of considerable geomorphic contrast, he quite rightly became interested in the variable content of thermophilous elements (which was formerly explained as due to age differences) as a reflection of vertical zoning. This was a new idea in paleoecology, and, as we shall see, it was to bear much fruit.

Reid and Chandler (1926) were among the first to show that in paleoclimatology it is communities, not individual species that count. In a study of the Paleogene flora of the London Clays they were confronted with an extraordinary diversity of burial conditions and analyzed not only the paleoclimatic aspect of successions but also the sum effect of climatic changes and transgressions and regressions in combination with a vegetation that was distinctly zoned owing to edaphic and topographic factors. Reid and Chandler stimulated paleobotanists to evaluate the transportability of plant organs (primarily fruits and seeds) and to always take note of adaptations to various modes of transport.

In the 1920s research on paleoecology was conducted on an intensive scale in the USA (Berry, Chaney, Dorf, Axelrod, MacGinitie, Mason, Elias, Condit, Wolfe, and others). Numerous factors contributed to the advances made in this science: the rich diversity of beds and forms of paleorelief, the definite connection that was discovered between changes of climate and syngenesis and mountain formation, the good degree of preservation of paleorelief structures, and also the combination, on a relatively small territory, of different zonal types of recent vegetation and relict groups. The methods of reconstructing and paleoecologically evaluating ancient cenoses that had been indicated by the founders of paleoecology were greatly improved upon thanks to the use of the latest achievements in geology and phytocenology and were placed on a firmer footing with the aid of statistical methods, composite climatic indexes (effective temperature), detailed studies of various burial situations, and so on.

Much attention was paid to questions of florogenesis (the history of the Madro-Tertiary flora, masterfully outlined by Axelrod, is an example of such a type of investigation: Axelrod, 1958), to migrations, and to an analysis of the hypsometric level to which the Tertiary biotopes were confined (the basis for this last had already been laid by Nathorst). One of the most important theories to come out of this school was the concept of **isoflors**, which was derived from the principle of homotaxis. Restating Hooker's notion (Hooker, 1891), Chaney writes that "floral and faunal resemblance across a wide range of latitude may be an indicator not of contemporaneity but of discrepancies in age" (Chaney, 1940, p. 476).

Since the altitudinal zonality roughly corresponds to the latitudinal zonality, a climatogenic shift of latitudinal zones takes place primarily by means of vertical migrations ("upland-lowland"). However much we may be averse to the idea of a "Miocene" forest on the Eocene uplands of Oregon, it does, according to Chaney (1936), reflect the actual situation. The concept of evolution by way of successive migrations from the north, which was initiated by Buffon (1778) and perpetuated by Gardner, Asa Gray, Krishtofovich, and others, found its culmination in Chaney's works. Engler, unlike Asa Gray, showed the importance of autochthonous processes in the evolution of the Cenozoic flora of North America and warned against Gardner's overhasty stratigraphic conclusions. Chaney's florogenetic constructions, like those of his predecessors, suffer from excessive schematization of natural-historical processes.

Another tenet of the Chaney school is expressed in the concept of a geoflora as a complex floristic system that preserves its structure during millions of years. Independently of Chaney, Krishtofovich introduced the notion of a **polychronous flora**, illustrating floristic homeostasis on a global scale.

Since the lateral variability of ancient vegetation, as we have seen, stimulates paleoecological investigations, Soviet paleobotanists have proved to be in an advantageous position: the territory of the USSR embraces entire paleofloristic regions, even kingdoms (Vakhrameev et al., 1970). On the other hand, the wide distribution of enormously thick continental deposits promoted the development of phytostratigraphy. Recently the paleoecological aspects of both phytogeography and phytostratigraphy have been arousing increasing interest in the Soviet Union. Meien (1968), studying the phytogeography of the Carboniferous and Permian, demonstrated that a natural classification of the paleofloras of that time requires that not only the floristic but also the paleoecological characteristics be taken into account. He originated the idea that signs of the Namurian floristic "leap" established by Gothan in Europe, may be found in Siberia: although the plant communities have a different habit here, so that the actual "leap" looks different too, the reason is apparently the same. Such an approach to stratigraphic correlation may be termed ecogenetic. I have tried to show that plant communities may take on the role of operative units of both a phytogeographic and a phytostratigraphic classification, and have indicated a system of subdivisions for a sedimentary series: ecostratons, which correspond to different episodes in the evolution of paleoecosystems (Krasilov, 1970). The aggregate of such units corresponds to a certain stage in the development of the biosphere, or the paleobiosphere. Consequently, the stratigraphic boundaries of universal significance reflect relatively brief episodes in the reorganization of the overall structure of the biosphere.

In speaking of vertical variability related to climate as being a most important source of paleoecology, it must be acknowledged that the Quaternary period is the most suitable model for studying this process. In essence, it is only due to the considerable advances made in Quaternary palynology that plant paleoecology came to be an independent science. The first spore-pollen analyses of peat bogs were carried out at the end of the 19th century (Früh, 1885; Trybom, 1888; etc.; see also Erdtman, 1954). Weber, the initiator of peat bog stratigraphy, specified the percentage contents of pollen grains of various species (Weber, 1893). Lagerheim considered palynology as a most important method of learning the history of a flora (in particular the immigration of tree species after glacial retreat). His student von Post became the father of modern palynology (von Post, 1916; this admirable report was published in English in 1967). Von Post showed that samples from different rocks variously characterize vegetation: some contain "regional" pollen of dominant trees, others "local" pollen and spores of swamp and aquatic vegetation. Attention should be focused on the regional spectra, as these reflect changes of climate. Linking palynology with paleoclimatology, von Post made this an essential branch of Quaternary geology. In addition, he supplied palynology with a rather rough but effective methodology of vegetation reconstruction that remains of significance to this day. Thousands of sections have been studied with the use of his method, and hundreds of analyses of subfossil, surface, and atmospheric samples have been conducted with the aim of perfecting his procedure. The purpose of these studies was to eliminate distortions arising due to disparities in the productivity and (or) the transportability of pollen, in the modes of its distribution (anemophily versus entomophily), in the burial conditions, etc. Many palynologists have directed their efforts toward laying down a statistical theory of interpreting spore-pollen assemblages (see Davis, 1963; Livingstone, 1968; Tauber, 1967; and others). Recently, more and more attention has been paid also to "local" assemblages of marsh plants. Their relation to climate is more complicated and is realized mainly through edaphic factors. A more precise method of interpretation is required here.

Palynology has found wide application in petroleum geology, in connection with which its stratigraphic scope has broadened considerably. Chaloner mentions that oil-field palynology has proved to be a link between the palynology of the Quaternary and that of the Carboniferous (Chaloner and Muir, 1968) and has enabled the methodology to be integrated.

Coal geology provided the second stimulus for the development of plant paleoecology, one just as significant as the spatial and temporal differentiation of vegetation. Coal is phytogenic rock and at the same time a form of burial of plant remains. Thus, its genesis is a paleoecological problem. Back in the 1830s two hypotheses were formulated on the origin of coal: from stems carried by a stream and from peat bog (see H. Potonié, 1907a). Brongniart, Unger, and Heer supported the peat-bog hypothesis. Gümbel (1883) divided coals into allochthonous and autochthonous, and later these terms began to be applied to beds in coal-free rocks as well. The peat-bog hypothesis stumbled upon a contradiction: the life forms of plants of the European Carboniferous seemed to point to a tropical climate, while peat bogs are developed chiefly in the high latitudes. This contradiction contributed to a more intelligible application of the principle of uniformity, but on the other hand, it lent impetus to efforts to further pursue ecological-morphological investigations, to render the reconstruction of the Carboniferous landscape more accurate, and to seek analogues of the Paleozoic swamp forest (mangrove swamps, the Everglades, the tropical bogs of Sumatra: H. Potonié, 1907b). Two fundamental differences have been determined concerning the conditions of coal formation and plant burial: paralic and intracontinental ("limnic"). Zeiler, H. Potonié, and others noticed that many Carboniferous plants had a local distribution. Similar observations yielded two phytocenological generalizations: the division of the Paleozoic communities into 1) "external" (paralic) and "internal" and 2) into anthracophilic and anthracophobic. For the first of these rules we are indebted to Gothan (1907). He assumed that external and internal groups were distributed on opposite slopes of Hercynian mountain structures. It is my belief that paleobotanists whose prime interest has lain in Gothan's phytogeographical constructions have underestimated his conclusion about the connection between community differentiation and relief. Yet this conclusion opens up wide possibilities for paleomorphological theories. The second generalization is based on the remarkably scrupulous investigations of Davies (1921) and on later data of Gothan and Gimm (1930).

Thus, while Heer proceeded from the similarity between the Tertiary and recent communities in his reconstructions, the phytocenology of the Carboniferous rested primarily on the facies limitations of beds. The **palynology of coals** gave a more accurate picture of phytofacies differentiation (see Ibrahim, 1933; R. Potonié, 1952; Kremp, 1952; etc.). At first it developed as a practical method of **identification** ("synonymization") of beds (founders: Lyuber and Val'ts in the USSR, R. Potonié in Western Europe, and Reistrick, Thiessen, and Shopf in the USA). R. Potonié brought together numerous observations in his **"spore-petrographic" rule** (Sporen-Petrographische Regel), which affirmed that all petrographically distinguishable units in a coal bed and in the enclosing rocks are palynologically identical. It was he who was behind the notion on the palynocenosis (or even the facies as a whole) as an "index fossil."

Potonié's idea has a direct bearing on what we have said above about eco-stratigraphic schemes. Reistrick and Thiessen used not only the composition of spore-pollen assemblages but also the quantitative ratios for correlating coal beds. This came close to von Post's paleoecological method. An essential difference, however, lies in the fact that the palynology of coals studied mainly bog associations, not sod vegetation. Paleosuccession in coals is associated first of all with fluctuations of edaphic conditions. Whence the frequent reversions, on the background of which it is very hard to determine the overall trend of syngenesis. Nevertheless, new methodological achievements in phytofacies analysis make it possible to reconstruct Paleozoic communities in a considerable amount of detail (see, for example, Habib et al., 1966).

Thus, we can speak of three trends in paleoecology, the sources of which were phytogeography and paleoclimatology (Heer and the schools of Chaney and Krishtofovich), Quaternary geology (the von Post school), and coal geology (the Potonié school). Although the methods used in these different trends and schools are in many cases the same, still, the first and second pay more attention to paleophytocenology, while the third concentrates on phytofacies analysis.

These days we are witnessing a merging and interpenetration of the trends and methods. The way is open for a paleoecological synthesis.

Part II

BURIAL

... a thick fall of leaves was there, enough to shelter two or three men in winter time ....

The Odyssey

In order to understand the origin of plant beds, we must know first of all the source of the material, then the manner in which it reached the place in which it is found embedded in sediment, the process by which it became fossilized, and finally, the nature of the landscape which formed the background for all these happenings. We then get a rough idea of how a mass of buried remains relates to the continental vegetation. There are burial sites from which fossils have been extracted now for more than a hundred and fifty years, and no end is in sight. Taking into account that only a fraction of the annual "rain" becomes fossilized, we may assume that the phytomass and productivity of the original communities were tremendous. The vegetation here had such a strong effect on the sedimentation process, the chemical composition of the sediments, etc. that we may say it prepared its own burial ground.

Two variants of burial can be visualized: aerial organs must somehow or other become enclosed in sediment, whereas subterranean organs are already embedded while alive. In the latter case burial consists merely in a further sinking of the root zone (true, roots are found in a secondary burial much more often than in a primary one). Burial takes place both in clastic sediments and in more or less pure biogenic sediments, consisting of remains of actual plants. There may be a wide disparity between the age of a buried plant and that of the enclosing sediment: a vertical stem buried in the position it had in life is to be considered an element of the bedding surface, whereas plant remains transported by water or wind along with clastic material assume the role of sedimentation particles. Remains that are confined to the top of a layer or that form a biogenic stratum were deposited during a pause in sedimentation. Finally, downwashed remains are usually younger than the stratum in which they are found (in exceptional cases, which will be discussed below, they may be older than this stratum).

1. MATERIAL

Sometimes a plant is buried whole (this occurs relatively rarely and only with small plants) and sometimes in parts due to disintegration during life or after death. Disintegration during life is a feature specific to plants, being related to the fact that plants are less "all of a piece" than animals, and it is concomitant with stepwise ontogenesis. Two forms of disintegration during life are to be distinguished: **traumatic** (caused by wind, rain, or pests, for instance, the shedding of needles attacked by fungi: Hartigan, 1968) and **physiological** or **functional**. The following are associated with the latter:

a) Shedding of assimilating organs. In a climate with seasons the loss of leaves, annual shoots, leafless green shoots, and bark in xerophytes (Vasilevskaya, 1954; Fahn, 1964) protects the plant against drying. The leaves of subtropical evergreens usually live 3–4 years (from 1 to 15–17 years). The regular shedding of brachyblasts (e.g. branch abscission in poplar or in kauri pine, *Agathis australis*: Licitis-Lindbergs, 1956) may be for the purpose of preserving the optimal geometric structure of the crown under fluctuating conditions of photosynthesis.

b) Detachment of organs related to reproduction. Examples are vegetative organs (the branches of some trees, the brittle leaves of mosses, etc.), specialized reproductive organs (the proliferous bodies of Bryophyta, isospores, megaspores, seeds, fruits, and analogous structures), and organs responsible for fertilization (microspores, pollen). Some plants scatter more complicated formations, for instance, tetrads of spores (aquatic liverworts) or the contents of an entire sporangium (pollinia in insect-pollinated angiosperms), or, finally, whole flowers (Hydrocharitaceae). In the liverwort *Riella* not individual spores but entire sporangia are scattered. Some observations, to which we shall return later, show that ancient Lycopodiales (*Lepidocarpon*, *Pleuromeia*) also reproduced in this way.

c) Detachment of organs that have fulfilled their purpose. Shortly after pollination the corolla of the flower disintegrates. The empty strobili of gymnosperms still remain for some time on the branch. They are then cast off whole or break down into parts. The bracts and sometimes also the hairs that protected the juvenile organ fall off.

d) Detachment of organs that have not fulfilled their purpose. Nonpollinated ovules of *Gingko* cease to develop and fall off. A similar phenomenon apparently also occurred in many extinct gymnosperms. This explains the frequent finds of aborted ovules in the Paleozoic and Mesozoic (Taylor, 1965).

The organs found buried may be dead or alive (e.g. seeds that have not lost their capacity for germination).

In addition, as a result of posthumous disintegration, dispersed leaf cuticles, the inner membranes of ovules, antheridia of Bryophyta, and other small objects become fossilized.

2. CHANGES OF THE MATERIAL IN THE COURSE OF EVOLUTION

The probability of burial depends on the amount of remains, and this in turn depends on the size of the plant and on functional adaptations. If only whole plants or products of traumatic and posthumous disintegration had been preserved, we should have far fewer beds. However, functional disintegration associated with reproduction has been a property of land plants from the very beginning. As a rule, the more primitive the reproductive organs and dispersion mechanisms, the more dispersable elements are produced (the meadow mushroom *Agaricus campestris* produces as many as 40×10^6 spores per hour and the shield fern up to 5×10^6 in a season). Evolution toward perfection in both these cases is therefore a negative taphonomic factor. With the development of heterospory the reproductive function passes to the megaspores, the microspores now being necessary only for fertilization. A tendency is observed in many phyla toward a decreasing number of megaspores in the sporangium. While the function of dispersion goes over to the diploid membrane of the megaspore, the number of functioning megaspores is reduced to one. This is observed both in Paleozoic sporangia-dispersing Lycopsida and in seed plants. Chaloner (1968) notes that in the Carboniferous the number of megaspores steadily decreases upward in a section, and he says this is because there was a progressive transformation of spore plants into seed plants.

The number of seeds produced depends on many factors, chiefly the size of the plant, its phytocenotic status (plants in pioneer communities generally produce more seeds than those in climax communities: see Stebbins, 1956), the mode of pollination (anemophilous angiosperms usually have fewer ovules per flower than entomophilous forms; Faegri and Pijl (1966) consider a large number of ovules to be part of the melittophily syndrome – pollination by bees). In angiosperms the reduction in the number of ovules is well expressed in the series follicle–achene (Chute, 1930), and in pteridosperms no less distinctly in the series *Calathospermum–Gnethopsis* (Smith, 1954). This is apparently a general trend, which is often reversed in the case of zoochory.

According to Rempe's calculations (Rempe, 1937), there are between 1000 and 10,000 pollen grains per cm^2 inside the corolla during flowering.

In forested areas the pollen rain yields several thousand grains of tree pollen per cm^2 in a year. Trees are more productive than grasses in this respect (except for such forms as *Secale* and *Calluna*). In northern forests the average productivity of anemogamous trees is about 10,000 grains per anther and of zoogamous species 100–1000 grains. A ten-year-old branch of a beech tree gives about 28×10^6 grains annually, birch, spruce, and oak more than $100 \times \times 10^6$, and pine 350×10^6. Productivity depends on the canopy density, the periodicity of flowering, and on other factors (Faegri and Iversen, 1950). Plants constituting underbrush yield little pollen, but in open places their productivity increases.

As a rule, aquatic and zoogamous plants produce less pollen than anemogamous species (Table 1). In principle, biotic pollination (zoidophily) is more economical than physical pollination and leads to reduced productivity, but if the pollen itself is the means of attraction (rather than the nectar, or in addition to the nectar), a certain amount of surplus pollen is essential. This is why flowers pollinated by insects, for which pollen is the main source of protein, often produce as much pollen as wind-pollinated forms.

Table 1. Number of pollen grains per anther, thousands (after Pohl, 1937)

Species	Mode of pollination	Number of grains
Corylus avellana	Anemophily	2546
Fagus silvatica	Anemophily	637
Aesculus hippocastanum	Entomophily	452
Acer pseudoplatanus	Entomophily	94
Tilia cordata	Entomophily	44
Plantago lanceolata	Anemophily	15
Sanguisorba officinalis	Anemophily	11
Betula verrucosa	Anemophily	7
Polygonum bistorta	Entomophily	6

While the dispersal of organs for reproduction is a primary property, the discarding of assimilative organs is not an obligatory attribute of terrestrial life. We have no reason to suspect that the architelomes of *Rhynia* were shed. Shedding presupposes, on the one hand, quite a high degree of differentiation of the plant body and, on the other, continued growth. It seems that only with the appearance of a secondary thickening of the stem did leaf fall become a regular feature (like the shedding of assimilative cortex). Seasonal shedding of leaves is observed in flowering plants among the most different gymnosperms: larch, gingko, xerophilous cycads. We may therefore assume that extinct groups of seed plants also acquired this property under certain conditions. Seasonal leaf fall greatly increases the likelihood of burial. The huge accumulations of leaves of glossopterids in the Paleozoic of the southern hemisphere and of Cordaitales in the Paleozoic and Czekanowskiales in the Mesozoic of North Asia are probably due to the seasonal shedding of leaves and (in the Czekanowskiales) brachyblasts.

A reduction in the size of the leaf blade does not usually result in a smaller total photosynthesizing surface (in some cases this even increases, e.g. in conifers as compared with broad-leaved trees), because small-leaved trees accordingly have more leaves.

On the whole, the increased body size observed in many phyla can be considered to be a positive taphonomic factor, whereas oligomerization and perfection of the reproductive sphere, which lead to economy of material, are negative factors. "Regressive" development of the reproductive sphere: self-pollination and apomixis, also diminish the likelihood of burial (Chigyryaeva, 1970), although some apomicts produce a fair amount of pollen.

3. CHANGES OF THE MATERIAL AS A CAUSE OF PALEOSUCCESSION

Paleosuccessions related to qualitative and quantitative changes of the material in the course of plant evolution can be qualified as **mutational** (Krasilov, 1969b). Let us list the most important morphological and functional innovations that can bring them about.

a) Secondary thickening of the stem, increased size of the plant body; large trees with secondary wood are known as early as in the Devonian.

b) Seasonal shedding of organs. Deciduous Cordaitales apparently dominated the flora of the Angara Region beginning from the Namurian. Their antipode in the southern hemisphere was the deciduous Glossopterides. The abundance of stomata on the bark of arborescent Lycopsida (Part III, section 3) and the exceptional wealth of bark remains suggest that seasonal shedding of bark took place here. The shedding of abbreviated shoots was probably a feature of Paleozoic gymnosperms, but this phenomenon became most widespread in the Mesozoic plant formations dominated by Czekanowskiales, that shed brachyblasts with leaf bundles.

c) Switchover of the dispersal function from megaspores to sporangia or to seeds. Aborted ovules without an embryo are commonly found in the Paleozoic and Mesozoic. This may be partly explained by the casting of unpollinated ovules as in gingko, but finds of pollen in pollen chambers show that this explanation by no means always applies. Apparently, as Emberger and other authors repeatedly pointed out, the Pteridospermae, Cordaitales, and some Mesozoic groups of gymnosperms scattered ovules, not seeds. This mode of reproduction was more wasteful and, consequently, favorable for burial.

d) Evolution of pollination. Remembering that the appearance of bisexual flowers is related to insect pollination, this phenomenon can be dated back to the Triassic period. In the Mesozoic we have direct proof that plants were visited for their pollen – coprolites of a small animal with pollen of *Caytonia* (Harris, 1956). However, it is not to be considered proven that zoidophily took place here: the winged pollen grains of *Caytonia* could also have been wind-dispersed. More likely is insect pollination in Mesozoic conifers with pollen belonging to the morphological group Operculati (*Classopollis* and related form genera). This pollen is often found in tetrads or conglomerated into a compact mass. Apart from this, the formation of pseudopores may be somehow related to the sucking out of the protoplast by insects. Whatever the case, the manner in which pollination was carried out did not affect pollen productivity: pollen of *Classopollis* is found very frequently. It may be that its abundance is due to the proximity of maternal plants to the place of burial.

In the opinion of some authors (Delevoryas, 1963), the flowers of Bennettitales of the phylum Cycadeoidea did not open, which inevitably led to self-pollination (autogamy). The large size of the pollen grains and stomata in some Bennettitales may point to polyplody, a common concomitant of apomixis. In fact, the amount of pollen in the spore-pollen spectra in Bennettitales as a rule does not match the content of their organs in large taphocenoses (it is much lower). Still, the reliability of these data should not be overestimated, since it is easy to confuse the pollen of Bennettitales with the single-furrowed pollen of other gymnosperms.

It is to be noted that changes of the material may be superimposed on other factors causing succession. For instance, a change in vegetation brought about by a worsening of the climate goes along with seasonal leaf fall and the transition from biotic to abiotic pollination and, in some cases, to autogamy. On the other hand, an improvement of the climate results in the formation of phytocenoses with a more complicated structure. Under the canopy of a tropical forest the air is almost motionless and wind pollination is ineffective. Moreover, there are always enough animals here to act as pollinators.

Apart from mutational paleosuccessions, there are paleosuccessions (much rarer than Heer (1859) thought) related to seasonal changes of the material (phenological: Krasilov, 1969b). Seasonal lamination is encountered most often in lacustrine sediments. We may cite as an example the well-studied sediments of the Oligocene lake Florissant (McLeroy and Anderson, 1966), where individual members consist of alternating layers of diatomite 0.01–2 mm thick without large plant remains, sapropelite (0.05–2.5 mm) with leaves and insects, and tuff (8 mm) with plants but without insects. Diatomite is a springtime formation (corresponding to the springtime maximum of diatom productivity) and sapropel a fall-winter formation. The pollen content of conifers in the former reaches 62%, in the latter 48%, while that of flowering plants is, respectively, 38% and 12%. Both types of plant flower in spring, but some species found here of the genera *Myrica* and *Ulmus* blossom in the fall.

Among the phenomena masking seasonal lamination and the phenological succession of taphocenoses that is associated with this, the most important are the shifting of pollen in fresh sediment by mud eaters and the mixing of sediment by waves, gas bubbles, near-bottom ice, groups of animals, etc. (Nichols, 1967), and also the reflotation of pollen (see Part II, section 4). Traverse and Ginsburg (1967) noted a high concentration of pine pollen in the waters of the Bahama Bank one month after flowering; this pollen was also found near islands where the pines do not grow.

Studying the content of pollen in suspension in the Delaware delta, Groot (1966) established that the time of pollination is scarcely reflected in the quantitative participation of particular species. This is connected with the fact that the transport of pollen by the river is accompanied by repeated deposition and renewed transport and extends over a long period.

4. IN SITU BURIAL

There are two essentially different possibilities for burial in the place of growth. The first is associated with soil formation and analogous forms of phytogenic rock formation. The second takes place under conditions of a catastrophically rapid accumulation of clastic sediments.

Subterranean organs are generally preserved in the root zones of mangrove swamps, and also freshwater bogs and marshes. In the estuaries of tropical rivers tree trunks are sometimes seen floating vertically because the roots are weighed down with stones. In the large majority of cases, however, a vertical position of a trunk is a sure sign of burial in situ. Catastrophic sedimentation may be related to mudflow, volcanic eruptions, and landslides. Chaney (1938) described an area of a forest that had been buried when volcanic ash slid down the valley. Landslips of this kind are usually formed during the underground tremors that accompany an eruption. In Tertiary basalts of

Iceland vertical trunks are preserved the tissues of which are charred or replaced by chalcedony (Friedrich, 1968).

A more commonplace source of in situ burial is inundation. Floodplain trees often develop protective adaptations such as negative geotropism of the roots, or else form adventitious roots on the part of the trunk filled up with alluvium (sequoia, for example: Stone and Vasey, 1968). Thanks to these, a large tree will not die even when a large part of its trunk is submerged. Owing to the asymmetrical thickening, equilibrium is maintained when the trunk leans.

Figure 3. Strata with horsetail roots in the Jurassic of England (after Batten, 1969)

The famous burial site of Devonian Psilophytales in the Rhynie chert, where a layer 40 cm thick contains countless remains of these very ancient terrestrial plants, is considered to be a fossil peat bog. In Devonian delta sediments of the Appalachians (Barrel, 1913) stems of plants that apparently inhabited mangrove swamps have been found preserved in the vertical position, broader toward the base. Quite frequently found in sections of Late Paleozoic coal-bearing strata are fossil soils and peat with stigmariae, roots of *Calamites, Pecopteris* (Remy, 1952), and Cordaitales (Cridland, 1962), and in countries of Gondwana with *Vertebraria* (Schopf, 1970). For instance, in the Döhlen basin beds of coal are usually overlain by coarse arkoses, above which occur fine sandstone and siltstone with vertical stems, and above these soil with roots of *Calamites.* The soil precedes a layer of burnt shales and a new coal bed. Burnt rocks sometimes pass into fine green clay with roots of *Pecopteris* and remains of *Pantelosaurus* (Reichel, 1968).

In the European Carboniferous autochthonous stems in places form true "fossil forests" (Klusemann and Teichmüller, 1954).

Bog soils with roots of Equisetaceae are very common in Mesozoic sediments. Also frequent are massive burials of vertical stems of horsetails (Figure 3) (what are known as "*Equisetites* beds": Lock, 1953; Batten, 1969, and many other publications). Horsetail thickets were as a rule confined to littoral bogs and marshes. Autochthonous beds of them mark out the position of the shoreline.

Six horizons of fossil soils were described in detail by Ortlam (1967) from the Triassic of the Black Forest. Similar horizons with roots of ferns and horsetails and with remains of vertebrates occur in a section of variegated sandstone in the Vosges (Gall and Grauvogel, 1967). In fossil soils of a certain locality of Rhaetic-Liassic plants in the vicinity of Beirut fern roots are preserved that in some cases have been successfully matched up with leaf remains (Weber, 1969). Similar burials in fossil soils have been described from the Rhaetic-Liassic of Sweden (Vossmerbäumer, 1969).

The question of how the famous "petrified forest" of Arizona came to be buried provoked a long controversy. It was thought that the countless stems of *Araucarioxylon* that make up most of the forest are in an allochthonous bed. However, Daugherty (1941) described trunks in which roots that could be traced for more than 3 m were preserved. Ash (1970) described an almost vertical trunk of *Araucarioxylon* about 1 m in diameter with roots embedded several feet. He considered that the tree had been buried in situ, although the trunk might have been slightly displaced as a result of erosion. An accumulation of detritus was observed near the base of the trunk in a layer of carbonaceous sandstones and siltstones. This layer is hypothetically considered to be a fossil soil. Preserved here are twigs of *Pagiophyllum* that possibly belong to the same plant as the stems.

In the USSR fossil Mesozoic forests consisting in the main of conifers have been described from Central Asia (Sikstel' and Khudaiberdyev, 1968). In Middle Jurassic sediments of the Yagnob river valley six horizons with vertical trunks have been discovered that were probably submerged by mudflows. The tree stems are up to a meter thick in some horizons. In the upper horizon the youngest forest is preserved, in which the average stem diameter is not more than 0.25 m. Fifty trunks were counted here, spaced 1.5–5 m apart. A Cretaceous forest on the Amu-Darya consists of conifers, the stems of which are covered with ferruginous sheaths (Shul'ts, 1967). In the Seattle area in the west of the USA there is a "fossil gingko forest" preservation. Daber (1968, etc.) described interesting autochthonous beds with *Weichselia* in the Neocomian of Quedlinburg.

Fossil soils are a common study object in Quaternary geology. They contain varied information on climate and vegetation.

Autochthonous beds are most valuable for reconstructing the forms of growth of extinct plants and their associations during life. It must be borne in mind, however, that not all burials of subterranean organs can be considered autochthonous. For instance, Rupke (1969) found stigmariae to be in an allochthonous bed in some horizons of a Carboniferous section of Nova

Scotia. This is proved by the fact that their orientation coincides with the predominant direction of the currents. Some of the stigmariae are tilted and deformed. They are distributed at various levels, which with an autochthonous burial would denote several successive horizons of fossil soils. In the same section there are vertical stems of *Lepidodendron* and Calamitaceae up to several meters tall that thread through whole members of interbanded sandstones and siltstones. These stems are evidence of high sedimentation tempos. Under such conditions it is unlikely that several horizons of soils were formed (for a criticism of Rupke's hypothesis see Ferguson, 1970).

5. TRANSPORT OF MATERIAL

Only hydrophytes and plants of accumulative lowlands have some chance of being buried autochthonously. The large majority of beds are formed of plant material that was transported.

Modes of transport

Only the spermatozoids of spore plants and the lower gymnosperms have the property of active movement. All other organs move only passively. The following modes of transport are distinguished:

a) **Mechanical dispersion** realized by special adaptations which include elaters, twisted valves (in Anthoceratales), peristome (in mosses), explosive dehiscence of fruits, ejection of pollen through a pore in the anther (*Rhododendron,* Melastomaceae).

b) **Dispersal by wind.** Plant organs are carried by the wind either in a suspended state or by tumbling. In the first case the distance they are borne depends on their weight and on the ratio of surface to volume, as in mineral particles. Some measurements of ranges are available for the latter: grains 1–8 mm in size can be carried several meters, those of 0.05–0.125 mm for several kilometers, and those of 0.015 mm and less for thousands of kilometers (Shvetsov, 1958). Ridley (1930) presents data showing that quartz grains weighing 0.05 mg are carried 600 miles by the wind, those weighing 0.04 mg, 700 miles. Much heavier plant remains can probably be transported the same distance, since as a rule they have a higher surface to volume ratio than mineral particles. The speed at which pollen falls in still air averages around 40 cm/sec for heavy grains to 2 m/sec for light ones. The weight of pollen grains is directly related to their size, although there are exceptions: the pollen of *Juniperus, Taxus,* and other genera is relatively light for its size (Dyakowska, 1959). In the case of sacculate pollen the rate of deposition depends on the ratio of the size of the sacs to the body of the grain. The distance of transport is determined from Schmidt's formula

$$F = \frac{V}{C}2 \cdot \frac{A}{Q}\eta,$$

where V is the wind velocity (about 10 cm/sec), Q is the air density ($1293 \times \times 10^{-3}$), and A, C, and η are constants (see Dyakowska, 1937). The distance that more than 1% of all objects reaches is taken to be the average range limit, and here the maximum distance is usually not more than double the average. In practice, the range deviates in some way from that calculated by Schmidt's formula, depending on temperature, humidity, wind force, the type of vegetation, and on repeated transport. Faegri and Iversen assume that the distance cumuli move (50–100 km) determines the limit of dispersion of heavy pollen grains, which manage to settle during the night, whereas light grains are picked up again the next day by ascending air streams and are borne much farther. Numerous observations of spore and pollen transport (as well as of aerosols, radioactive dust, etc.) have shown that, contrary to former concepts on pollen "rain," these particles move horizontally almost all the time (Tauber, 1965; Schmidt, 1967; Manten, 1967). The following categories are distinguished: 1) transport above the forest canopy (range from several to many kilometers; the pollen falls only into large water bodies); 2) filtration between trees (the wind velocity here changes from 0.75 m/sec to 1.5 m/sec depending on the state of the foliage; range from several hundred meters for large grains to 1000 m for small ones); 3) rain wash (yields about 10% of the pollen in samples). On the average, according to Tauber, samples from lake sediments give a picture of the vegetation within the limits of several kilometers (from several meters in small lakes to several tens of kilometers in large lakes). Andersen (1967) showed that in soil samples from forested areas there is a predominance of pollen from trees growing within a radius of 20–30 m. Tauber (1937) demonstrated the significance of reflotation – rain wash and the redeposition (reworking) of pollen that settled on leaves and branches (Table 2).

The small percentage of pine and oak pollen that can be refloated shows that reflotation is of importance mainly for trees that grow close together; it therefore increases the content of local pollen. The reflotation maximum is observed in November, when the leaves filter pollen more easily. Some idea of the possibilities of reflotation is given by the following figure: 40,000 pollen grains were counted on one willow branch, and all the local species were represented. Obviously, such factors as canopy height and the distance between trunks considerably influence the ratio between local and regional and also large and small pollen (small pollen filtrates more easily).

Table 2. Changes in pollen content during the year (after Tauber, 1967)

Genus	% of grains in March–August	% of grains in August–November (reflotation)	% of total content	% of pollen on account of reflotation
Fagus	2.1	0.9	1.7	18
Quercus	6.0	0.7	4.2	6
Betula	45.5	93.1	61.9	52
Picea	0.3	0.1	0.2	17
Pinus	22.2	0.7	14.8	2

Tauber (1965) advanced an interesting hypothesis explaining the reduction in the amount of elm pollen at the boundary between the Atlantic and Subboreal periods of the Holocene in Europe (this phenomenon, which is observed everywhere, is usually linked to a cooling of the climate or to human activity): the somewhat reduced atmospheric precipitation at this time caused the marginal areas of water bodies to dry up, and in their place appeared dense thickets of alder and willow that hindered the transport of large pollen of *Ulmus* (and also of birch and linden). As a result, the false notion grew up about the degradation of the broadleaf forest.

All authors without exception mention an increased content of regional pollen on forest-free areas and in clearings. The significance of long-range transport markedly increases over open spaces of tundra and ocean. The dispersal limit of 50–100 km usually quoted in textbooks on palynology (Erdtman, 1943) may here be greatly surpassed. A large amount of exotic pollen was discovered in atmospheric traps above arctic Canada (Ritchie and Lichti-Fedorovich, 1967). The pollen of broad-leaved trees (oak, elm, *Pterocarya*) is carried a distance of 1500–2000 km. At a height of 3300 m in the Alps, pollen of African fir was observed, borne along by air currents (Bortenschlagen, 1965). The possibility of far-flung transport over forest-free spaces must be reckoned with when interpreting Pleistocene spectra. Ice-covered level surfaces are particularly favorable in this respect. In winter, ice connects up areas of arctic land, permitting the broad scattering of disseminules. This explains some geographical features of plants from the high latitudes. Remains carried to the ice surface from surrounding territories entered glacial sediments during the thaw. This is why a mixture of pollen from different biotopes is often observed in moraine (Grave et al., 1969).

If we climb up above a seaside town, we sense the smell of the sea more strongly. In the same manner, the significance of regional pollen, usually masked by local influences, is progressively augmented with increasing distance from the shore (Chaloner, 1969a). In atmospheric samples pollen of birch and pine, main constituents of forests in the North Atlantic region, was found about 1000 km from the shore. There are a good number of examples of transport on such a scale. In the Pacific Ocean alder and pine pollen has been noted 7000–10,000 km from the nearest source (Koreneva, 1964). In general, however, deepwater muds in the central parts of oceanic depressions hardly contain any pollen.

Meteorological survey data give a more detailed picture of wind dispersal of spores and pollen. Hirst et al. (1967) established that long-range transport of spores and pollen over the sea is a common phenomenon in temperate latitudes. They studied the vertical section of a spore cloud over the North Sea, showing on diagrams the height, distance from shore (British Isles), and lines of uniform spore concentration ("isospores"). In good weather the maximum concentration of pollen and spores was noted hundreds of miles from shore. At the maximum distance the lower strata of the spore cloud were "eroded," and the majority of the grains were situated at a height of 500–1500 m. Two types of grains were identified: those dispersed primarily during the day and those dispersed primarily at night. The first group was tagged according to pollen (mainly of grasses) and conidia of *Cladosporium* and the second group according to different ascospores, basidiospores, etc. It was found that each type forms its own cloud. In a "daytime" cloud the pollen and conidia are at the same distance from shore but at different heights, the pollen being almost 500 m lower down (it settles more rapidly). "Daytime" and "nighttime" clouds alternated along the path of the airplane, which intersected the production of several diurnal cycles.

A rough idea of the range of wind transport is obtained from the distribution of spore plants on oceanic islands (in cases where other modes of dispersal, e.g. by means of continental bridges or drift, are unlikely). Bryophyta with very small spores are among the first colonizers (in spite of the a priori notions held by some bryologists who refuse to accept the possibility of their being dispersed via the sea) and are known on all oceanic islands. Some species inhabiting St. Helena Island have a South American origin. According to Ridley, the number of species of ferns on such islands is as follows: the Azores 30, St. Helena Island 25, the Bermudas 22, Tristan da Cunha 24, Norfolk Island 40, the Chatham Islands 40, Juan Fernández Island 44, the Galapagos 75, and on the Hawaiian Islands 128. Such Hawaiian species as *Schizacea robusta, Asplenium fragile,* and *A. aspidioides* originate from South America. Spores of *Psilotum* are also successfully dispersed by the wind: they are found on islands as far from the continent as the Galapagos, Bermudas, Pitcairn, Norfolk, and others. The rapid colonization of new islands or islands laid waste by volcanic eruptions proves that a considerable number of spores constantly come to rest a very long way from their source.

For example, three years after an eruption on Krakatau, 12 species of ferns were found there that had been carried at least 25–30 miles (Seward, 1910).

Very small and light seeds are dispersed just as far by the wind as are spores. For instance, the seeds of Orchidaceae are transported from South America to the Galapagos and Trinidad and from Africa to the Azores. Larger seeds and fruits are borne along due to their having special adaptations. In Ridley's excellent summary on this topic it is mentioned that winged seeds and fruits are usually carried more than 30 m in a strong but not hurricane-force wind, while the range in *Pinus silvestris* is up to 880 m. Pubescent seeds and fruits can soar in the air and, caught in a gust of wind, are borne large distances. Data on the dispersal of such plants on islands show that they overcome water barriers about 500 km wide or even much more extensive: the penetration of the North American species *Solidago sempervirens* to the Azores and of African Compositae to St. Helena Island are examples of very long-range wind transport.

It must be taken into consideration that the range of transport on land is not determined by the distance traveled during a single transit, since, having landed on the ground, seeds may be picked up by successive gusts of wind. It is therefore of great importance whether they lie flat on the ground, like the seeds of pine, or whether they are raised on various outgrowths. For other large organs wind dispersal apparently does not have great significance.

c) **River transport.** River currents carry along masses of remains of shore plants. Flood waters and surface runoff collect plant material from a very extensive area. Surface runoff as a rule shifts only fine detritus, but during tropical rains it can displace large objects as well. A large amount of plant material is carried down slopes into the valley with landslides. In the rainy season in the tropics whole areas of a mountain rain forest are moved by huge landslides several square kilometers in area. Such landslides are especially frequent in areas subject to earthquakes and volcanic activity. They promote not only the transport but also the rapid burial of plant remains. The following figures give an idea of the amount of various plant organs borne along by rivers: the number of pollen grains per gram of sediment in river mouths usually varies between several thousand to several tens of thousands (Cross et al., 1966; Groot and Groot, 1966; Muller, 1959). Calculations of the number of seeds transported by North American rivers through irrigation canals about 4 m wide in the course of twenty-four hours yielded figures from 4,000,000 to 9,000,000 (Ridley, 1930). The concentration of wood debris carried by rivers in the muddy sediments of the shelf of California reaches 53 g per liter (Uchupi and Jones, 1967). Masses of various detritus and whole trees swept away by the current float in the estuaries of tropical rivers. They sometimes take root on offshore islands. Aquatic plants borne along by tropical rivers form dense masses, or floating islands ("sudd"). In African rivers "islands" of this type consist of *Azolla, Cyperus, Trapa, Utricularia, Pistia, Vossia,* and other forms. For many aquatic plants, transport by rivers and flood waters is the principal mode of dispersal. Stolons of *Trapa* with leaf rosettes resume growth after falling into still waters.

The distance which rivers can transport plant organs depends both on the velocity of the current and on the buoyancy of the organs. The current velocity of plain rivers is about 0.20–0.50 m/sec, less often 1–2 m/sec. Mountain rivers have a current velocity up to 5–10 m/sec. The buoyancy of spores and pollen is apparently determined by the same factors as volatility: pine pollen holds first place with either mode of transport (Pennington, 1947). Microscopic remains are carried many hundreds of kilometers (up to 700 km: Fedorova, 1952) by rivers. It is clear that rivers and surface runoff play a very important role in the scattering of disseminules. But organs dispersed by this means, unlike seeds and fruits that are transportable by sea currents, usually do not have special adaptations enhancing buoyancy (with the exception of the Paleozoic megaspores: Potonié, 1953). Inhibiting adaptations are evidently more important to them: a ribbed or tuberous surface cover is typical for seeds borne by surface runoff (Ridley, 1930). According to Ridley, about 60% of the fruits and seeds of plants from the British Isles sink immediately or float for at most a few days, 25% can keep afloat for a month or more, and 15% for more than 6 months. Leaves as a rule cannot float for a long time, but if attached to a stem, they are carried far: the branches of most trees float for at least two weeks. Odzaki (1969) mentions that leaves of a beech growing at an altitude of 650–800 m are transported by rivers 600 m along the vertical to the foot of slopes, where they may become mixed in beds with leaves of lowland plants.

Axelrod (1938) reports finds of cones belonging to *Pinus sabiniana* some 25 miles downstream from the place of growth. Thus, the cones of this species may join up with remains of littoral species of pine. For the strobili and fruits transport is realized not only in the suspended state but also by means of tumbling along the bottom.

d) **Transport by currents.** River-transportable plant remains are found in marine sediments far from shore. For example, river-borne pollen is still observed some 80 km from the mouth of the Volga (Mal'gina and Maev, 1966). In a Japanese deep glacial trough the bathyscaph "Archimedes" found pollen of broad-leaved trees at depths of 4500 and 9200 m (Boulouard and Delauze, 1966). Twenty miles from the Hawaiian Islands the ocean floor is strewn all over with stems of sugar cane at a depth of up to 12,000 feet. Among the mechanisms responsible for such finds the leading role is played by currents and suspension flows. These arise when a large volume of sediment is deposited on a continental slope (often in deltas) and are particularly common in seismically active zones.

The effectiveness of transport by currents can be assessed from finds of remains of exotic plants on shores and also from the dispersion of species borne by currents. The Gulf Stream, which moves at a rate of about 9 km per hour, carries a mass of fruits and seeds to the coast of Florida (Larden, 1968) and farther, into the North Atlantic. Found on the beaches of Iceland are gigantic fruits of the tropical bean *Entada,* that grows at the edge of mangroves and along river banks (Ridley, 1930).

Almost every tropical family of dicotyledons includes riparian and littoral species that are apt to be transported by currents. Ranking first among these are the famous mangroves (Rhizophoraceae, Avicenniaceae, Combretaceae, some Anonaceae, Apocynaceae, and others), and among the monocotyledons the mangrove palm *Nipa,* the coconut palm *Cocos nucifera,* and some species of *Pandanus* that reproduce vegetatively, by means of shoots dispersable by currents. For us it is especially important that such a mode of dispersal is also present in gymnosperms: for its wide distribution on sandy and rocky shores of Japan, the Malay Peninsula, New Guinea, Australia, Polynesia, East Africa, and many oceanic islands, *Cycas* is obliged to the fact that its seeds are carried by currents and remain viable after a long stay in seawater. *Cycas* was one of the first woody plants to settle on the island of Krakatau after the volcanic eruption in 1883 which wiped out the vegetation. Let us note that the seeds of *Gingko* are also able to withstand quite a long sea voyage: in an experiment, out of twenty seeds six germinated after a forty-day period in saline water (Ridley, 1930).

Fruits and seeds of plants that are carried by currents often have adaptations that raise their resistance to seawater and enhance buoyancy: the air space between the embryo and seed coat in Leguminosae, the floatlike fruits of *Scaevola sericea,* the capsules of cotton, the polished surface of coconuts that safeguards them from corrosion, and so on (see Carlquist, 1965). The vegetation of low oceanic islands consists in the main of plants whose disseminules are borne along by currents. A study of the dispersion of cotton, coconut palm, and sugar cane yields many examples of long-distance transport by currents. A coconut will keep in seawater for more than 110 days, and during this time can cover 3000 miles (Purseglove, 1968). At any rate, the coconut palms growing on islands 700 miles southeast of Java arrived here via the sea. Also found here are stems of sugar cane brought from Java. The penetration of cotton to Pacific islands is associated with dispersal by currents (Stephens, 1966). It is thought that *Anona* reached Africa from South America via the Equatorial Current – a journey that must have taken 2–3 months (Ridley, 1930).

Organs that do not possess a high degree of buoyancy can be carried by currents far from their native shores bolstered by such "floating aids" as tree trunks, floating islands on which palms sometimes grow (Carlquist, 1965), icebergs with plant detritus that has landed on them, and pieces of pumice with small seeds trapped in the pores.

Long-distance transport by currents raises two main questions in paleoecology: about the mixture of different floristic elements and concerning the position of the shoreline. Finds of *Nipa* fruits in the Eocene of the British Isles were formerly tentatively attributed to the action of drift currents from the Equatorial Region. However, Chandler established that *Nipa* was not an alien element in the flora of the London Clay formation and that the localities where its fruits are discovered accurately contour the Eocene shoreline. Many places of occurrence of land plants are known that are situated far from the present shoreline, apparently indicating that this has moved. Of course, due to currents and suspension streams, large aggregations of fine plant detritus may be formed far from the shore. For instance, in one sample of recent bottom sediments taken 180 miles off the east coast of the USA, the concentration of pollen reached 1900 grains per gram (Stanley, 1965). This is probably the maximum amount for sediments lying outside shelf limits. Habib mentions that in deepwater sediments not only the concentration of the pollen but also its diversity diminish. Yet in deepwater samples from Cretaceous sediments north of the Bahamas he discovered a concentration of 42,000 to 42,300 grains per gram, which approximately equals the concentrations in same-aged continental deposits of the Atlantic shore of the USA. The considerable species diversity also confirms the proximity of the shoreline (Habib, 1968).

e) **Transport by animals.** Animals transport plant organs that happen to attach themselves to their bodies and are used to feed the young, and also in the form of excreted food remains. Pollen grains or entire pollinia of zoogamous plants stick to the body of pollinators, the most important of which are beetles, ants, bees, birds, and bats. The bee *Bombus* carries up to 10 mg (about 100,000 grains: Faegri and Pijl, 1966) in corbiculae (pollen baskets) specially adapted for the purpose. This material has naturally been lost to the geological record. It is now known that the seeds of many plants (*Celtis, Convulvulus, Malva, Rhus,* etc.) may be preserved in the stomachs of birds for up to half a month (Proctor, 1968). Seeds more than 1 mm in size with a hard membrane are regurgitated by birds (*Charadrus vociferis, Erola minuta*) only 100 hours later. During this time the birds may cover a considerable distance, but it has been noted that before taking flight, they clean the digestive tract. Seeds and fruits sticking to animal feathers or fur can endure long transport. The vegetation of many oceanic islands consists chiefly of species dispersed by birds, and their seeds and fruits have special adaptations for being so transported. Taylor (1954) described an interesting case of fruits and seeds being carried along stuck to the feet of albatrosses. Like many other sea birds, albatrosses in building their nests regurgitate an oily fluid that turns into cement when it dries. Trapped in this fluid and adhering to the feet of the birds, the seeds are held fast when the birds are in flight or even when they dive into water, but on land the cement crumbles and they fall off. Considering the size of colonies of marine birds, we may assume such a mode of transport to have played an important part in the formation of beds of plant remains.

In some plant formations almost all the dominant species are zoochorous. For example, African bushland plants have seeds with different kinds of holdfast devices. Up to 810 seeds and fruits have been counted on hares living here, and the animals clean themselves at least once a day. Some seeds cling too firmly to hare fur and are probably adapted to being transported by the more fleet-of-foot antelope (Agnew and Flux, 1970).

6. CHANGES IN ADAPTATIONS TO TRANSPORT IN THE COURSE OF PLANT EVOLUTION

The factors that affect the amount of material (Part II, section 3) also alter the conditions of transport. An increase in the size of a plant has a positive significance in this respect, because organs falling from a considerable height are carried far by the wind. It is not by chance that tall trees are usually anemochorous. Even in the tropics, where there would seem to be no lack of living agents of dispersal, the largest trees – Dipterocarpaceae – have winged fruits, *Dyera* and *Koompassia* winged seeds, and *Scaphium affine* boat-shaped carpels up to 15 cm long (Poore, 1968). Andersen (1967) notes that horizontal transport of pollen in a mixed deciduous forest in many cases does not exceed 20–30 m, i.e. more or less equivalent to the height of the trees (around 20 m). The seasonal shedding of assimilating organs can evidently be either positive or negative with regard to transport. If the leaves fall in the dry season of the year, there is no possibility of their being borne away by surface runoff.

Biotic pollination may be considered a negative taphonomic factor both from the point of view of the amount of material (Part II, section 2) and in respect of the conditions of transport. True, an increase in the amount of pollen in cases where the pollen itself serves as the means of attraction promotes anemophily, and some of it is in fact apparently carried away by the wind. The pollen grains of anemophilous species generally have a smooth, dry surface and are typically between 20 and 30 μ in size. Larger grains are often equipped with adaptive devices enhancing volatility. The large pollen of pines (50–150 μ, weighing 30–300 $\times 10^{-6}$ mg) is extremely transportable owing to the presence of air sacs. The size of pollen grains of entomophilous plants varies between 10 and 300 μ. Their surface is as a rule sculptured and sticky.

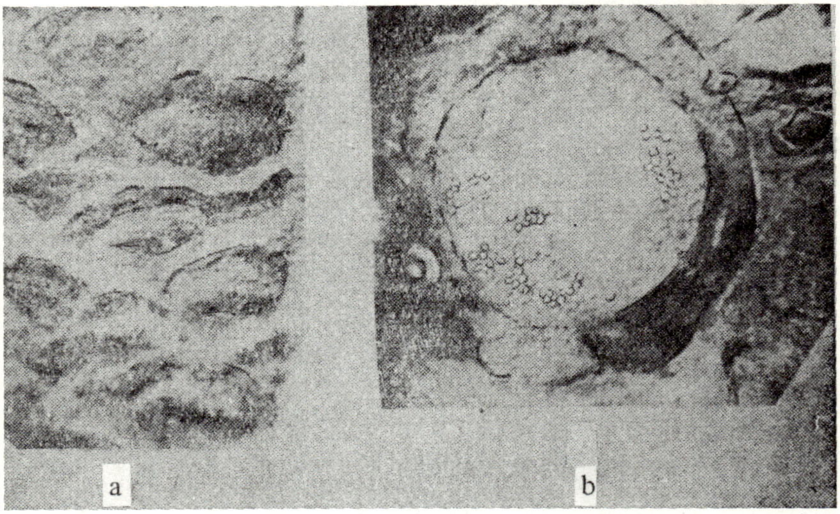

Figure 4. Dispersal of sporophylls in *Pleuromeia*:

a) part of strobile, $\times 3$; b) sporophyll with megasporangium among remains of marine mollusks, $\times 7$; Lower Triassic of Russkii Island, collection of Yu. D. Zakharov.

The most ancient isosporous plants were apparently dispersed by the wind. From the point of view of transport, the development of heterospory may be considered as an adaptation to hydrochory, as the megaspores of Paleozoic plants were very large (on the average about one-and-a-half times as large as recent plants, and some attained gigantic proportions, e.g. *Cystosporites giganteus* Zerndt, with a diameter up to 11 mm: Potonié, 1953). Many of them had adaptations for enhancing buoyancy (for example, the equatorial zone of the anastomosing appendages in *Triletes ramosus* Arnold, that double the surface area of the spore and retain air bubbles: Potonié, 1953) or for promoting anchorage (the anchor-shaped penetration apparatus in *Lagenicula*).

Some of the ancient Lycopodiales apparently scattered not megaspores but whole sporangia, which were transported by water. For instance, the halophyte *Pleuromeia,* which formed pure thickets along the shores of Triassic seas, had large, almost spherical megasporophylls with a smooth surface that were admirably adapted for transport by currents. A multitude of megaspore-filled sporangia has been found in beds with Ammonoidea and other marine organisms (Figure 4). The ancient ovules were also apparently too heavy to be effectively carried by wind. The angular form and ribbed surface of many Paleozoic ovules, the air chambers in the cover of *Aetheotesta elliptica,* the pubescence of *Physostoma elegans,* etc. (see Potonié, 1953) are evidence of hydrochory. Wing-shaped excrescences and pubescence increasing buoyancy can at the same time be considered as a preadaptation to anemochory and zoochory. It is interesting that the anemochorous seeds of Paleozoic plants were generally symmetrically winged (*Samaropsis, Cordaicarpus,* etc.), and the width of the wing or winglike edge was not greater than the diameter of the central body. Only the Permian seeds of *Sylvella* (Sukhov, 1969) had a unilateral wing, and it was markedly larger than the nucleus. Recent seeds are predominantly unilaterally winged.

The dispersal of symmetrically-winged seeds by the wind does not seem to be too effective. This is indicated if only by the fact that the Taxodiaceae, which have such seeds, cannot overcome water barriers (there are no sequoias on the islands situated close to the coast of California). In beds seeds of Taxodiaceae are as a rule encountered together with shoots and cones, which points to an absence of differentiation in terms of transportability. The unilaterally-winged seeds of Pinaceae are much more often found isolated from other organs, among remains that have been transported a good distance. Nevertheless, some Pinaceae with winged seeds scatter the cone scales with the seeds. The purpose the wing serves is not clear in this case. The scattering of cone scales (elementary strobili) was a feature of some Paleozoic conifers. This phenomenon is most clearly expressed in the Araucariaceae. Their winged scales (*Araucarites*) are found more plentifully in Mesozoic beds than any other organ of the same category.

The most ancient parachute mechanisms of dispersal belonged to the Late Jurassic seeds of *Problematospermum* (Turutanova-Ketova, 1930). In general, though, such objects are rarely found in the fossil state, since they are much more easily transported than sedimentation particles apt to be buried, and they are borne far beyond the zone of active sediment accumulation.

Zoochory apparently achieved wide distribution only with the appearance of birds and mammals. It is known, though (Ridley, 1930), that fish eat the fruits of *Pandanus,* and crocodiles and iguanas may play some role in spreading the "alligator apple" (*Anona palustris*). The Permian rhynchocephalians ate the strobili of conifers (Weigelt, 1930), but it is not known whether zoochory actually took place here. Conifers with arils appeared in the Jurassic period (*Palaeotaxus*: Florin, 1958), possibly in connection with the spread of ancient reptiles and archeopterices.

Seeds with a hair cover are found from the Devonian (*Archaeosperma*), but this feature is by no means always related to zoochory. Seward (1933) assumed that the hairs of the Carboniferous *Thysanosperma* were true glochids. This is probably the way to interpret the outgrowths of the Triassic *Carpolithes chinleana* (Daugherty, 1941) and the Late Jurassic *Tyrmocarpus*, which I described from the Tyrma River (Figure 14). It is not known when a hair cover appeared in mammalian reptiles or the primitive mammals. At any rate, fleas that lived in the fur of animals have been described from the Lower Cretaceous (Riek, 1970).

The fruits of angiosperms far surpass the analogous structures of gymnosperms in terms of diversity and degree of adaptation to transport. Often participating in the formation of the flight apparatus here are gynoecial organs (styles), perianth, extrafloral bracts, etc. (see Ridley, 1930). Originally, however, the fruit arose as an adaptation to zoochory. The widespread occurrence of inedible seeds inside edible fruits determined a trend toward more numerous and smaller seeds. The seeds of many angiosperms (Orchidaceae, Gesneriaceae, etc.) are much smaller than the seeds of gymnosperms. True, a tendency toward a reduced size of seeds is also manifested in the Mesozoic Caytoniales and Czekanowskiales with polyspermous capsules. I have found fully formed ovules less than 0.25 mm in size in Upper Jurassic deposits (Krasilov, 1971b). At the same time, a reduced size and weight of seeds may be regarded as a preadaptation to anemochory. In fact, the anemochorous seeds of angiosperms (in contrast to gymnosperms) often have no adaptations of any kind to wind transport apart from their exceptionally small size.

7. CHANGES IN THE CONDITIONS OF TRANSPORT AS A CAUSE OF PALEOSUCCESSION

The appearance in taphocenoses of organs which became increasingly transportable in the course of evolution (for instance, the winged cone scales of conifers) or the disappearance of organs due to the acquisition of taphonomically negative adaptations (zoochory, zoidophily) can be considered mutational paleosuccession (see Part II, section 3). Taphogenic paleosuccessions (taphosuccessions) brought about by a change in the conditions of transport are encountered much more frequently in actual sections (Krasilov, 1969b). Weber was one of the first to draw attention to this circumstance, noting that heavy objects (stems, cones) brought by water, predominate in some beds and light ones (leaves), carried by wind, in others (Weber, 1852;

of course, the burial of leaves is certainly not always related to wind transport). Reid and Chandler (1926), who always laid the accent on adaptations to transport when analyzing taphocenoses, showed that anemochorous organs are most often found in sediments of small lakes and lagoons. This observation is borne out by data on taphocenoses of the most different age. One of the richest beds of anemochorous organs was discovered in sediments of the Late Jurassic lake Karatau (Turutanova-Ketova, 1950). The part played by the wind is especially important in the burial of spores and pollen of anemophilous plants.

Koreneva (1964) mentions that samples from deepwater muds east of New Zealand proved to be more saturated than muds lying to the west, this being due to the prevalence of westerly winds. Changes in the concentration of pollen related to the prevailing wind direction are also observed in lake sediments. It is clear that under these conditions a change in the force or direction of the wind may result in a succession of spore-pollen assemblages in a section. However, prevailing wind directions, conditioned by general regularities of the planetary system of atmospheric circulation, are extremely stable. After a study of the effect of meteorological conditions on the distribution of pollen in arctic Canada, Terasmae (1967) believes that the prevailing direction of winds has not altered here for at least 2000 years.

The nature of vegetation greatly influences the wind transport of pollen. We have already pointed out (Part II, section 5) that the significance of such factors as canopy height and the distance between tree trunks has only recently become clear following the changed concept of "pollen rain" (Manten, 1967). The lowered content of tree pollen (of *Ulmus* and others) in the spore-pollen spectra at the boundary between the Atlantic and Subboreal periods is ascribed by Tauber (1965) to the fact that the marginal zones of water bodies dried up and belts of willow and alder formed here, hindering the transport of tree pollen.

On the other hand, the nature of the vegetation is to no small extent determined by the wind (here we are speaking only of the direct action of the wind, not of its influence on the amount of atmospheric precipitation and other ecological factors). The distribution of anemophilous species displays a definite connection with wind force: on the open sandy shores and islands of the North Sea such species constitute about 47% of the total, whereas in the inland regions of Central Europe they account for 21% (Faegri and Pijl, 1966). Large populations are favorable for anemophily. In a tropical forest, where the specimens of a particular species generally grow at a good distance from each other, anemochorous plants form more distinct groups (Poore, 1968). These patterns are, of course, reflected in the composition of taphocenoses. In the case of river transport, a change in the drainage system may lead to a paleosuccession that in some cases simulates a change of vegetation (Davis, 1967). An analysis of spore-pollen spectra from delta sediments of the Nile showed (Soad and Sami, 1967) that their composition (the content of pollen of species growing outside the bounds of the delta) depends on the fluctuations of the river flow and on the annexation of new channels to its drainage basin.

It should be pointed out that since transport is realized by agents that are closely bound up with the climate (wind, surface runoff, rivers), the effect of a change in the conditions of transport is superimposed on climatogenic succession.

With combined river and wind transport we observe interesting patterns in the distribution of plant detritus. For instance, in the sediments of Lake Kinneret (the Sea of Galilee) pollen of plants belonging to maquis and plain groupings (batha), brought by the wind, is concentrated some distance from the shore, whereas the pollen of hydrophilous species and spores of *Thelypteris,* which arrive with the Jordan River, settle immediately they fall into the lake. The pollen of ruderal species also collects near the shore (Horowitz, 1969).

The differentiation of plant material in the case of transport by inshore currents and the associated alternation of assemblages in a section are particularly well illustrated in the celebrated flora of the London Clay formation. Taphocenoses that have undergone considerable transport consist mainly of fruits of mangrove and marsh plants: *Nipa,* various Anonaceae, Apocynaceae, Icacynaceae, and others (Chandler, 1961). The distribution of pollen in marine sediments also depends on currents. Highly transportable pollen (*Pinus*) is carried beyond the zone of psammitic sediments and concentrates in thinner muds (Traverse and Ginsburg, 1966).

8. FOSSILIZATION

Plant remains undergo microbiological decomposition and are transformed into humus (they are humified). Kononova (1970) states that any component of plant tissue may serve as a prime source of the structural units of humus in the form of breakdown products (phenol compounds), metabolism (metabolites forming upon the utilization of carbohydrates by microorganisms), and also decay and resynthesis (amino acids and peptides). Humic substances are formed as a result of condensation and polymerization of the structural units. About one fifth of the humus formed from the annual deposit of vegetable matter is preserved in the soil (the coefficient of humification is

0.3 according to experimental data). Val'ts (1968) observed that the parenchymous tissues of herbaceous plants decompose rapidly and pass into structureless light-orange substance. The leaves of woody plants become humified just as readily. Wood passes into attrital substance of a reddish brown color with signs of jellification. The wood of conifers is preserved longer owing to the large amount of resin and dense structure. The cork layer is preserved longer than the other stem tissues. The attritus, jellified tissues, and the coagulation products of the humic acids (dopplerite substance) form the bulk of coal. Depending on conditions, woody remains may give rise to various components of coal (Spackman and Barghoorn, 1966).

Another form of transformation of plant matter is fusainization, which is especially characteristic for Paleozoic coals. Fusainization may take place very rapidly, as during a forest fire, or slowly, when oxidation occurs under anaerobic conditions. The results are practically identical in both cases.

Remy (1954), having studied "leaf fusite" (Laubfuzit) from leaves of pteridosperms with very prominent veins, concluded that such a form of preservation arises when fallen leaves are burned. Charred masses of fallen debris in a present-day forest strongly resemble Paleozoic leaf fusite. Teichmüller (1952) also connects the formation of fusite with fires, but at the same time he reports find of bark of *Lyginopteris* where the parenchyma was preserved in the form of vitrain while the sclerenchymous tissues were fusained. Obviously, in this case fire is not a suitable explanation.

As rock becomes more compact, organic remains are compressed. In Paleozoic coals plant remains are compressed about 7 times (in comparison with the situation of identical remains in concretions: Teichmüller, 1952), while in durain coals they are compressed 12–15 times (Robertson, 1952). Compression in clastic sediments is illustrated by an interesting find of the flattened body of an ichthyosaur 2 cm thick, the original thickness being about 40 cm (Kräusel, 1968).

In sediment biochemical decomposition takes place under conditions of high temperature and pressure. This causes carbon enrichment on account of a loss of hydrogen, oxygen, and nitrogen (coalification). If the process advances considerably, only the most stable components – exine, cuticle, etc. – are preserved. However, even cell nuclei are preserved if burial occurs quickly. Remains of plants in bituminous Eocene deposits of the Green River formation are distinguished by an exceptionally good state of preservation (Bradley, 1931, 1970). The leaves here are variously colored from white to dark brown, with occasional greenish ones. The use of chromatography revealed in them the pigments pheophytin A and chlorophyll A, that is, the first breakdown products of chlorophyll (Dilcher, 1967). Cell nuclei are occasionally preserved in older sediments (Eisenack, 1965). Taylor (1965) gave a detailed description of Paleozoic ovules with the gametophyte and archegonia, in which the coalified cytoplasm of the egg cell was preserved.

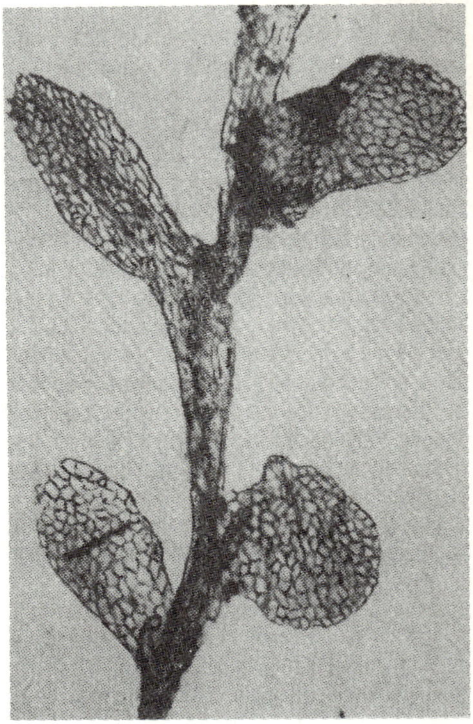

Figure 5. Preservation of soft tissues: phytoleuma of a liverwort from the Upper Jurassic of the Bureya basin, × 150, collection of the author

Coalified remains of plants (phytoleumas: Figure 5) may be more or less strongly mineralized. In the case of strong mineralization they form a sort of transition between a phytoleuma and a fossil. Fossils are formed when the plant tissues are replaced by fossilizing substances. Deposited on the surface of the plant remains, these substances form incrustations or impressions (the latter are essentially a very thin incrustation). In most cases an impression cannot be considered to be a mechanical imprint. As a rule, it differs from the surrounding rock in structure (which may be very fine even on coarse rock) and color. A "corona" is often formed around the impression, that is, a border of varying width that is of the same color but of a less intensive shade. It is not hard to understand that an "impression" arises as a result of complex chemical interaction between the buried plant and the sediment.

The most different minerals may act as the fossilizing substances, but usually these are ferric oxides, ferrosulfides, manganese oxides, silica, calcium, and dolomite. Clay minerals often replace the original incrusting mineral when weathering takes place. Khaitser (1970) described kaolinization of shells from Permian deposits. Here authigenic kaolinite replaced calcite mainly on the inside of the valve, and consequently was not formed upon weathering. Apparently, kaolinization plays an important role in the formation of plant "impressions" as well. Rarer fossilizing substances are minerals of the apatite group (phosphatized stems with the cellular structure preserved have been described from the Carboniferous of Czechoslovakia: Skoček, 1969), copper sulfides (shoots of *Ulmannia bronni* in the Zechstein show the same degree of preservation), and others.

The distribution of iron in a sediment is directly related to the presence and concentration of organic matter (Huber, 1958). In organically enriched layers of mud, reducing processes develop and free hydrogen sulfide is formed which binds iron, manganese, calcium, and other cations present in muddy waters. The result is a concentration gradient of these cations in the mud solution, and they migrate from the reduced layers without organic matter to the layers with it; the latter thus become a trap (Sevast'yanov, 1968). The fossilizing substances are deposited on the surface of the plant remain (incrustation) or replace the plant tissues (mummification). It is impossible to draw a clear-cut boundary between these processes. In a thin incrustation creating an "impression" a certain amount of powdered plant substance is preserved, which determines the coloration. Such incrustations are associated by more or less gradual transitions with true mummifications that preserve the original volume of the tissue. Mummifications are often encountered in peat-dolomites (coal balls), which are dolomitic or sideritic concretionary bodies with a diameter of up to 20 m. Coal balls are generally formed in paralic peat bogs. Iron is supplied to them together with products of erosion, while calcium and magnesium are brought from seawater (Teichmüller, 1952). According to Teichmüller, plant tissues are partly replaced by metasomatism, partly pass into a colloid state, and, upon the crystallization of fossilizing minerals, are mechanically forced out into the spaces between the crystals.

A no less important mode of mummification is the replacement of tissues by silica. A study of siliceous wood with an electron microscope gives a more precise idea of the mummification process. Eicke states that part of a fossil consists of amorphous silica, while another part is formed of crystalline silica. In this case the crystals were formed not by crystallization of amorphous SiO_2 but directly from the solution, since the lines of their coalescence are visible. There were apparently two phases to crystallization: in some samples only the cell walls had been replaced while in others the cavity was also filled. Larger crystals were deposited on smooth-walled cells. The tori were replaced by very small crystals (Eicke, 1954). Interestingly, individual silicified cells, what are known as phytolitharia (Hajós, 1968), often pass into the fossil state. Phosphatization (Skoček, 1969) and pyritization (which often combines with calcitization: Brzyski, 1969) apparently proceed in the same way as silification.

A curious case of fossilization was reported from the Hawaiian Islands: a cast of a fern frond was excellently preserved in olivine basalt; even the hairs of the sori could be distinguished (Palmer, 1947, Am. J. Sci., No. 245, p. 320). The moist frond was evidently caught up in lava that had already cooled.

9. CHANGES IN THE CONDITIONS OF FOSSILIZATION AS A CAUSE OF PALEOSUCCESSION

With animals it is usually the hard skeletal tissues that become fossilized, but it does happen that soft tissue does too (for example, in the Middle Cambrian of British Columbia or the Eocene of Geistthal: Kräusel, 1968). In the case of plants, preservation does not depend on mechanical durability. Darrah said that there is practically no tissue or organ "too delicate" to become fossilized. If a certain type of tissue has not been preserved, the explanation must be sought in the conditions of burial (these ideas were expressed in reference to the gametophyte of *Rhynia,* which was not found for a long time; it subsequently became clear that the reason was not, in fact, that the gametophytes were not durable enough, but simply that no one had been able to recognize them).

The fossilization of plants is governed by the following:

a) **Sedimentation rate.** We have already noted that under conditions of rapid burial even the most unstable components become fossilized (cell nuclei, the first breakdown products of chlorophyll, etc.). The concentration of spores and pollen in samples (the absolute number of grains) depends on the rate of sedimentation and on the degree of decomposition of peat; the latter is inversely proportional to the growth tempos of peat bogs, so that the connection is rather complex (Aario, 1932; Tsukada, 1967). Let us give one more example of the changes occurring in the composition of a taphocenosis in dependence on the promptness of burial. There are generally no sphagnums in the peats of the Atlantic period; with the warming that occurred later their remains prove to be fully decomposed. However, although a peat bog will be quickly overlain and "conserved" by a layer of sand when a transgression takes place, sphagnum remains will be preserved (Donseloar and Jonker, 1952).

b) **Climate.** The sedimentation rate and chemism of muddy waters are bound up with the climate, hence the overall influence of climate on fossilization. Its specific effect is manifested in the state (degree of decomposition) of the material prior to burial. Val'ts (1968) mentions that wood in a box grove in Khosta was found to be more markedly altered after 10–12 years of aerobic decay than wood in Tertiary lignites. Dying parts of plants rot much faster in the tropics than in temperate latitudes. Particular note is to be taken of the activity of termites, which contribute to the exceptionally rapid decomposition of wood (Hopkins, 1966). In a dry climate, on the other hand, wood rots slowly and is found buried more frequently than leaves and pollen, which soon become desiccated and therefore less liable to be fossilized. According to Brush (1966), desiccated pollen grains decompose faster than fresh ones in sediment.

c) **Organic content of sediment.** A high concentration and uneven distribution of organic matter produce intensive migration of elements in mud solutions, ensure a sufficient supply of fossilizing substances, and consequently promote burial. If there are few organic remains or the remains are scattered evenly in the sediment, the likelihood of burial diminishes, since the influx of fossilizing substances may prove inadequate. Hence, in a rich bed even sparse remains have more chance of becoming fossilized.

From the above we see that the probability of a plant's being buried depends more on chemical composition than on mechanical durability. The conditions of burial alter the content of the relatively unstable components and thereby change the composition of taphocenoses. Nikitin (1970) mentions that accumulations of plant detritus yield both dead and live (still viable) seeds (so that such an accumulation cannot, strictly speaking, be called a thanatocenosis). Live disseminules are more resistant than dead ones. In unconsolidated sediment gravitational separation takes place, and small, light seeds become concentrated in the upper layer, which is more intensively decomposed. Selective burial of heavy objects may thus occur.

Special studies have been conducted on the durability of pollen grain casings. Firbas (1949) states that the pollen grains of *Populus, Carpinus, Fraxinus, Taxus,* and *Juniperus* are not durable and readily collapse, making determination difficult. Fortunately (so he says), these trees were not important components of the Holocene forests of Europe, and selective fossilization had relatively little effect on the composition of palynocenoses. In Canada, on the other hand, *Populus* was apparently one of the chief dominants, and the lack of resistance of its pollen gives rise to serious distortions.

Von Post earlier noted that the casings of spores and pollen are best preserved in sapropelic muds (see also Berezina and Tyuremnov, 1967). In a lowland peat bog pollen grains show a markedly selective disintegration, toward a concentration of conifer pollen (Hallik, 1968). The pollen of conifers is evidently more durable in marine muds as well. According to some observations, here oak pollen suffers most from corrosion, linden pollen less so, and alder and conifer pollen still less (Schubert, 1933; Vorman, 1952). The spores of ferns are less subject to corrosion and are found in deepwater muds. Here we come up against a complex problem, one to which we will return later: are the differences in the composition of freshwater and marine formations primary or secondary? Apparently, primary differences still have the main significance, while selective fossilization only intensifies the sum effect.

The form in which remains are preserved gives us a certain amount of information on the diagenesis of the sediment. With an increased degree of diagenesis (which often corresponds to greater geological age) the content of cellulose and hemicellulose diminishes. Miocene wood of Japan does not contain lignin (Kagemori, 1969), but lignin is present in wood of the same age from the Senckenberg lignites (Morey and Morey, 1969). Also found here were structurally preserved polysaccharides forming cellulose. Preservation of these substances in a highly organized state depends on the stage of decomposition at which mineralization set in and, consequently, on the rate of diagenesis. Bacteria decomposing cellulose are a key factor in coalification.

Gijzel et al. (1967) discovered that the color of pollen under a fluorescent microscope depends on its age and changes from reddish brown in Paleozoic bisaccate forms to light blue in subfossil pine pollen. The intensity ratio between blue-green and orange-red luminescence formed the basis of a table for age determination. Val'ts (1969) showed that the optical anisotropy of cellulose depends on the degree of humification.

The biochemical evolution of plants probably lay behind the varying probability of fossilization and, accordingly, mutational paleosuccession. The most important event was the appearance of lignin (which apparently occurred independently in Bryophyta and Pteridophyta). In the cuticle of the Permian conifer *Voltzia* n-octosane was found which is not a component of the cuticular wax of recent conifers (Knoche et al., 1968). In a study of organic compounds in the stems of Paleozoic *Calamites* in comparison with those in recent *Equisetum* it was found that they have a different galactose/glucose ratio. The importance of cellulose in the phylogeny of the Arthrophyta has apparently increased (Swain et al., 1969). There are still few data of this nature.

To the above it should be added that the activity of benthos has a marked influence on the mineralization of bottom sediments and on the preservation of small plant remains. Because the benthos mixes up the sediment, the age of pollen grains fossilized together may differ by a hundred years and more (R. B. Davis, 1967).

The changes occurring in the chemical composition of microfossils give evidence of the conditions of metamorphism of the rock (Correia, 1968; Burman, 1969). Brooks and Shaw (1971) showed that the exine is gradually carbonized at temperatures from 100° to 250°. More intensive carbonization and changes of sculpture proceed on heating above 300°. At a temperature of 350–450° a spore casing rapidly loses its form and sculpture and becomes amorphous. It is to be noted that determinable plant remains are sometimes preserved even in graphite (Hamilton et al., 1970).

10. REWORKING AND DOWNWASH

Remains that we usually call impressions are not apt to be redeposited (reworked) (they may be redeposited in debris of a plant-bearing layer, but such cases are extremely rare). On the other hand, petrifactions and phytoleumas, especially small phytoleumas – spores, scraps of cuticle – are quite often reworked. Two variants of reworking are to be distinguished: a) reworking in as yet unconsolidated sediment; b) reworking as a result of scouring of consolidated rock. Migration in a fluid sediment activated by the organs' own weight is of definite significance for the burial of fruits and seeds (carpocenoses). Nikitin (1969) noted that small seeds are concentrated in upper layers of formations, which may be the cause of selective burial of heavy seeds. For pollen, gravitational migration is usually of minor importance, although in peat heavily intergrown with roots it may be considerable (see Firbas, 1949). In unconsolidated muds spores and other small remains are moved by bottom currents, waves, lifting of near-bottom ice, and gas bubbles liberated upon putrefaction (Nichols, 1967). The inhabitants of a water body and animals fording it stir up the mud, but the main work is performed by mud-eaters, which are especially active in the upper 8 cm of mud. The shifting of spores by worms was observed in an experiment (Walch et al., 1970). Tagged grains in a box with soil were raised 25–55 cm during a rainy six-week period. In a control box without worms the grains, on the contrary, were pushed down under the action of the rainwater. Leaves present in the soil retarded the sinking process, and some of the grains were held up. Grains brought to the surface by the worms dried out and were blown away (see also R. B. Davis, 1967).

As a result of the combined action of all the agents displacing spores and pollen grains, the original differentiation of the latter is leveled out to a large extent. The composition of spore-pollen spectra from different parts of a water body proves to be much more uniform than might be expected on the basis of the differentiation of the shore vegetation.

Aggregations of cuticle scraps that have undergone natural maceration are formed, in the opinion of Harris (1958), when unconsolidated beds are washed out. Among the spores and pollen transported along with clastic particles, reworked ones usually constitute a negligible proportion. For instance, in the Delaware delta, with a sporopollen concentration averaging 2000 grains/gram, less than 1% comprised reworked grains (Groot and Groot, 1966). Nevertheless, in some situations the number of reworked microfossils markedly increases. Thus, North Sea muds show a high content of reworked pollen, brought here upon the erosion of Holocene peat bogs (Firbas, 1949). A large number of reworked palynomorphs is observed in the Triassic and Jurassic of the Donets Basin (from the Carboniferous: Gluzbar, 1968), in the Miocene of Europe (from the Triassic: Kedves and Karácsonyi, 1967), in the Paleocene of Alabama (McLean, 1968), etc. Clarke (1970) noted that the content of reworked Mesozoic pollen of *Classopollis* reached 30% in Quaternary deposits in Devon. The source of this pollen was apparently submarine outcrops of Jurassic deposits, eroded by currents when the sea level rose in the Holocene.

Reworked remains have sometimes led to taxonomic errors. For example, large megaspores are quite frequently found in the Eocene London Clay formation. Potonié surmised that they might have been reworked, but Reid and Chandler were certain that the megaspores belonged to some Eocene Lycopodiales. The question remained controversial for a long time, until finally, the "Eocene" megaspores were accurately identified and their Paleozoic age proved (Dijkstra, 1950). Megaspores of Carboniferous plants have been found quite often in Western Europe in various-aged deposits (Jurassic, Cretaceous, Tertiary, Quaternary). Some of them were reworked during the erosion of Carboniferous sediments or were accumulated together with pieces ("pebbles") of coal. But a major source of them was apparently downwash from coal dispersed during many years of working and transport (Dijkstra, 1950; Harris, 1958). This is one of the rare instances when downwashed remains are older than the rock in which they are found. Generally, they refer to a much later period (e.g. Mesozoic plants in fissures of Carboniferous limestones: Lewarne and Pallot, 1957). Apart from passive downwash, active penetration sometimes occurs of blue-green algae, fungal mycelia, or roots in fissures (Harris, 1958). Downwash may change the composition of a bed appreciably. Chaloner named this phenomenon the "Dimbleby effect" (Chaloner and Muir, 1968).

Reworked and downwashed spores may be distinguished according to their degree of preservation, nature of mineralization, staining with safranine, autofluorescence, etc. (Gijzel, 1967; Kedves et al., 1966; Stanley, 1966). They are often looked on as a troublesome hindrance that distorts the composition of spore-pollen spectra and leads to wrong dating (downwashed Paleozoic pollen in the Precambrian of Krivyi Rog is a known example). But they can become a study object themselves and a source of various information. For instance, with their aid we can:

a) obtain data on plants whose original beds have been wiped out by erosion;

b) determine the upper age boundary of deposits that do not contain any other fossils; for example, downwashed spores and pollen of Jurassic plants in primary kaolins of the Ukrainian Shield attest to the pre-Jurassic age of the weathering crust (El'yanov and Teteryuk, 1970);

c) establish the direction of currents and the paths of eroded sediments. Interesting studies have been performed by American palynologists on the dispersal of eroded Devonian red sediments traced by reworked spores. Reworked red sediments have been found on the continental slope from Newfoundland to the Bahamas. A high content of reworked spores is evidence of intensified erosion during glaciations (Needham et al., 1969; Stanley, 1966). In Cretaceous-Paleogene flysch in the Carpathians five zones were identified on the basis of palynomorphs from reworked fragments of Paleozoic coal. The distribution of Carboniferous outcrops during flysch formation and some paleogeomorphological features are established according to these zones (Turnau, 1970).

11. STRATA WITHOUT FOSSILS

The absence of fossil plants in continental and coastal deposits would be most simply explained by the absence of vegetation and a bare, desert landscape. This is apparently how many conceive of the conditions under which series of red beds were formed which very rarely yield either large remains or microfossils. Clapham (1970) rightly assumes that such a picture has little to do with the real situation. In red beds one often sees intercalations of gray and greenish rocks with a wealth of plant remains. The rich burials of dinosaurs and other large reptiles in red beds attest to conditions quite other than desert ones. On the other hand, spores, pollen, bits of cuticle, and wood often fall even into deposits of rock salt (Rudau, 1966), which is considered (not always correctly, however) an indicator of an arid climate. The absence of plant remains can probably be attributed to the following: rapid decomposition of dying parts of plants; even distribution of organic matter within the sediment, unfavorable for redistribution and concentration of fossilizing materials; formation of the sediment in an oxidizing environment, above the groundwater table. It is the last factor that seems to play the chief role. In formations such as weathering crusts plant remains are as a rule absent. Fragments of sideritized tissue are very rarely found in bauxites (Kaye, 1962); if they are, it is assumed that they were formed below the groundwater table.

Red beds are thought to have a various origin: redeposition of laterites or primary red sediments, depositional or post-depositional formation of red pigment (Van Houten, 1964; Schlegelmilch, 1968). A lateritic profile arises in a humid climate (at least 800 mm rainfall a year according to Brooks, 1949) with a definite combination of seasonal humidity and mean-annual temperatures. However, with erosion of laterites red soils are not formed, since the red color is masked by the abundance of gray and brown (Walker, 1967). Much has been written on the conditions of formation of red beds. The main question here is, can primary red beds be considered a sign of aridity? Sochava (1968, 1970) mentions that the ratio of ferric (comprising the red pigment) and ferrous iron depends on the Eh of the sediment during the time of formation and diagenesis. This in turn is connected with the climate and the organic content. In periods of overall aridization the illumination intensity increased, and so did phytoplankton productivity – the main resource of free oxygen. The content of free oxygen in the atmosphere was augmented, speeding up oxidative reactions. This is what governed the wide distribution of red beds, the climatic range of which was much broader in such epochs.

Sochava's interesting hypothesis stumbled up against a number of difficulties (Verzilin, 1970). A paucity of plant remains in general and an absence of phytoleumas are evidence, as we have said, of red-bed formation above the groundwater table, and consequently of a certain dryness of the climate, that was still apparently humid enough to support tree vegetation. Direct indications to this effect are given by the Triassic petrified forest of Arizona. The red calcareous sediments of the Chinle formation with bentonites and gypsum contain a tremendous number of *Araucarioxylon* stems. It was assumed that the stems were allochthonous, but at least in some of them roots were preserved which went down 10 feet into fossil soil with small plant detritus (Daugherty, 1941; Ash, 1970). Daugherty believes there was a sparse forest here, similar to the savanna of the Brazilian uplands. The splitting of the terminal zones in the annual rings bears witness to a dry period that intruded into the summer rainy season (as in Texas). The fallen trunks had been dry for a long time – as shown by the work of bark beetles. Meanwhile, in Chinle we see autochthonous beds of vertical stems of *Neocalamites* and other plants of swamps and marshes.

12. PLACE OF BURIAL

The conditions of burial are governed by the tectonic position of the area, its geological history, and paleogeographic situation. For example, large deltas, in which the richest formations are found, are generally formed on shores of the Atlantic or Sea of Japan type, whereas they are absent on littorals of the Andes type (Mitchell and Reading, 1969). Land deltas often develop along linear tectonic structures in zones of accelerated vertical movements of the Earth's crust. In seismic zones suspension flows often arise which move plant remains beyond the shelf limits. Along rifts lie chains of lakes, each with its own specific formations. Coal formation proceeds differently in mobile zones than on platforms (Timofeev and Bogolyubova, 1962). In mature areas of land chemical erosion predominates, and sediments are formed that are characterized by a high monomineral coefficient and a low content of unstable components (Kazarinov, 1969), including such ones as phytoleumas. Finally, volcanicity is a vital factor. It follows that in principle, plant beds may be classified in terms of their being confined to certain tectonic structures (geosynclines, platforms, aulacogens,* etc.). However, there are not enough specific observations to do this as yet. We shall therefore restrict ourselves to a landscape characterization of various conditions of burial, paying attention to the trend of facies changes observable in each type that leads to a sequence of taphocenoses in a section.

A. Limnic complex

Strictly speaking, the term "limnic" refers to lake-marsh sediments, but in the wider sense it is used also to denote the whole facies complex of an alluvial plain.

a) River valleys

Burial sites confined to certain morphological elements of a river valley differ in richness and composition. The coarsest material (shingle, gravel, coarse and medium-grained sand) is deposited in the active watercourse. Fine sand and silty material appear in sediments of the scroll meander and levee. Silts and siltstones are distributed on the floodplain, and lenses of pelitic sediments are formed in oxbows and floodplain lakes (Lunev, 1967).

The facial features of a river valley combine with the characteristics of the soil-forming processes and of the distribution of the vegetation. Floodplain vegetation shows a general pattern of change from pioneer groups on new alluvia to near-sod communities of the second height interval of the central floodplain (Mirkin, 1969). The distribution of the communities is governed by the amount and composition of alluvium and by changes in mineral alimentation (replacement of eutrophic organisms by oligotrophic organisms

* [Graben syneclise.]

with the transition from the channel area to the central floodplain) and in soil moisture. In an arid climate a xerophilous and halophilous vegetation develops on the central floodplain. Vegetation near the river channel develops under conditions of "catastrophic alluvion" and is subject to inundations. Some plants (black poplar, for instance) settle only on new alluvium. The facial features of a river valley are combined with soil-forming processes and vegetation characteristics. Meanders divide a floodplain into segments of rectangular, oval, or ellipsoidal form (Shilov, 1969; Figure 6). In each of these geobotanists distinguish longitudinal strips (channel, central, terrace) and three transverse zones: a xerozone with the highest relief and well-drained soil, a mesozone with loamy and sandy-loamy soils and the most differentiated vegetation, and a hygrozone, which is characterized by an annual deposit of fine alluvium, clayey fertile soils, poor drainage and swampiness. In elliptical segments the zonation is not clear-cut. With a break in the meander neck the segment is transformed into an island segment. Here we now have another, concentric distribution of zones: in the center loams, with meadow forbs, and along the periphery weakly developed sandy-loamy soils with willow groves.

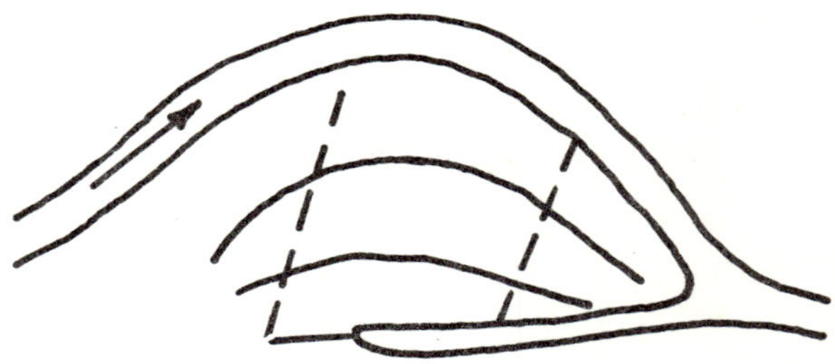

Figure 6. Oval segment of a river floodplain. Longitudinal zones: channel, central, and terrace; transverse zones: xerozone (beginning), mesozone, and hygrozone (after Shilov, 1969).

Longitudinal zonation is well expressed in valleys of tropical rivers, which are probably more similar to the ancient alluvial lowlands where the limnic coals of the Carboniferous were formed. Some of them have a very gently sloping longitudinal profile. For instance, in the Amazon valley there is an elevation of only 150 m at a distance of 3200 m from the mouth. The valley is 1.5 km wide in some places. A boggy strip (igapó) stretches continuously along the channel; it is submerged for a long time during floods.

Knowing the distribution of the lithofacies and of the vegetation along the horizontal, we can go on to interpret their succession in a vertical section.

Alluvial deposits are characterized by a rhythmical structure caused by a certain repetition of facies in the section. A rhythm, or cyclothem, is a series of strata united by a common direction of change in granulometric composition (from coarse to fine) that is repeated in the following rhythm. At the same time, consecutive cyclothems have a varying thickness. A change of granulometric composition within a cyclothem may be gradual (graded) or nongraded, without transitions between granulometric varieties (Bishop and Force, 1969). The coarse-grained part of a cyclothem is usually characterized by a marked variability and complex combination of different types of lamination, whereas the fine-grained upper part has a simpler structure.

The rhythmicity of alluvial deposits was for a long time attributed to vertical accumulation with different time periods in connection with the periodicity of crustal movements and (or) of climate. It was thought that the sequence of coarse to fine sediments was related to a leveling of the relief, a reduced tempo of river flow, or a drop in river level. It was only relatively recently that Shantser (1951) and then Beerbower (1964, 1969) demonstrated cyclothem repetition in a section: an autocyclical process governed by stream migration. At first it was established that the main part is played not by vertical but by lateral accumulation of alluvium (Allen, 1965, 1970; arguing with Moody-Stuart, Allen contended that lateral sedimentation predominates not only in meandering streams but also in streams with insignificant sinuosity). Consequently, a change along the vertical is the result of lateral migration of the stream. Migration occurs due to the deposition of sediments at one bank and the erosion of the opposite bank. As the curvature of meanders increases, the necks become narrower and narrower and are burst through by the stream.

The coarse-grained element of a cyclothem is usually interpreted as channel alluvium and the fine-grained constituent as floodplain alluvium. The formation of a cyclothem presupposes a temporary stabilization of the channel and later its subsidence under floodplain sediments. "Cyclicity means that channel and floodplain have followed each other many times at a place subject to subsidence" (Allen, 1970, p. 320).

Floods are an important factor in the formation of plant beds. It often happens that during one flood up to 1.5 m of mud is deposited, submerging shrubs and uprooted trees. A formation in a river valley with superimposed volcanic activity is a most interesting feature. Chaney made a special study of this phenomenon, investigating subfossil beds formed following the eruption of the Katmai volcano in 1912 (Chaney, 1938). A stratified layer of ash 10 and more feet thick was deposited within an 8–10-mile radius of the crater. The underground shocks attending the eruption caused the still-hot mass of ash to slide down the valley slopes. A nonstratified series 200 feet thick was formed here, with charred tree trunks (Chaney's book shows a photograph of poplars submerged by the landslide during the eruption). No leaves are found in these layers. The Tertiary tuff beds in Oregon were apparently formed under somewhat different conditions. Here leaves are concentrated at the base of the series, directly above the layer of lapilli. Probably, after the throwout of lapilli, heavy leaf fall took place and leaves that had dropped into small water bodies were buried by the ash that slid down the slopes. The material is well preserved because it was buried rapidly.

A hundred miles from Katmai, on Kodiak Island, a thin layer of ash was deposited that still covered most of the island in 1913. Twenty years later an ash film 6–7 inches thick remained only in the low-lying parts and was by now covered with a thin layer of humus. There are hardly any plant remains (just spruce needles) in this subaerial layer. The ash brought by rivers settled near the mouths and in lakes near beaver-built dams. Here there are many remains of plants and insects, concentrated mainly in the lower half of a layer consisting of poorly sorted and nonrounded material which was formed right at the time the ash was deposited. The branches and leaves of alder, willow, and spruce that fell here were ripped off the trees by the ash or by the heavy rain and wind that accompanied the eruption. It is interesting that fallen leaves of the same plants are strewn along the lake shore, but there are no plant beds outside the ash layer. The upper, darker part of the layer was formed by finer, well-rounded and sorted material that was washed out and redeposited by the river after the eruption took place. There are few plant remains here. Chaney correlates this layer with the Tertiary Bridge Creek formation, where plant remains in a 2–3-foot layer of tuff are concentrated in an intercalation several inches thick of angular, nonrounded grains.

Schwarzbach (1968) tried to establish the prevailing wind direction from the ash content in Rhineland brown coals and from the distribution of tuffs. One of the deposits (Rott) 1 km^2 in area, where thinly banded coals overlie trachytic tuff and alternate with siliceous rocks, is classed as sediments of a crater lake. There are fewer remains of thermophilous plants here than in other formations.

A good number of well-studied alluvial formations are known, but there is hardly any need to cite numerous examples. The reconstruction of the burial conditions of the Chalk Bluffs Eocene flora performed by MacGinitie (1941) is interesting in that he succeeded in sketching the outlines and morphology of a V-shaped valley 2 km wide with steep terrace cusps, filled with gravelly material with clay lenses in which plant remains were preserved in the form of phytoleumas (reducing environment, burial below the groundwater table). Overlying this is lighter-colored, better rounded terrace gravel with numerous intercalations of light-colored and chocolate-brown clays.

The plant remains enclosed in these are stained with iron oxides. These sediments were formed when the control of erosion was raised, when the valley was wider, material was transported slowly, chemical erosion was intensive, and when the products of erosion of arkose and biotite rocks were transformed into quartz-anauxite sands and pure clays. Changes in mineral composition of this type are often observed in sections of alluvial series and serve as a basis for lithostratigraphic zonation. For example, in a Cretaceous section of the Suifun Basin (Maritime Territory) arkosic sandstones of the Neocomian and Aptian are succeeded by polymictic rocks of Albian age, whereas in the Bureya Basin the same kind of succession takes place in the Neocomian at the boundary between the Urgal and Chagdamyn suites. Differences in the original material probably play a certain (probably minor) role. Of main importance, apparently, is a morphological change in the river valley, on which, in the cases described above, a change of climate is superimposed.

On the other hand, alluvial strata are known where chemical "maturity" of the rocks is preserved for very long intervals. The Cretaceous and Tertiary sediments of the Zeya-Bureya basin are an example. The upper horizons of the Cretaceous here are composed of shining sands and gravel with silicified wood and lenses of light-gray, yellowish, and brownish clays to which the beds of the well-known Tsagayan flora are confined. These lenses, about 20 m in diameter and several meters thick, contain several plant-bearing intercalations. Phytoleumas are practically absent here; impressions are stained with iron oxides. Foliage is often found together with fruits and seeds of the same plants, which indicates that the formation is close to autochthonous. The roof is almost everywhere overlain by a layer of sandy siltstone that is crammed with haphazardly oriented rhizomes, seeds, and cuticles of *Nymphaea*. This layer was probably formed when an oxbow was abandoned by floodplain alluvium (Figure 7). Above it often lie sediments of scroll meander or levee, containing typical sandy "sandwiches" with thin clayey and carbonaceous intercalations overflowing with plant detritus.

The Paleogene deposits of this area are characterized by an abrupt lateral succession of facies and variable thickness, reflecting an uneven base relief. On elevated areas thin, rather coarse facies of bench placers predominate, in which redeposited bones of dinosaurs are sometimes found. In depressions there are several coal beds quite persistent along the strike; these were probably formed in longitudinal channel zones of flooded swamps of the Amazon "igapós" type. The white clays at the roof of the coal beds are sediments of water bodies that developed when the swamps were flooded; these water bodies had almost standing water and open aquatic vegetation, and they contain rich beds of carbonitized and sideritized plants. The picturesque bluffs of the Zeya River are composed of white Miocene sands of exceptionally

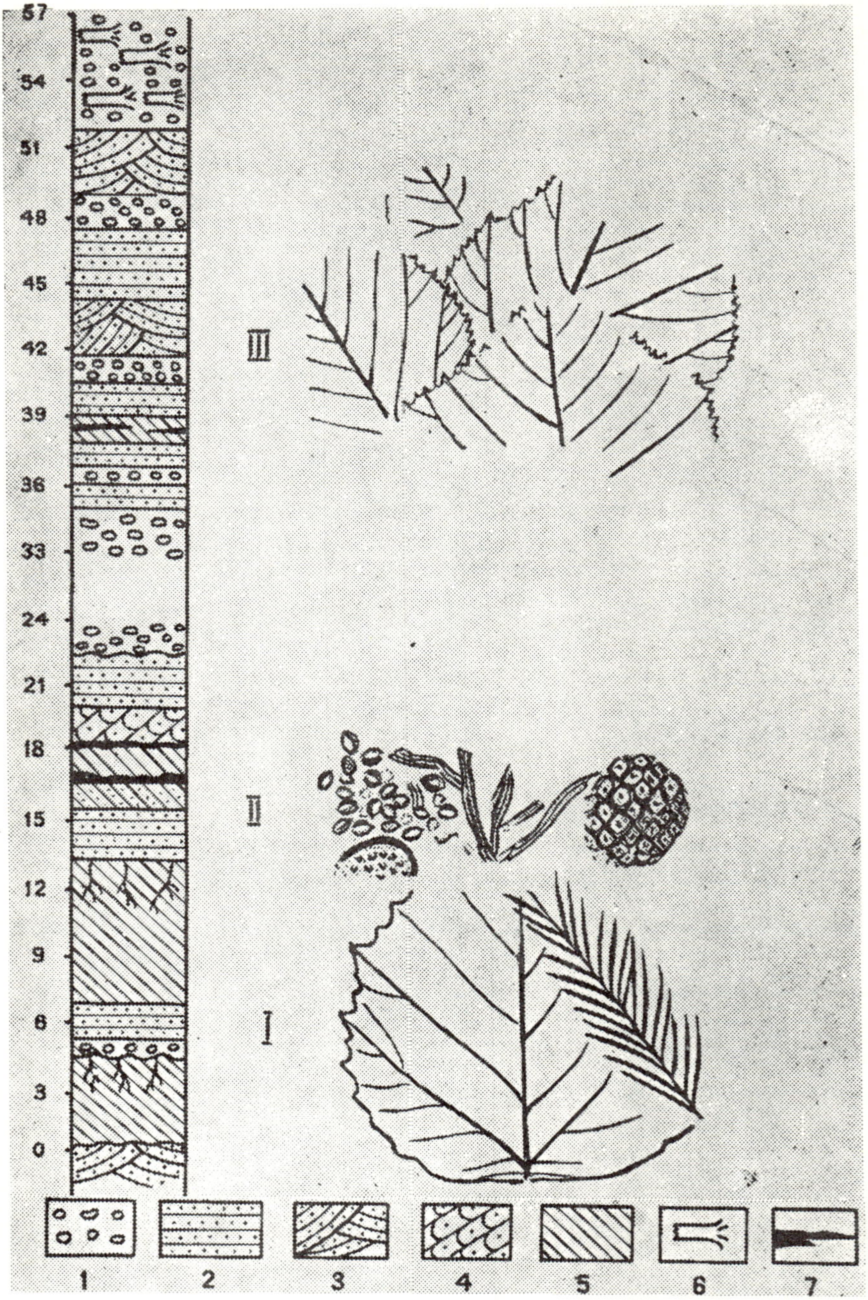

Figure 7. Section of alluvial deposits from the Upper Cretaceous of Bureya Tsagayan, and the nature of preservation of vegetation:

I) clay lens (sediments of floodplain water body) with *Platanus* and *Taxodium*; II) root layer with seeds and rhizomes of *Nymphaea*; III) platy sandstones (channel facies) with *Tiliaephyllum*. Notation: 1) shingle, gritstones; 2) platy sandstones; 3) cross-stratified sandstones, lamination of the large ripple type; 4) sandstones and siltstones, lamination of the small migration ripple type; 5) clays and argillites; 6) tree trunks; 7) coal.

homogeneous composition, in which lignites and small lenses of plant-bearing clays are only occasionally encountered.

b) Lakes

When we speak of fossil plant beds, the picture that is usually conjured up is one of a calm lake, where the rustle of falling leaves is heard. Old lakes in river valleys (see above) are in fact usual burial sites. But in the case of lakes that are not situated in river valleys, conditions are generally not favorable for burial, or more exactly, they depend heavily on such factors as the size of the lake, the shore relief, and the nature of sediment runoff. Lacustrine sediments can be recognized both from their sedimentological features and from the characteristic limnofauna, diatoms, and higher aquatic plants. Miki (1941) described a Tertiary lake with *Brasenia* the sediments of which were enriched with aluminum oxide. Here the type of sedimentation meets the environmental needs of *Brasenia*, but it must be remembered that the remains of aquatic plants found in lacustrine facies do not always belong to natives of the lake in question: they may have been brought from other water bodies in a time of floods.

Lakes of the "gobi" type (Florensov, 1968) can serve as a present-day model of water bodies in which lake formations most often occur. These are wide but small bodies of water with a variable level, and they are aliment-ed by small rivers that are for the most part drainless or with an irregular flow and are brackish to a varying degree. They were formed in wide, flat inter-montane troughs under dry-climate conditions. The small lakes of Transbai-kalia (depth from 5 to 25 m, with a fairly extensive area) are similar to this type (Lut and Agafonov, 1968). These are drainage lakes, but from time to time they lose their link with rivers and become salinized. Their slopes are gentle and even (sometimes with small ellipsoidal elevations), and their bot-toms are covered with viscous, organically enriched mud. Sandy bottoms are distributed only near the shore. The shores are sometimes boggy.

The Mesozoic lakes of Transbaikalia were probably of this type. Skoblo (1968) described lacustrine lenses up to 1 km^2 in area that were formed either in small water bodies that ceased to exist and filled up with sediment (in this case coarse rocks occur in the upper part of the lenses) or when sub-merged by lava sheets. The lake sediments – fine-grained silty sandstones and siltstones with siliceous cement, sometimes silicites – have a complex rhythmical structure brought about by a combination of seasonal rhythms and periodic changes of the accumulation regime; the latter are due in the main to pulsations of volcanic activity which affected the rate of sedimentation and the silica content. Here are found numerous remains of Lamellibranchia,

Phyllopoda, Ostracoda, and especially insects, which are confined to the finest sediments with horizontal lamination, signs of fine rippling, and sun cracks. The ribbonlike lamination is emphasized by the content of clayey material, detritus minerals, and plant detritus in argillites and carbonaceous layers, sandy and micaceous-clayey siltstones, etc.; the stratified member splits into sheets from several millimeters to tenths of a millimeter thick. Layers with insects would seem to be a more appropriate place for plants to collect, but they yield hardly any well-preserved remains; only fine, charred detritus. Large plant remains are found only in a few beds, where detritus is absent.

Sediments of much deeper Late Mesozoic lakes are distributed in Eastern Transbaikalia. These are siltstones, bituminous argillites, clay dolomites, and dolomitic limestones with very fine horizontal stratification ("paper shales"). The thickness of the deepwater lacustrine facies reaches 1000 meters. The lakes were probably a few hundred meters deep. Some of them may have had a zone of near-bottom hydrogen sulfide contamination, as is evidenced by the abundance and excellent preservation of fish remains. Plants are found very rarely.

From Transbaikalia Jurassic and Cretaceous sediments of this type continue in a broad belt across the Zeya-Bureya basin and Manchuria to the Pacific shore. In the Maritime Territory they are represented by a series of black siltstones of Albian age with a characteristic limnofauna. A wide lake, 70 km across from southwest to northeast, was formed after a brief marine ingression (*Trigonia* beds) that interrupted paralic coal formation. At the base of the lake sediments lie inequigranular cross-stratified sandstones with plant remains. The depth of the lake basin and the volume of sediment runoff varied widely: the black siltstones and argillites are relatively deepwater facies with remains of phyllopods and almost no plants (only very sparse remains of conifers), and they alternate with gastropod beds, which are fairly coarse sandstones with large, well-rounded shells. Plants are as a rule absent in the gastropod beds. A few formations are present in fine greenish sandstones and siltstones, which probably should be considered to be deposits of a small lake delta. Along the periphery of the basin conglomerates and breccias – sediments of proluvial debris cones and slope landslides – irrupt into the lake facies. Neither dolomites nor "paper shales" were deposited in this lake. The climate was less arid than in Transbaikalia, but humidity was evidently still insufficient. Red beds occur above the series of black siltstones.

Much richer formations (among the richest) are associated with sediments of Oligocene and Neogene lakes in the USA that are situated in intermontane basins of an extensive volcanic plateau. Best known among these is the Oligocene lake Florissant, that according to the calculations of McLeroy and

Anderson (1966) existed for between 2500 and 5000 years. The volcano slopes bordering the lake were covered with conifer forest, chaparral, and thorny shrubs along dried-up river channels. The climate was quite dry (MacGinitie, 1953).

According to MacGinitie and Axelrod, on the northern slopes of the volcanic mountains surrounding the lake were coniferous and deciduous forests, while the drier southern slopes were covered with woodland or savanna (*Celttis, Arbutus, Sapindus*), chaparral (*Ceanothus, Cercocarpus,* etc.), and thorny shrubs (*Zizyphus, Caesalpinia*) along the dried-up channels of streams. A seasonal stratification is observed in the lake sediments: an alternation of diatom layers (that formed in spring during the reproduction peak of diatoms in cold, transparent water) and sapropelite, deposited during the dry summer, fall, and winter in a reducing environment. The lake was probably chemically stratified (meromictic) with alternating acid and alkaline vertical zones. The bulk of the foliage and insect remains is confined to sapropelites. Large plant remains are also found in intercalations of coarse-grained tuffite that formed when ash material brought by spring waters from the slopes was deposited in the near-shore part of the lake.

The Miocene Blue Mountains taphoflora beds on the Columbia plateau were formed under similar conditions (Chaney, 1959). These are confined to deposits of a lake situated near an active volcano. The Pliocene Chalk Hills taphoflora (Axelrod, 1962) is contained in lake diatomites that seem to have been deposited after the eruption of one of the many eruptive centers of the andesite plateau. A lake was situated northwest of the volcanic plateau, and the colder northern and eastern slopes, covered with broadleaf forest, faced it. For the Pliocene Mount Eden taphoflora (as, moreover, for other components of the Madro-Tertiary geoflora), which is preserved in sediments of small lakes with a rapid and sharp sequence of facies, Axelrod (1938) reconstructs arid conditions.

With that, apparently, we may end our survey of formations in relatively small, short-lived lakes. The distribution of plant remains here is related to the alternation of littoral and more deepwater facies in the section and also to seasonal stratification and volcanic activity.

Let us now discuss large lakes with a long life-span. Many investigators believe that it is just such water bodies that occupied the central areas of both the Paleozoic and Mesozoic superimposed basins and intermontane troughs on the territory of North Asia. The best known among these is the Kuznetsk basin. Its Late Paleozoic flora is one of the richest in the world. Early Carboniferous formations are contained in delta facies with marine fauna. Later on, apparently in the Namurian, both the conditions of burial and the vegetation underwent a marked change. The fauna above this borderline is a freshwater one. The isometric configuration of the distribution area of coal-bearing sediments, the confinement of fine-grained granulometric varieties to the central part, the complete rhythms, that are well sustained along the strike, and the freshwater conchiofauna call to mind an extensive intermontane basin with a lake in the center.

Belenko (1970) postulates that during the time of maximal downwarpings the lake was similar to the Aral Sea in terms of its size and hydrodynamic activity. When it became shallower and its hydrodynamic activity declined, mudstones came to occupy an increasingly larger area (Belenko follows Strakhov's scheme, according to which the ratio of granulometric varieties is directly related to the size and depth of a basin: in Lake Balkhash there are practically no sandy sediments, in the Aral Sea such sediments form a narrow coastal zone separated from a broad region of mudstones by a strip of silts, in the Caspian the zones of silts and mudstones are more or less equal, etc.). Finally, shoaling and filling up with sediment led to swamping. Coal beds are persistent along the strike (a very extensive bog covers the entire area of the former lake) and indiscriminately overlie bar sandstones, lagoonal sediments, and so on. While this scheme fits the Kuznetsk basin well, it could actually be applied to other Asiatic coal-bearing basins of matching size. Here, however, some difficulties arise in choosing a present-day model. In fact, the Aral Sea is a very unsuitable model, since in terms of climatic conditions and sedimentation type it is rather the antipode of coal-bearing basins. Lake Baikal, which is a lake comparable in size with ample river alimentation, has a long, narrow basin, whose configuration is related to the tectonic nature of the depression. Here there is a narrow (0.2–8 km) zone of sands that were formed at a depth of 100–150 m. Disseminated in it are patches of coarse-fragmented sediments: boulders, pebble, and gravel, occurring in the zone of maximum hydrodynamic activity at depths of 20–30 m (Goldyrev and Vykhristyuk, 1968). The area of sands becomes expanded in the inlets and at the mouths of rivers. Coarse siltstones form a narrow zone on the periphery of the sands at depths from 50 to 400 m. These are gray and dark-gray viscous silts that contain an abundance of plant detritus. Almost the whole area of the bottom in the northern part of the lake is taken up by fine silts that are fine, almost black muds. Only in the center is there a strip of mudstones 6–12 km wide. In the middle of Lake Baikal mudstones, on the contrary, cover a large part of the bottom, forming gray, bluish, or greenish silts. A section of bottom sediments has a rhythmical structure in places, the rhythms being formed by the alternation of sand, pelitic mud, and diatomaceous ooze.

Lakes of the Baikal type are genetically related to deformation structures in rift zones (Transbaikalia, East Africa). Among water bodies of a different origin they resemble only lake depressions formed by a glacier. Long-lived lakes of either type were formed under conditions of sediment runoff that was insufficient for basin flooding. Lake Baikal has been in existence at least since the Pliocene. Its water level is subject to only negligible periodic fluctuations, whereas for basins of the Kuznetsk type we should assume frequent shoaling and even total drying out, followed by an "inherited" new lake in the same place. A present-day model of a huge and, judging from the standpoint of hydrodynamic activity, a fairly deep lake with a long existence and at the same time sharp fluctuations of level, apparently does not exist. Before we assume that the conditions of formation of limnic complexes in the past could have been substantially different from those today, we must examine the possibility of interpreting the Permian paleolandscape of the Kuznetsk basin as a river valley; this, of course, does not mean that more or less stable floodplain lakes with diverse conchiofauna could not have developed in the central part.

B. **Paralic complex**

In mobile zones on the continental margin an extremely intricate complex is formed by sediments of above-water and underwater delta, sandy shoals, spits, and bars, mud flats, lagoons, salt marshes, and mangroves. A more or less gradual transition relates it on the one hand to an alluvial terrestrial facies complex and on the other to a littoral-marine complex. More than half of the major formations are confined to paralic facies.

a) Deltas

Large deltas are formed on Atlantic-type shores (Amazon, Niger, etc.). The morphology of a delta reflects the characteristics of the tectonic regime (Tageeva, 1968). Generally, when a delta begins to form, river-borne material is deposited on the shelf, creating a prodelta. As the deltaic cone grows, material is discharged primarily in the upper part, and the delta acquires a fan shape. Several classifications of deltaic elements and the corresponding sediments have been proposed: subaerial part (intradelta), with primarily freshwater sediments, delta platform, formed by tides, and submarine part (delta front), which includes the prodelta shelf, the continental slope, the abyssal cone or submarine fan and other morphological structures.

The well-studied Mississippi delta will give us some idea of the distribution and thickness of deltaic sediments. It occupies an area of 110,000 km^2, of which 28,500 km^2 refer to the delta plain, 45,000 km^2 to the prodelta shelf (to a depth of 180 m), 22,000 km^2 to the continental slope (to a depth of

1800 m), and 18,000 km^2 to the submarine fan. The average thickness of the sediments that have been forming continuously since the middle of the Quaternary is 70–120 m within the limits of the delta plain and 260–600 m in the submarine part (Fisk and McFarlan, 1955). Each structural element of a delta is characterized by a certain assortment of lithofacies, organic remains, and types of cross-bedding (Botvinkina, 1962; Allen, 1962; Bouma, 1963; Jopling, 1963; Kukal, 1964; McKee, 1964; Taylor, 1963; van Straaten, 1960).

Autochthonous and hypoautochthonous formations are established within the intradelta and deltaic platform and allochthonous formations in the prodelta. Sections of deltaic sediments, like alluvial sediments, have a rhythmic structure, although the cyclothem pattern may be slightly different here. The rhythmicity is due to the migration of morphological elements of the delta, the recurrent succession of prodelta and intradelta sediments, etc. The structure of a cyclothem is usually complicated by a migration sequence of second-order elements: delta arms, bars, shallows, etc. The rapidly expanding Terek delta illustrates this process (Baidin, 1970). The sediment runoff of the Terek amounts to about 16×10^6 tons annually. This material at first forms a submarine delta, bars, and estuarine spits in the mouth. The rate of accumulation depends on the depth of the offshore zone and the flow of detritus along the shore. As the estuarine part becomes longer, the current velocity decreases and the material is dumped into the channel, the bottom of which is raised. The branches of a delta are elevated above the delta plain. Water often sweeps in through levees. A good deal of the runoff pours into the rush, and the old bed silts up and disappears. In the next stage of development the hydrographic network of the delta becomes a system of drainage lakes that regulate runoff. The elevated sectors that gradually divide the lakes are washed out, and instead of a lake system develops at first a multibranched and then a one-branched delta. The material is no longer discharged in the intradelta lakes but passes in transit to the mouth, where a prodelta once again builds up. A new cycle is begun.

Autocyclical processes of this kind develop against a background of evolutionary changes related to tectonic movements and eustatic fluctuations of the sea level. Clapham (1970) described an alternation of palynocenoses of two types in Permian deltaic sediments of Oklahoma: the first type is dominated by pollen of gymnosperms and the second by fern spores. The first type is confined to prodelta sediments, where there are also remains of Acritarcha, the second to sediments of the delta plain or platform with a fauna of mud-eaters. Bogs and marshes with ferns are thought to have developed on detritus during regressions that were quite short-lived. The succession of the palynocenoses (and the corresponding facies) can probably be attributed also to an autogenetic migration of the deltaic elements.

As an example let us examine the deltaic formations in the Wealden of England, which though not among the largest, have been studied in detail in recent years by lithologists Allen and Taylor. The lower part of the series is composed mainly of sandstones (Hastings Sands) with a freshwater fauna and the upper part chiefly of clays (Wealden Clays) with oysters and brackish-water mollusks. Taylor (1963) thinks that the deltaic complex was formed by several rivers issuing from the London-Belgian Ridge and the Devon and Cornwall heights. The thick clay series of Wealden Clays, Fairlight Clays, and others, from where a good number of the collections described by Seward originate, are interpreted as prodelta sediments. Variegated and gray, pyritized clays and siltstones alternate with paludous sandstones, coquinas, and ossiferous strata (bones of reptiles and fish), that were formed during pauses in sedimentation. There are several horizons of fossil soils with roots of conifers – the basin was very shallow. Typically there is an alternation of siltstones and limestones or a more complex interbedding of clays, calcareous siltstones, ostracod-bearing limestones, and argillaceous iron ores. The stratification has been destroyed by abundant fossorial organisms.

The basin was strongly freshened. A transitory salinization has been established according to bursts of microplankton. A high content of *Classopollis* (possibly mangrove pollen) is observed in fine sediments, and of floating bisaccate pollen in coarse ones (Hughes and Moody-Stuart, 1967). Intradelta strata are developed mainly in the northern part of the area. Here are found thin ribbonlike clays (*Tempskya* Clays) with rich remains of plants, siltstones with horsetails (*Equisetites* beds) and large and small fill structures with irregular bedding and plant remains distributed in different planes – signs of a turbulent current. Sun cracks, wind-transported angular sand grains in the clays, layers with roots – these are features of intradelta sediments, related to the shallowness and frequent drying out of water bodies. Root layers and fill structures are also characteristic for sediments of a delta platform. In contrast to the situation in the prodelta, here there are hardly any fossorial organisms and the thin original bedding is preserved. Intercalations of shingle correspond to brief transgressions that abraded the platform, while thin layers of lignites indicate shoaling of the water body and swamping.

We find approximately the same facies spectrum in other plant-bearing delta sediments, beginning with the well-known Devonian delta of the Appalachians (Barrel, 1913) and ending with the no less famous Eocene deltas of England (Chandler, 1964).

b) Marshes and mangroves

Sediments of marshes and mangrove swamps as a rule form part of a deltaic complex, but they should be studied in more detail, since they are a prime source of plant beds and coal formation. Furthermore, there is reason to think that the first land plants were inhabitants of salt marshes. Typical for **salt marsh** facies are: minor thickness, a wealth of root remains of halophytes (at present primarily *Salicornia* and *Spartina*), slightly sinuous parallel bedding, alternating thin sheets of clayey sand and clayey (or sandy) silt, poorly sorted material, and an absence of gradual transitions between granulometric varieties (Bouma, 1963). These sediments are generally dark-colored and have practically no faunistic remains. Due to the high salinity and frequent drying, there are few fossorial organisms here, so that the original lamination is preserved. In a section of paralic strata marsh facies can often be recognized by horizons of fossil soils. Salt marshes sometimes combine with mangroves (in Sydney, Florida: Clarke and Hannon, 1967). Mangroves grow best in the equatorial zone, but they also occur in places far from the tropics: as far as Japan and the northern end of the Red Sea (30–32°N) in the northern hemisphere and to New Zealand (44°S) in the southern hemisphere. They develop chiefly on gently sloping shores, in river mouths, where abundant terrigenous material is deposited, and around lagoons (Walter, 1971; Ridley, 1930; van Steenis, 1962).

The width of a mangrove belt is governed to a large extent by tidal amplitude. The frequency and duration of flooding determine the salinity and aeration of the soil and the development of sprouts growing between pneumatophores. On the land side limiting factors are dryness and a soil salinity that is insufficient for halophytes. Mangroves are characterized by a distinctly expressed zonality that is related mainly to fluctuations of tidal amplitude. In Australian mangroves, according to the data of Clarke and Hannon (1968), the vertical amplitude of the water does not exceed 1–1.2 m. On the shore of the Gulf of Mexico the spread of mangroves is controlled by the seasonal rains and storms raising the water level in the gulf. Mangroves fringe small (2–3 m deep) coastal lagoons with oyster banks on the flat bottom and channels of streams whose depth reaches 20 m. During the winter rains black streams, saturated with organic matter and bearing water hyacinth, fill the channels. The lagoons burst their banks, flooding the swamps and marshes. The violent winter hurricanes, lifting up the water in the gulf, prevent the runoff of surface waters. From April to September the mangrove area is free of surface waters, and at this time the soil salinity is higher (Thom, 1967). In the dry climate off the East African coast the salt content in mangrove soil is greater with increasing distance from the sea. In the inland *Avicennia* zone the osmotic pressure of the soil solution reaches 40 atm (Walter, 1971). Behind the *Avicennia* zone lies a sandy strip devoid of vegetation that is flooded only twice a year, during syzygial tides. Plants cannot adapt to the sharply fluctuating salt concentration in this strip.

The most complicated structure and richest floristic composition are seen in the mangroves of the Indo-Malay Region, that are framed by thickets of the palm *Nipa fruticans* and spread far inland along river valleys. The fern *Acrostichum aureum* often grows at their inner edge. The Eocene mangroves of Europe often had more or less the same structure. Finds of *Nipa* fruits mark the position of the mangrove front, while *Acrostichum* marks the inner margin. From the Lower Bagshot series Chandler (1962) described a member of siltstones with black intercalations crammed with pinnules of *Acrostichum* and monocot leaf cuticles. Fragments of *Anemia* leaves were also found here. All this probably denotes a tropical salt marsh vegetation. Beds with *Acrostichum lanzeanum* have also been described from the Eocene of Hungary (Pálvfalvy, 1966). But where more ancient mangroves are concerned, it is far more difficult to make a reconstruction, and we have to draw our assumptions mainly on facies boundaries of distribution. Some clues are provided by remains of algae and animals, especially Isopoda and Limulida, as well as traces of their life activity. Small water bodies and mud flats above a delta are commonly inhabited by the young of these crustaceans. When they become adult, they swim away beyond the tidal zone (Bandel, 1967). Limulida are often found in Paleozoic plant beds and much less often in Mesozoic beds (we were lucky to find one specimen in the plant-remain-filled Upper Jurassic siltstones of the Bureya basin that overlie marine strata).

From the Jurassic of Yorkshire Black (1929) presented a detailed description of allochthonous formations ("drifted plant beds") with numerous remains of Gingkoales in delta-front sediments. Harris distinguishes here autochthonous horsetail beds, thin mudstone sediments of lagoons and meandering river beds with a slow current, containing very well-preserved remains (the well-known Gristhorpe Bed), sandy sediments of streams with a swifter current (the Whitby formation, from which Brongniart had described excellently preserved leaves of ferns), drifted plant beds with siltstone intercalations among delta-front sandstones, allochthonous beds with tree trunks, and reworked beds with dispersed cuticles. He noted that in those beds of the Lower Deltaic series of Yorkshire where *Pachypteris* dominates (about 25 localities), Hystrichosphaeridaceae and *Tasmanites* are also encountered. He also refers *Stenopteris, Ptilophyllum, Zamites gigas,* and *Brachyphyllum* to possible mangroves. Concerning *Pachypteris, Stenopteris,* and similar Mesozoic Pteridospermae with thick pinnules, other authors (Wesley, 1965) share Harris' opinion. Some consider the species *Lepidopteris ottonis,* also characteristic for the Rhaetic of Europe, to be a halophyte. Judging from its distribution, *Ptilophyllum* is one of the most likely halophytes among the Bennettitales. From Jurassic deposits of the Donets basin (Kamenka) I described a bed of cemented clay that was stuffed with leaves of *Ptilophyllum.* No other

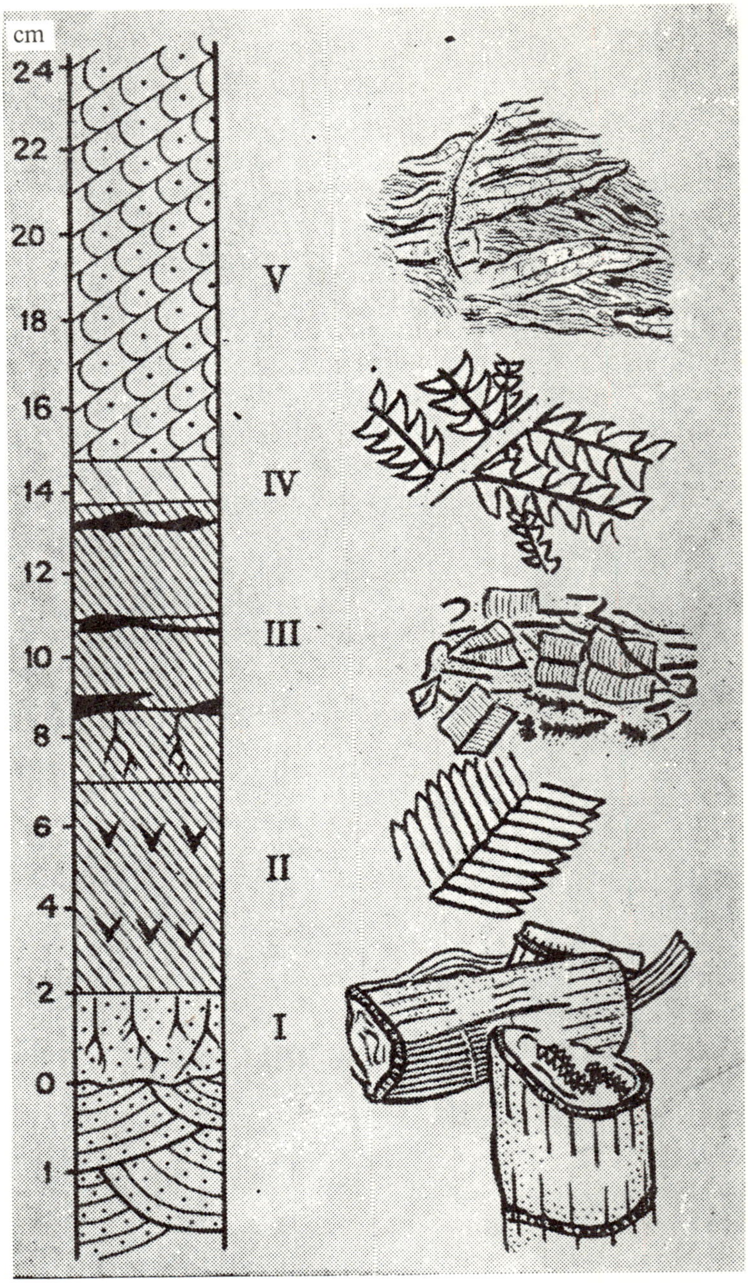

Figure 8. Section of intradelta sediments of the Jurassic of Kamenka in the Donets basin and the nature of preservation of plants:

I) ferruginous sandstone with *Equisetites*; II) cemented clay with *Ptilophyllum*; III) coal clay with *Ptilophyllum, Elatides,* and *Taeniopteris*; IV) white clay with *Todites*; V) cross-stratified sandy siltstone with *Dictyophyllum*. Notation as in Figure 7.

plants are in evidence here. *Ptilophyllum* and *Williamsonia* occur also in intercalations of limonite filled with horsetail stems (some of them oriented vertically) and leaves of the fern *Klukia.* Cemented clays and iron ores are usually confined to a zone of transition from cross-bedded glauconitic sandstones of a delta front with marine mollusks and Ophiuroidea to intradelta sediments with plant-bearing lenses of clays (Figure 8). Iron enrichment here was associated with marsh sediments.

The most splendid specimens of *Ptilophyllum* and of the related genus *Otozamites* that I have ever seen were found in the Upper Jurassic of Karatau in sediments of a water body with a high salinity. All this obliges us to reckon with the possibility that the "*Ptilophyllum* flora" of India is of the nature of a mangrove. It seems to me that the famous Triassic *Pleuromeia* was also a halophyte. For disseminules it had large spherical sporangia crammed with megaspores; these are often found in marine deposits together with ammonites.

Of course, these guesses can be pitted against data on finds of both *Ptilophyllum* and *Pleuromeia* in sediments known to be freshwater. It may be, however, that these Mesozoic plants were only facultative halophytes and were able to grow in nonsalinized soils as well.

Although a mangrove belt is not wide, it is situated in exceptionally favorable conditions for burial. This is probably why the role of mangroves in fossil floras has been greatly exaggerated. Their remains form the basis of the paralic formations of the Euramerican Carboniferous. American palynologists have established that the Paleozoic mangroves had in general the same zonality as recent mangroves. Peat beds in Florida are now being formed in freshwater marshes with *Mariscus, Eleocharis,* and *Scirpus* and stretch across a salt mangrove swamp for 1.5 miles inside the bay. Surface runoff carries the pollen of freshwater marsh plants into the mangrove peat, but the mangrove pollen does not fall into the freshwater peat. In the Paleozoic Kittanning coal series palynocenoses with *Densosporites* occur only in that part of the coal bed that is overlain by marine strata with *Mesolobus* or by brackish-water strata with *Lingula.* This facies corresponds to a mangrove peat bog with pollen of *Rhizophora,* buried under marine sediments of the Gulf of Mexico. Another pollen assemblage with *Lycospora* is contained in coals the roof of which is formed by sediments with a freshwater fauna that correspond to freshwater marshes. Sporomorphs of the complex with Lycospora are encountered together with *Densosporites* but not vice versa (Habib et al., 1966).

c) Lagoons

Plant beds associated with terrigenous lagoonal sediments are common in a deltaic complex. Plant remains occur much more rarely in chemogenic sediments.

Often found in rock salts are pollen, wood fragments, and other small remains that are of interest to crenopaleontologists (Rudau, 1966). As a rule these remains belong to sod plants and have been brought from afar (as, for example, in Oligocene sediments with potash strata: Kirchheimer, 1946). It seems that the shores of lagoons in which salt settled were devoid of vegetation. It was under less arid conditions that the gray Jurassic limestones of Veneto were formed that Wesley (1965) considers deposits of lagoons framed by mangroves. In the Bahamas he observed lagoons of this type that were protected from waves by a belt of mangroves, where almost pure aragonite was deposited in the shallow, calm water. Thick, leathery leaves of pteridosperms are admirably preserved and are found in association with families of the same plants, indicating that this is more or less an autochthonous bed. Remains of conifers proved also to have undergone little transport. On the other hand, the Veneto beds do not contain many plants that are usual for deltaic Jurassic taphofloras (e.g. *Nilssonia*). Formed under similar conditions were the Solnhofen "paper shales" that have become famous due to the finds of archeopterices, flying reptiles, and other animals that pushed the more humble though quite numerous plants out of the limelight here. The Upper Jurassic plant-bearing beds of Karatau are similar to this group of lagoonal formations, although Gekker assumes that they have a lacustrine origin.

The Triassic lagoonal formations in the Vosges were formed under conditions of a dry climate. They are lenses of clay in sandstones with *Voltzia*. The clays are finely stratified, which underlines the seasonal distribution of plant detritus. Conifer and horsetail remains commingle with insects, scorpions, xiphosurans, and fish. The accumulations of aquatic faunal remains at the roof of the strata attest to periodic drying out of water bodies (Gall and Grauvogel, 1968).

d) Flysch

A genetic link exists between river mouths and submarine canyons on the shelf and continental slope (Moore, 1970). Flysch often passes along the strike into deltaic coal-bearing deposits (Nalivkin, 1961). Allen notes a paragenetic connection between paralic coal-bearing facies and flysch in the Carboniferous of England. The same type of relationship may be traced in the Senonian deposits of Sakhalin (Zhonkier suite). The terrigenous Cretaceous series of the West Sakhalin Mountains was formed at the back of island arcs, with which siliceous-effusive deposits of the same age, developed on the eastern shore of the island, are connected. The coal-bearing members of the Zhonkier suite, with abundant remains of ferns, *Nilssonia, Sequoia,* and (more rarely) angiosperms, often contain remains of *Inoceramus* and *Trigonia*. In some sections (for instance, in exposures on the picturesque Cape Zhonkier) they border directly with flysch strata formed by the regular alternation of thin layers of black argillites, siltstones, and lighter, often ferruginized sandstones about 5 mm thick. There is a remarkable display here of diverse bioglyphs, small festooned filling structures, and other flysch attributes. Large plant remains are very rarely found, as a rule only in concretions. They were probably brought in by suspension streams.

Plant remains occur much less often in typical flysch than in flysch-type strata. Berger (1965) reports finds of the wood of conifers and Lauraceae in Cretaceous alpine flysch. Wood of *Cedroxylon* and pieces of resin have been found in Paleogene flysch. This emphasizes that there are no remains of mangrove plants here. Note is to be taken of the predominance of stable remains in flysch beds, especially resins, in which small fruits and seeds are sometimes enclosed (e.g. fruits of *Myrica* in resin from Paleogene flysch: Bachmayer, 1968; see also Knobloch, 1970a).

C. Bog

We have talked about beds in bog facies when describing both paralic and limnic complexes. But the part played by bogs in the formation of beds is so important and so unique that the subject deserves a few more pages. The formation of peat may be considered a special variety of nonclastic burial, that in the given case coincides with rock formation.

a) **Present-day model.** The study of Carboniferous deposits and their taphofloras in the second half of the 19th century posed before paleobotanists and coal petrographists a number of problems, the solution of which demanded an investigation into recent conditions of peat accumulation. The bog masses of the temperate latitudes that were well known by this time were little suited to this purpose, because they had developed under different climatic conditions, in many cases under the influence of Quaternary glaciation.

The factors leading to bogging in coastal areas were eustatic fluctuations of the sea level resulting from the succession of glacial and interglacial zones. A section of Quaternary peat bogs often begins with reed-sedge peat or gyttja with remains of aquatic plants, succeeded above by peat with *Eriophorum* and pollen of woody plants and, finally, sphagnum peat (Birks, 1970; Donseolar and Jonker, 1952; Hallik, 1968; Vorman, 1952; and others). A good example is the Avaste bog in the Baltic area, that developed in place of the Littorina Sea (Kask, 1965). At the base of the peat deposits occurs gyttja with remains of aquatic plants, corresponding to a limnic and a telmatic stage. This is followed by a semiterrestrial stage of bog forests (wood peat) and sedge bog with the gradual development of a moss cover on the hummocks and a decline of hydrophytes in the space between hummocks. This is succeeded by a *Myrica–Schoenus* bog that directly precedes drainage, the spread of *Molinia*, and the elimination of hydrophytes. Foci of transition to upland bog appear in the form of sphagnum hummocks and bog forest developing on the hummock islets.

In Western Siberia the factors promoting bog formation were plain relief, stagnation of soil-ground waters, poor drainage, shallow-lying impervious beds with concave roof, and also permafrost (Glebov, 1969; Sergeev, 1969). Impervious clays isolate the peat from the water-bearing horizons, making the bog oligotrophic. These clays are thought to be deposits of wide, shallow lakes from the warm Atlantic period of the Holocene. At the boundary between the Atlantic and Subboreal periods the lake site was occupied by swampy forest, and intensive bogging set in (Khotinskii et al., 1970).

Thus, in the temperate latitudes a fairly long succession series leads to an oligotrophic bog climax. But where the lignites and hard coals of Europe are concerned, judging from plant remains, they were formed under climatic conditions more or less similar to tropical ones. The tropics had been minimally studied in H. Potonié's time. It was thought that bogs, especially upland bogs, were absent here. The discovery of extensive tropical bogs in Sumatra brought some enlightenment. According to Potonié's description, these are forest bogs with a peat deposit more than 6 m thick. The trees here have aerial roots, a tall, straight trunk, and a small canopy. The underwood contains many lianas, palms, and candle-shaped trees. A grass cover is not developed, and ferns are rare (Potonié, 1907).

It later became clear that the Sumatra bog is far from an ideal model for reconstructing the bog landscapes of the past. In most cases paralic deposits were formed from marsh peat bogs and mangrove swamps. True, there are those who assert that here, too, the principal role was played by upland moors (Afanas'ev, 1968). However, the frequent interbedding of coal seams with limestones containing remains of marine organisms, the absence of any signs whatsoever of erosions or any kind of subaerial processes in the soil of the beds (e.g. in the Donets basin: Tkachenko, 1968), the peat dolomites, and many other features contradict this hypothesis. We discussed mangrove landscapes in the previous chapter. Naturally, it is only in very rare cases that we can prove that Paleozoic and Mesozoic plants whose remains are found in clays belonged to mangroves (such proof is furnished by finds of roots with aerenchyma, as in *Amyelon*). But as early as in Eocene clays we find remains that unquestionably relate to mangroves. For instance, in the Lower Eocene Cherra series (Assam) coal beds alternate in the section with carbonaceous clays. The beds are dominated by pollen of Rhizophoraceae and palms, which are also represented by macroscopic remains (Dutta and Sah, 1970).

River channel bogs of the "igapó" type in the Amazon valley (see the section "River valleys") give an idea of the conditions under which many intracontinental deposits were formed. Timofeev (1970) compares the Jurassic coals of the Angara-Chulym trough with the peat bogs of the Rioni River delta. The Colchis lowland grew up in place of the bay in which the mountain rivers Rioni, Inguri, Kodori, and others emptied. With shoaling submarine ridges – "lido" – were created in the bay. The part of the bay demarcated by them was freshened and transformed into a lagoon, which gradually filled up with sediment and developed marsh vegetation. The repeated transgressions and regressions of the Black Sea were expressed in an alternation of peat bogs 2–3 m thick with marine sediments. The peat bogs of the deep-lying horizons are very densely compacted and have more or less become transformed into lignites. They now occupy a narrow strip of the Colchis lowland along the Black Sea shore, from which they are separated by a levee. The form of the peat deposits is determined by the hydrography of the coastal strip, that is threaded by the channels of large and small rivers. The deposits are lenslike bodies drawn out along channels that peter out rapidly in the direction of the sea and more gradually toward the edges of the plain. According to Timofeev, the Jurassic coal bed Moshchnyi is made up of two such deposits, separated by the channel of the Angara.

There is thought to be a marked similarity between the "lignite" landscapes of Europe and the subtropical bogs of Florida, in which Monk (1966) identifies several types (mainly on the basis of floristic characters). Apart from the well-known *Taxodium* bogs fringing lagoons and lakes, sedge bogs and mixed bog forest are distributed along the river banks and channels, while on the poorer and more acid, less frequently flooded soils there are bogs with evergreen trees, *Myrica,* and *Woodwardia* in the lower horizons. It is to this last type that the peat deposits are confined. In the Rhineland Oligocene brown coals dark layers (dunkle Becken) and light layers (helle Schichten) are distinguished. Thomson (1949) correlated the first with the peat of forest bogs and the latter with treeless bogs in Florida. However, on the basis of a detailed analysis of all climatic indicators, Schwarzbach (1968) rejected the comparison with Florida and postulated a Virginia-type climate (defined by the symbols Cfa according to the Köppen classification) for the Rhineland lignite deposits. Jux (1968) performed a detailed comparison of the lignite sections with contemporary coastal deposits in various climatic zones, from Canada to Mexico. On the coast north of Boston dark forest peat occurs on the Boston Clays and above this light marsh peat with *Spartina* up to 6 m thick;

this differs from "hellen Schichten" in its high ash content. In Tamiahua lagoon (Mexico) *Rhizophora mangle* is the main peat-former. Its roots form traps filled with mud with plant detritus and coquina. A rich clayey wood peat of minor thickness accumulates here. In river deltas develops a flourishing aquatic vegetation and sedge thickets producing black muddy peat in which there is hardly any tree pollen (on the other hand, pine pollen predominates in the lagoonal sediments). These peat bogs cannot be correlated with the lignite peat bogs, where mangroves are totally absent. The closest resemblance is to the peat bogs of the Mississippi delta, which are characterized by a rapid succession of facies, a wide diversity of plant groupings, cyclicity, erosion surfaces, and seasonal stratification due to hurricanes.

On the basis of the correlation with the Mississippi delta the following "lignite" landscapes are reconstructed: 1) nonproductive surfaces with stumps of *Sequoia* and *Pinus*; 2) broadleaf forest with oak, magnolia sweet-gum tree, and others (nonproductive); 3) bog forest with *Taxodium* and *Nyssa* (weakly productive; thin, dark forest peat); 4) scrub bog with *Myrica–Salix* (highest productivity; dark peat); 5) reed swamps and salt marshes (light peat). Referred to a special type are beds with *Marcoduria,* which Schwarzbach considers allochthonous, or from eel-grass that was cast onto the shore during a storm. Schneider (1969), on the basis of dispersed cuticles, notes in these deposits two types of boggy forest: with *Pinus* and with broad-leaved trees and *Sequoia* and also grassy bogs with *Marcoduria,* etc. He stresses the absence of *Taxodium–Nyssa* associations in many formations. Sequoia cuticles associate with remains of Lauraceae, pine cuticles with *Myrica* and palms.

According to Teichmüller (1952), Paleozoic coal accumulation is in many ways similar to Cenozoic coal accumulation: also observed here are facies of forest peat (lepidophyte coal), of treeless fern bog, and "reed" marshes (coal with cuticles, bark, roots, and peat dolomites). Durain coal is formed from sludge in lagoons and bays. In contrast to the preceding types, this type contains many sporomorphs. It has been experimentally shown that microlamination may be obtained upon compression of allochthonous peat but not autochthonous nonlaminated mangrove peat (Cohen, 1970).

Analogues of the Gondwana Paleozoic coals should probably be sought among the bogs of the cold-temperate zone of the southern hemisphere, where such types as sphagnum bogs with Gleicheniaceae, that cover the bog relief with dwarf tree ferns, and others have been identified (Kats, 1966).

Dynamics of peat accumulation. Peat bogs display a rhythmical alternation of broad light and narrow, undulating dark bands denoting a succession of relatively moisture-loving and xerophilous communities, with an increased proportion of the latter in the upper part of the deposit (Faegri and Iversen, 1950). The rhythmicity may be associated with alluvial cycles, irregular growth of a delta (as in the peat bogs of the Mississippi delta described by Jux, 1968), eustatic fluctuations of sea level, and changes of climate. It goes along with autogenetic development of a bog from eutrophic to oligotrophic, related to peat bog growth. Changes in mineral alimentation and chemism in the course of a bog's evolution lead to altered conditions of burial. In an upland bog mineralization, pH, and the rate of decomposition of vegetable material are lower, while the tempos of peat accumulation (and accordingly the thickness of the peat deposit) are higher: about 0.5–1.2 mm a year in the oligotrophic bogs of Vasyugan (Khotinskii et al., 1970). For the Holocene bogs of the Urals and Russian Plain Neishtadt presents similar figures. In more decayed lowland peat stable components become concentrated (e.g. pollen grains of conifers).

The degree of interference of external factors depends on the age and size of a bog. On poor substrates bogs are oligotrophic from the very beginning. Minor fluctuations in the amount of atmospheric precipitation and of the groundwater level have a marked influence on vegetation and on the conditions for burial. However, large bogs show themselves to be more stable systems. The periodicity in their development does not coincide with dry-valley cycles (Glebov, 1969).

As peat accumulates, not only the trophic characteristics but also the surface relief of a bog change. This is another aspect of development conditioned by internal factors. The surface of the bog becomes convex. Bottom subsidence (expressed variously depending on the nature of the underlying rocks) alters the surface slope. An increase in the gradient may result in afforestation of a bog.

The morphology of a peat bed depends on the geomorphological position of the bog. In river valleys a deposit has its greatest thickness at some distance from the channel (Kryukov, 1956). As it approaches the channel, the bed becomes thinner or else the thickness is split up, in which case the overall thickness may even increase due to rock intercalations. Morphological changes of a bed are often accompanied by a change in ash and sulfur content and other qualitative characteristics (Kizel'shtein and Serebryakova, 1970). A mangrove peat deposit thins out toward the sea. The morphological structure of a peat bog is also governed by the system of streams, lakes, and fens inside the bog. A multitude of channels run through mangrove swamps. The ash content of peat increases near a stream inside the bog. The bed here often shows an increased thickness and complicated structure. In zones of such complicated structure the content of jellified tissues often increases due to the better aeration, but, on the other hand, the content may diminish on account of a more rapid accumulation of peat (Kizel'shtein and Serebryakova, 1970; Fisk, 1966), so that the connection here is complex and ambiguous.

In just the same way, the relationship between mineral alimentation, pH, ash content, and swamp vegetation is so complex (a plant association reflects the conditions of peat accumulation more minutely than the aggregate of numerical indexes, and yet under certain conditions differences in vegetation still do not adequately reflect the changes in the pH, ash content, etc.) that it is very hard to adopt a unified principle of classification that would permit all the key characteristics of a bog to be taken into account. Therefore, parallel classifications are drawn up on the basis of peatiness, chemical parameters, morphological characters of the peat deposit, and plant communities. Attempts are also being made to combine several classification techniques in a single system (Glebov, 1969).

Plant associations are also considered when classifying hard coals (for instance, one speaks of lepidophyte coal or coal with *Lyginopteris*). This mode of classification may eventually become more widespread.

Many investigators believe that the petrographic type of coal reflects both the original type of material and the conditions of burial, particularly the degree of allochthonousness. H. Potonié had shown that durains were formed in water bodies with an open surface, primarily from allochthonous material. According to Teichmüller (1952), as plant material moves farther away from the original source, finely banded vitro-clarite is succeeded by durain and then by cuticular and spore clarite. At the same time, some coal beds consisting almost entirely of stable components – horsetail cuticles or megaspore casings (see Harris, 1952; Dulhunty and Dulhunty, 1949) – apparently do not always prove to be an allochthonous formation. Zhemchuzhnikov (1948) points out that the petrographic type of a coal must not be considered to give an accurate picture of conditions, since it is largely connected with secondary changes. Timofeev (1970) proposed a genetic classification based on the structure of the microcomponents of coals, which reflects the degree of decomposition of plant material. He specifies five genetic types: telinite (cellular structure of tissues preserved), posttelinite (the cellular structure is preserved in places, but the tissue is strongly pulverized and pressed into lens-like planes corresponding to individual groups of cells), percollinite (the tissues disintegrate into granulated particles corresponding to gelatinized cells), collinite (structureless substance or floccular structure), and leiptinite (durable microcomponents – cuticles, sporomorphs, etc. – predominate). The first two groups, according to his data, are formed in mobile bogs of alluvial and coastal plains, while the last three arise in stable limnic and paralic bogs.

Still, it is to be noted that in many cases a good correlation is observed between certain palynocenoses (assemblages with *Lycospora, Densosporites, Laevigatosporites, Torispora-Crassosporites,* etc.) and petrographic varieties or microlithotypes of coals (Smith, 1961; Smith and Butterworth, 1967; Navale, 1963; and other publications).

Teteryuk (1971) notes that the greatest diversity and variegation of palynocenoses are observed in the soil of coal beds. In the bed itself palynocenoses are more often oligodominants, reflecting bog vegetation in the main, while at the roof they are primarily allochthonous.

13. CLASSIFICATION OF BEDS AND TAPHOCENOSES

Each of the factors governing the formation of fossil plant beds that have been discussed in the preceding sections (material, transport, fossilization, reworking, facies distribution, paleogeographic conditions) can be used as a basis for their classification. We will call a **bed** a layer or part of a layer with remains of plants or animals. The bed is the main operative unit of taxonomic classification. The aggregate of plant remains contained in a bed is called a **taphocenosis**. Thus, we have two objects to be classified: the bed and the taphocenosis. The classification may be morphological or genetic in each case.

Morphologically, beds are distinguished according to the following signs:

1) Extent and form: plant-bearing strata, lenses, fissures, karst sinkholes, concretions, and combinations of these (e.g. a coal seam with peat dolomites).

2) Distribution of plant remains, which may occur inside the layer or on the bedding surface. I call beds of the latter type foliar roofs. They are formed during pauses in sedimentation.

3) Rock containing plant remains. The main types of beds in terms of this character are clastic, evaporitic, igneous, peat, and soil.

4) Thickness of the plant-bearing body. This ranges from zero (in foliar roofs) to 1–2 m (there seem to be no thicker plant-bearing strata that cannot be broken down into smaller components). Obviously, beds should not be designated as "thin" (as this term will be used to characterize the material) or "thick" (this could be confused with an estimation of the abundance of the remains). It is better to use the adjectives fine-bedded (0–5 mm), medium-bedded (5–20 mm), and thick-bedded (more than 20 mm).

5) Order in the arrangement of plant remains. In the absence of such order a bed is called disordered. Order occurs when the orientation the material showed during life is preserved or when the bed is situated in a mobile environment. Accordingly a bed may be residually ordered (with positive or negative geotropism of the elements) and dynamically ordered.

6) Concentration of the plant remains, determined by their amount per unit volume or weight of rock. In palynology the absolute pollen frequency

(APF) is calculated per unit weight; this is practically impossible in an evaluation of beds with large remains. Here the **saturation coefficient** of the bed is more conveniently calculated as the number of remains per unit volume, say per m^3.

Since the degree of saturation in beds of large organs depends not only on their amount but also on their size, in a preliminary evaluation of saturation we can consider as **concentrated** those beds in which the distance between adjacent remains does not exceed (or only slightly exceeds) their average length. In the opposite case a bed is to be qualified as **dispersed.**

The genetic classification of beds reflects the conditions of formation. In terms of paleogeographic conditions beds are divided into two major groups: paralic and limnic, with further subdivisions into deltaic, riverine, lacustrine, etc. Each of these types includes more fractionated groups of beds corresponding to individual geomorphological elements, superposition of volcanism, and other factors: e. g. prodelta, channel, shallow-lake, eruptive-lacustrine, etc. The classification may also reflect the overall geotectonic situation. The differences between stable and mobile zones in sedimentation rates, distribution of facies types, rhythm thicknesses, and volcanism leave a definite stamp on a bed. In particular, coals of mobile zones are distinguished by the weak extent of decomposition of the original material (Timofeev and Bogolyubova, 1962). In the Donets basin coal beds are seen to be confined to rhythms extending over 300,000–400,000 years, whereas the span of rhythms of the coal-free *Araucarites* suite is determined at 1,200,000 years (Balukhovskii, 1966).

Let us now go on to classify taphocenoses, basing ourselves on the following characteristics:

1) Predominance of certain organs. The terms leaf floras, seed floras, and so on are commonly used. Such terms are inappropriate, since the concept of "flora" does not marry successfully with a specific aggregate of organs. Spore-pollen assemblages are singled out in palynology, but these should be considered stratigraphic rather than taxonomic units. R. Potonié (1952) calls an aggregate of spores and pollen from one bed a **palynocoenum** (the more usual "cenosis" may probably be substituted for "coenum"). Correspondingly, an aggregate of fossil leaves may be designated a **phyllocenosis**, of fruits and seeds a **carpocenosis**, of wood a **xylocenosis**, of roots a **rhizocenosis**, etc. A territorial or stratigraphic aggregate of taphocenoses will be called a **taphoflora.** If it is desired to indicate the category of organs predominating in all the beds, phyllofloras, carpofloras, etc. may be distinguished.

2) Nature of fossilization. Above we mentioned that there are two main forms of fossilization: phytoleumas and petrifactions, and it was noted that there is sometimes a fine line between them. The name "ichnoflora" was proposed for taphocenoses consisting of impressions (Zhilin, 1969), but

word-building using the prefix "ichno-" has already been monopolized by the science dealing with traces of life activity – ichnology (see Radwanski et al., 1970). Furthermore, the form of preservation called an impression is a special variety of fossil rather than a mechanical imprint (trace) on rock.

3) Size of remains. Classification according to this sign makes sense, since remains of different size are studied not merely by different methods but, at present, by different specialists. Taphocenoses consisting of remains less than 200μ in size will be called here **microtaphocenoses.** Maceration and centrifugation are generally employed in their study. Entering into this class are spores, pollen grains, remains of algae and fungi, and other small objects. The second group is comprised of **mesotaphocenoses**, with remains measuring from 200μ to 2mm. This group includes almost all megaspores, small seeds, leaves and gametangia of Bryophyta, and scraps of cuticle. They can be distinguished on the surface of the layer, but can usually be brought to light by bulk maceration of the rock and individual scrutiny under a binocular microscope. Finally, **macrotaphocenoses** consist of remains more than 2 mm in size that can be seen with the naked eye. Of course, a single taphocenosis may include remains of all sizes. They are then called **heteromeric** as opposed to **isomeric**, where all the remains belong to one category. These signs reflect the degree of differentiation of the material.

4) Information losses. In animal taphonomy it is accepted that the biocenosis serves as the original material for a bed. If a taphocenosis corresponds to a group of organisms that were alive when buried (for instance, when part of the sea bottom is abruptly buried with its entire population), then we speak of a paleobiocenosis or paleocenosis (Merklin, 1969). A group formed posthumously is called a thanatocenosis (Boucot, 1953; Fagestrom, 1964). And finally, to show information losses in the course of diagenesis or the difference between what is buried and what falls into the hands of the investigator, Efremov's term (Efremov, 1950) oryctocenosis is used. Thus, these designations may be applied, on the one hand, as names for stages in the formation of a taphocenosis with progressive information loss (biocenosis – thanatocenosis – oryctocenosis) and, on the other, as names for types of taphocenoses in a case where only one of the stages had a decisive influence on their composition (of course, information losses are inevitable in the case of diagenesis, but I do not see any point in calling all taphocenoses oryctocenoses, since this term thereby loses its definitive meaning).

Many paleobotanists employ such a terminology, but here apparently we must take account of a feature that we have already mentioned: the burial of plants is comparatively rarely accompanied by their death. Most taphocenoses consist of organs that were detached from the plant during life. Besides, accumulations of plant detritus often include live (i.e. still viable) seeds.

Can we speak of a thanatocenosis in such a case? On the other hand, lithification losses depend not so much on the durability of remains as on the conditions of burial. Not only are they inevitable but in every case they are too considerable to be ignored. The scheme of progressive loss of information with the formation of a thanatocenosis has more or less the following appearance: 1) a phytomass of plants on a territory from which material is supplied to a bed; 2) the sum of mechanical and physiological deposit; 3) the part of the latter that gets buried, which, following Fagerlind (1952), we shall call the primary product Fp; Fp = Pr De%, where Pr is productivity, and De% the transport factor; 4) the part of the primary product which is conserved constitutes the secondary product Fs = Fp Co/100, where Co% is the conservation factor; 5) the part of the secondary product falling into the hands of the collector Ft = Fp Co Ma/1000, where Ma% is the manipulation factor, symbolizing the losses upon extraction of the remains.

5) Distance of the bed from the source of the material; this affects the value De%, which varies from 100% (in situ burial) to 0% (a bed is not formed). A taphocenosis that was formed in the place where the original phytocenosis grew is called autochthonous, while if it arose as a result of transport it is called allochthonous. Gümbel (1883) originally proposed these terms for coals. R. Potonié (1958) introduced the designation hypoautochthonous for taphocenoses that were formed without any considerable transport (as distinct from euautochthonous taphocenoses, formed in situ). In the paleoecology of terrestrial animals the relation between the composition of a taphocenosis and the distance of the place of habitat from the burial site is now also taken into consideration ("proximity method": Shotwell, 1964).

6) Species diversity and number of dominants. We call the dominants of a taphocenosis the species that dominate in terms of numbers: they do not necessarily correspond to the chief dominants in the original plant community. According to these characters taphocenoses vary from mono- to polyspecific and from mono- to polydominant.

7) Synusial composition of a taphocenosis. In many cases it is possible to determine quite reliably whether a taphocenosis corresponds to only one group of life forms of plants (e.g. only aquatic plants) or to different groups (hydrophytes, herbaceous plants of coastal marshes, mangrove trees, etc.). Merklin (1969) called the first type an isopaleocenosis. We may also speak of a **monosynusial** taphocenosis (by synusia is usually meant a certain aggregate of life forms, although the term is sometimes used in another sense). Accordingly, taphocenoses in which two, three, or more synusias are represented may be named oligo- and polysynusial. Il'inskaya (1958) classifies taphocenoses into monotopic (corresponding to one biotope) and polytopic (corresponding to many biotopes). However, biotopes are reconstructed

(with various degrees of arbitrariness) by means of taphocenosis analysis. If we then classify taphocenoses in accordance with this reconstruction, we come up against a vicious circle. Breaking down Il'inskaya's classification, Radchenko (1963) identified six types of taphocenoses: autochthonous monotopic (well-preserved remains belonging to a coastal phytocenosis), plesiopolytopic (with a small admixture of plants from more distant phytocenoses), hypsopolytopic (with an admixture of deformed remains of upland plants), allopolytopic (deformed remains belonging to various phytocenoses), and teletopic (sparse remains brought from afar). Here the distance of the bed from the source of the material is combined with the presumed biotopic composition and the nature of preservation. However, there is no direct connection between these two characteristics. An autochthonous bed may be polytopic and an allochthonous one monotopic. As for degree of preservation, it depends not only, and not so much, on distance, but more on the means of transport (in suspension or by rolling along the bottom), the state of the organs prior to transport, and on their displacement in nonsolidified sediment.

Part III

LIFE FORMS

And just as when I looked on that I marveled long within, since never before sprang such a stalk from earth.

The Odyssey

1. CLASSIFICATION

The concept "life form" (in the modern literature also "ecobiomorph") was introduced by Humboldt (1806), who distinguished 16 (or 19) forms: palms, bananas, lianas, cactuses, beefwoods, etc. Unlike phylogenetic systems, the classification of life forms was utilitarian from the outset, adapted to the needs of plant geography and geared to the description and definition of plant formations. No matter what the approach to the classification of life forms this objective remains in force ("Wenn wir nach den Grundlagen einer natürlichen Gliederung ökologischer Pflanzenformen suchen, haben wir zugleich an diese Zusammensetzung der Formationen zu denken, die ihrerseits das Fundament der 'ökologischen Pflanzengeographie' haben": Drude, 1913, p. 4). At the same time we see that some of the forms in Humboldt's system coincide with units of the phylogenetic system (palms, beefwoods). Grisebach (1872), who brought the number of life forms up to 54 and assembled them into groups of a higher order: trees, grasses, succulents, epiphytes, etc., tried to show the influence of climate on vegetation. In so doing, he laid the basis for a classification according to adaptive (epharmonic) signs, laying particular emphasis on leaf morphology. In the main the aim of the classification remained as before, since climate is the principal factor controlling the spread of plant formations. Different trends have been noted in the classification of life forms: some investigators have attempted to isolate groups that could help characterize environmental conditions (primarily climate), while others worked on a "natural" system of life forms. To the first group belong classifications in terms of temperature adaptations (de Candolle's mega-, meso-, micro-, and hecistothermic plants), aquatic adaptations (Schimper's hygrophytes and xerophytes), or in terms of adaptations for surviving an unfavorable season (Raunkiaer's system, based on the arrangement of perennating buds: phanerophytes, chamaephytes, hemicryptophytes, therophytes). Raunkiaer considered his classification to be natural, as it reflects the historical sequence of appearance of life forms (which is less questionable). Kerner, Reuter, Drude, and their followers understood naturalness of classification differently – as the expression of the ecological status of a plant, its role in the biosphere, that is, its ecological niche (ihre Rolle, ihre Besiedelungskraft in der Biosphäre: Drude, 1913, p. 5). This group comprises classifications:

a) in terms of the place occupied in the succession series (pioneer and climax plants; Ramenskii's "eksplerenty" and "patienty");

b) in terms of the overall role in the phytocenosis (ecological status). Of basic importance here are: division, introduced by Warming (1908), into autotrophs (autophytes) and heterotrophs, autonomous plants, parasites, epiphytes, etc.; subdivision according to the arrangement of the assimilative organs in the aerial or aquatic environment (aerophytes and hydrophytes: Drude); subdivision according to the life span (Serebryakov, 1955; Walter, 1971).

These groups are broken down into further subdivisions according to the combination of various epharmonic signs (with consideration of adaptations of both the vegetative and the reproductive sphere) and constitutional signs, between which, of course, there is no clear-cut boundary. The result is a physiognomic classification (Drude, 1913; Shmitkhyuzen, 1968), whose roots go back to Humboldt's system. However, as Shmitkhyuzen notes, a physiognomic classification always contains an ecological element as well, and Raunkiaer's purely ecological groups would seem to partly coincide with the physiognomic forms of growth.

Some physiognomic groups (in Humboldt's and Warming's systems) match systematic groups. Drude and other investigators (Bykov, 1970) consider the higher taxa of the phylogenetic system at the same time major subdivisions of life forms. In Drude's opinion, it is wrong to lump into one group of life forms, let's say, coniferous and deciduous trees or arborescent ferns and palms merely on the basis of a similar leaf arrangement, since they have different modes of reproduction, leaf organization, root-system formation, etc. At first glance, such a hybrid of classifications of species and life forms seems nonviable, because the first is based on a **causal idea** (affinity as the cause of similarity) and the second on a **teleological idea** (similarity as the result of adaptation to the same conditions). However, the so-called phylogenetic system must not be considered strictly causal. Many investigators feel that taxa of higher rank have a polyphyletic nature and are levels of organization corresponding to certain stages of biological progress, in other words, they are nothing other than teleological levels. As for the lower taxa, in only rare cases do we observe here a correspondence to the subdivisions of life forms. Nonrelated species often fall into the same group of life forms and, conversely, one species may be represented (in the adult stage) by various forms: for example, some species of *Ficus* and *Microsideros* grow as lianas in the forest but as shrubs in open places.

Let us note that in Hutchinson's classification (Hutchinson, 1969) herbaceous and woody dicotyledons are considered taxa of higher rank, corresponding to two main trends of evolution.

2. RESTORATION

Transition to the fossil state often goes along with disintegration and deformation. Before we go on to reconstruct life forms, we must piece together the original links between scattered organs and remove the effect of deformation. If the organic connection between organs cannot be established, the restoration will inevitably be hypothetical, resting on indirect proofs, the most important of which are the following:

a) **Common burial.** Organs of different categories (stems, leaves, strobili, pollen grains, etc.) have different weight to volume and surface to volume ratios, and therefore their capacity to be transported also differs. They thus become separated during transit and as a rule (though there are a few exceptions) end up in different beds (Black, 1929). Reid and Chandler (1926) divide the beds of the Oligocene Bembridge taphoflora into "leaf beds" and "seed beds," noting that in no case were leaves and seeds (fruits) of the same plant found buried together. Naturally, such instances are most unfavorable for restoration. The shedding of assimilative and reproductive organs at different times of the year may be an additional cause of taphonomic separation. Nevertheless, different organs of the same plant are often found together in taphocenoses that have not undergone any appreciable transport (hypoautochthonous). In the case of such beds Harris (1951) was right in saying that if reproductive organs are found, there must also be leaves (of which there are many more). Now everything depends on the state of the taphocenosis. The problem is greatly simplified if it consists of just a few species (which is more often the case in beds of the temperate climatic zone). In the Tsagayan deposits of the Bureya River I described several beds containing only *Trochodendroides* leaves and compound fruits assigned to *Trochodendrocarpus.* The connection is obvious here. When more complex taphocenoses are analyzed, we have to go by the method of elimination, which by no means always leads to an unambiguous decision. For instance, in the same Tsagayan beds there are the peculiar *Nordenskioldia* inflorescences associated with leaves of *Platanites, Papilionaceophyllum, Trochodendroides, Magnoliphyllum,* and some monocots. *Papilionaceophyllum* seems to be ruled out, as *Nordenskioldia* has little in common with the Leguminosae where gynecial morphology is concerned. *Magnoliphyllum* can hardly be considered a serious claimant: it is represented by single specimens, whereas there are hundreds of specimens of *Nordenskioldia.* In quantitative terms *Platanites* and *Trochodendroides* are the most probable, but for the first a connection is presumed with capitate inflorescences of the *Steinhauera* type and for the second a connection with *Trochodendrocarpus.* In the end, the leaves fit no genus. Krishtofovich suggested that plants with *Trochodendroides* leaves may have inflorescences of two types – *Trochodendrocarpus* and *Nordenskioldia.*

Such a complicated case is, of course, very hard to recognize and confirm. In other cases, on the contrary, there prove to be two or more candidates with more or less equal qualifications. For example, according to the data of Prinada (1962), the Jurassic microstrobili *Ixostrobus* associate with leaves of *Czekanowskia* and *Sphenobaiera* with equal frequency. Expectations of recurrent association are not justified in this case, since in the majority of Asian beds the leaves of both genera are found together. However, finds of microstrobili of another type in association with *Sphenobaiera* in the Triassic of Europe seem to tip the balance somewhat in favor of *Czekanowskia.*

The *Cycadocarpidium* riddle familiar to paleobotanists illustrates yet another complicated situation: these fructifications constantly associate with *Podozamites* leaves in Early Mesozoic deposits, but in the Middle Jurassic they completely disappear, whereas *Podozamites* remains one of the most commonplace fossils right up to the middle of the Cretaceous period. Florin (1951) and other authors assumed that *Podozamites* is a collective group and that the Late Jurassic species bear only a superficial resemblance to the more ancient forms associated with *Cycadocarpidium.* This explanation is hardly correct, as the structure of the epidermis in *Podozamites* is very distinctive, and at the same time some Early Jurassic and Early Cretaceous specimens are identical in terms of epidermal characters. Could it be that here, too, as in the case with *Nordenskioldia,* plants with the same leaves have different strobili? Is it possible that in spite of the traditional concepts, the leaves proved more conservative than the reproductive organs? We still have no answers to these questions.

It is still harder to connect up dispersed spores and pollen with large organs. Common burial tells us little: the difference in transportability is too great here. Besides, the subdivision of taphocenoses into allochthonous and autochthonous has significance only for some definite mode of transportation. For large organs transport by water is most important, while for iso-(micro)spores and pollen the principal agent is wind. Therefore, hypoautochthonous leaves and fructifications may associate with allochthonous pollen. Only very rarely can we note associations by basing ourselves on taphonomy. Thus, Harris (1965) observed that *Pteroma* pollen associates with *Pachypteris* leaves; in fact, in one bed there are no other leaves at all. The affinities of the dominant types of Mesozoic pollen are still not known. In Jurassic and Early Cretaceous spore-pollen assemblages large grains, reminiscent of the spores of *Anemia, Mohria,* and *Lygodium* (they were, in fact, formerly determined as such), hit one in the eye, but leaves of these genera are absent in the corresponding deposits. The groups *Normapolles* and *Aquilapollenites* dominate in Upper Cretaceous beds. But where are their leaves? *Platanus, Aralia, Sassafras, Viburnum, Ficus,* etc., known from determinations by paleobotanists, do not seem to fit the bill. Zaklinskaya (1970) proposed that this pollen was produced by herbaceous plants whose large organs have not been preserved. However, all the Cretaceous taphocenoses attest to a forest type of vegetation, and the percentage of nontree pollen in the palynological spectra of a forest is low as a rule. It is difficult to imagine that trees whose leaves dominate in Cretaceous beds would not have left traces in the palynological record, while the dominant types of pollen were produced by hypothetical grasses.

With regard to the Early Cretaceous spores of Schizaeaceae, it has become clear that their counterparts may be the numerous leaves of *Adiantopteris,* which are traditionally referred to another family. I think that in a case with Late Jurassic pollen lack of correspondence is to be attributed to taxonomic errors. In the examples set out above the effect taxonomic preconceptions exert on restoration is keenly felt. It was a long time before paleobotanists succeeded in recognizing the connections between the ovules and the "ferny" leaves of pteridosperms, between the pycnoxylic stems of *Callixylon* and the foliage of *Archaeopteris,* etc., because these organs were referred to different families and series and, moreover, because uniting them upset the system of "usual correlations" (Meien, 1970a).

b) **Structural similarity**. However representative it is, common burial enables us only to advance a working hypothesis, that requires confirmation from other sources. One of the key sources is identity of the epidermal and subepidermal structures – cells, stomata, hairs, and secretory organs. In some cases epidermal similarity is so obvious (for instance, in *Czekanowskia* leaves and *Leptostrobus* strobili) that no other proofs of connection are needed. On the other hand, the nonidentity of epidermal structures still does not prove that there is no connection. Bhatnagar (Bhatnagar and Meenakshi, 1969) mentions three groups of flowering plants: 1) with the same structure (ontogenetic type) of stomata on different organs; 2) with a different structure of stomata on different organs; 3) with a different structure of stomata within one organ. Florin (1933) notes differences in the epidermal structure of the leaves and floral bracts of Mesozoic Bennettitales.

By now we know of many finds of ovules in organic connection with the fernlike foliage of pteridosperms, but in Oliver and Scott's time such a combination seemed fantastic. The first hypotheses were based on the similarity of the glands present on the leaves of *Lyginopteris* and the cupules of the *Lagenostoma* type. No one any longer doubts that these organs belong to the same plant.

The identity of the secretory organs in the integument of the peculiar Carboniferous *Callospermarion* ovules (that occupy an intermediate position between the Lyginopteridales and Medullosales) and in the pith and cortex of *Callistophyton* stems helped Stidd and Hall (1970) establish the connection between these organs and institute a new family, Callistophytaceae. I used the same criterion in restoring *Pseudotorellia,* the leaves of which have secretory ducts running between the veins from base to apex. In the scales (*Umaltolepis*) and in the seed integument (*Burejospermum*) the ducts are not so long with tapered endings, but the diameter and the covering are just the same as in the secretory organs of the leaf (Krasilov, 1969).

c) **Identity of components.** If the organs being compared are fairly complex, we often discover that some of their components are identical. For instance, the spur of *Phoenicopsis* is covered with small scalelike leaves. Scales of exactly the same form and with the same epidermal characters are present at the base of the megastrobili of *Leptostrobus crassipes.* The identity of cataphylls served as additional proof that the leaves of *Pseudotorellia* and the scales *Umaltolepis* belong to the same plant: the cylindrical brachyblasts of this plant and the bases of the strobili are clothed in scalelike leaves that have the same form, epidermal structure, and resin bodies (fillings of the secretory spaces).

A study of pollen grains preserved in pollen chambers of ovules often aids in restoring the link between male and female fructifications. For example, sporangia of *Callandrium* and pollen chambers of *Callospermarion* contain pollen grains *Vesicaspora* (Stidd and Hall, 1970). The value of this criterion came under doubt after Sahni (1915) discovered foreign pollen in ovules of *Gingko biloba.*

Meanwhile, Oliver (1915) showed in a special work that mixtures of different pollen grains are never found in Paleozoic ovules. He considers the case described by Sahni to be an anomaly related to the artificial cultivation of *Gingko.* Apart from this, many Paleozoic plants had highly specialized pollen-trapping structures that apparently guaranteed that foreign pollen would not fall into the pollen chamber. Pollen grains are often preserved stuck to leaf cuticles. If the leaves had already put forth during anthesis, the plant's own pollen fell onto the leaves more frequently and in greater quantity than foreign pollen. All the same, such finds do not have much significance for restoration purposes. Attack of different organs by fungi of one species may be of some help in restoration. For instance, I found fruiting bodies of the same species of Ascomycetes on leaves of *Sphenobaiera* and on the seeds *Karkenia* that are associated with this plant.

Kokawa (1960) described a procedure for reconstructing the original form of seeds that were deformed upon compaction of the sediment.

3. RECONSTRUCTION

In accordance with what we discussed above (Part III, section 1), to reconstruct a life form means to explain the role of a plant in a phytocenosis in terms of the adaptations of the vegetative reproductive sphere and the combination of characters physiognomically determining the growth habit. A few plants have been preserved so fully and a few restorations are so perfect that they give a direct picture of the external habit of the plant. But more often we have to fill in the blanks with academic guesses based on morphological, paleogeographic, and taphonomic data. Conclusions on the adaptive value of certain characters are also far from always unequivocal.

A. Size

In forest formations the size and life span of a life form to a large extent determine its place in the structure of the phytocenosis. Many fossil plants have been described from remains of stems that give an idea of size (according to the trunk diameter one can approximately figure out the volume of the crown and that of the whole tree: Attiwill, 1969). It is known, for example, that the Paleozoic arborescent Lycopsida grew to a height of 40 m. In the case of dominant plants of the Paleozoic and Mesozoic known from leaf remains, a link has often been restored with a certain type of stem. Thus, leaves of *Archaeopteris* have been linked up with *Callixylon* stems (Beck, 1962), those of *Glossopteris* with *Vertebraria* (Yoshida, 1966; Pant, 1967), and those of *Dicroidium* with *Rheoxylon* (Archangelsky, 1968).

Only evergreen trees attain truly gigantic proportions, whereas the height of deciduous trees usually does not exceed 20–30 m. Prokhanov (1965) attributes this to the fact that at the end of the unfavorable season water could not rise to a very great height in deciduous plants due to the absence of an "upper motor" – leaves. He indicates that there are certain correlations between leaf morphology and size. An increase in size and intensification of the struggle for light lead to the appearance of entire blades on long petioles. Conversely, a decrease in tree height and an accordingly reduced water deficiency institute a trend toward a sparser network of veins, dissection of the blade, the appearance of lobes, the predominance of apetalous growth, and the formation of sessile leaves. If the Paleozoic Cordaitales and Mesozoic Czekanowskiales and *Nilssonia* of Siberia were deciduous, it is most likely that the canopy height of the forest formed by them scarcely exceeded 30 m.

In the Paleozoic conifers the lateral branches had a verticillate arrangement and a characteristic "featherlike" structure owing to the regular twin-row arrangement of the smallest twigs that were spread in one plane. On the basis of these characters, Florin (1951) suggested for them the form of a low-growing tree.

A distinct correlation with size is not manifested in the structure of reproductive organs. In the opinion of Thomas and Harris (1960), the megastrobili *Beania,* associated with *Nilssonia* leaves, were pendent, which is more typical for trees than for shrubs. In some cases, but certainly not always, a correlation is observed between the size of a tree and the size of its seeds.

Certain assumptions may be formed on the basis of the frequency of occurrence of remains. Chaney (1959) established that, other conditions being equal, leaves of trees figure much more importantly in beds than leaves of shrubs. However, as noted by MacGinitie (1941), the proximity of shrubs to the place of burial in most cases levels out this difference. In terms of pollen productivity trees as a rule outdo herbaceous plants, although *Secale cereale* and *Calluna vulgaris* produce more pollen than trees do.

B. Adaptation to climatic and edaphic factors

a) **Form of the stem.** Hygromorphous trees generally have a small crown and broad base. Their roots are often stiltlike or discoid. Sequoias growing in river valleys possess a number of adaptations to withstand flooding and silting up of the lower part of the trunk: asymmetrical thickening of the trunk in the presence of leaning, creating counterweight, negative geotropism of the roots, the formation of adventitious roots above the part of the stem buried by debris, etc. (Stone and Vasey, 1968). Devonian woody plants buried in situ in deltaic deposits (Barrel, 1913) had stems that were thicker toward the bottom, as is very characteristic for hygrophytes. Stems of this type have been described from Triassic deposits of Arizona (Daugherty, 1941) and other localities. The stigmariae of the Paleozoic Lycopsida are also considered to be a hygromorphous adaptation.

Xeromorphous stems are often thickened in connection with the development of water-storing tissue or pith, where inner adventitious roots are put out in xeromorphous cycads. The Mesozoic Bennettitales had barrel-shaped stems, this being regarded by many as a xeromorphous sign. Others thought the peculiar form of the stem to be a constitutional rather than an epharmonic character here, since Bennettitales are found in coal-bearing facies. In fact, various groups of Bennettitales apparently existed: those that are commonly found in coal-bearing facies usually had thin, forked stems. On the other hand, Cycadeoidea with short, thick stems covered with leaf bases were probably true xerophytes. This is borne out by an analysis of taphocenoses, as will be discussed below, and also by a comparison with recent xeromorphous cycads. *Encephalartos barteri,* an inhabitant of the stony savanna of West Africa, has a short stem (about 40 cm), whose height almost equals its thickness. The trunk is covered with a dense sheath formed from the bases of abscised leaves, which acts as a shield against both desiccation and fire (Hall and Jenik, 1967).

b) **Roots.** Hygrophytic trees often have a superficial root system that extends horizontally, like the Paleozoic stigmariae. The stiltlike roots can at the same time function as respiratory roots, or pneumatophores, which are particularly characteristic of mangroves. These roots absorb air instead of water. During high tide the oxygen contained in intercellular spaces in the aeriferous root tissue is used for respiration. Aeriferous tissue has been discovered in the roots of Paleozoic Lycopsida (Potonié, 1951). Cridland (1964) made a detailed study of the root system of *Amyelon* belonging to the Cordaitales of the American Pennsylvanian (Figure 10) and showed that they have a siphonostelic structure and are supplied with aerenchyma formed by phloem and phylloderm. These cordaites were thus mangroves. Pant and Singh (1968) have recently confirmed the formerly expressed assumption that *Vertebraria indica* represents the roots of *Glossopteris.* In my opinion, it is not excluded that these roots were pneumatophores.

c) **Leaf size.** Raunkiaer (1916) demonstrated that leaf size is connected with climate and distinguished 6 size classes: leptophyll (25 mm^2), nanophyll (225 mm^2), microphyll (2025 mm^2), mesophyll (18,225 mm^2), macrophyll (164,025 mm^2), and megaphyll (more than 164,025 mm^2). According to Axelrod and Bailey (1969), the abundance of large and very large leaves distinguishes tropical vegetation from pseudotropical vegetation (temperate, frost-free climate), where the leaves are also smooth-edged but predominantly of medium and small size.

A decrease in size often goes along with an increase in the number of leaves, so that the overall surface of foliage remains the same. However, a small entire leaf is in a better situation, because it is compact and all its parts are situated a short distance from the main water canals. This is especially important if there is strong development of mechanical tissue, which impedes the lateral distribution of water in mesophyll tissues.

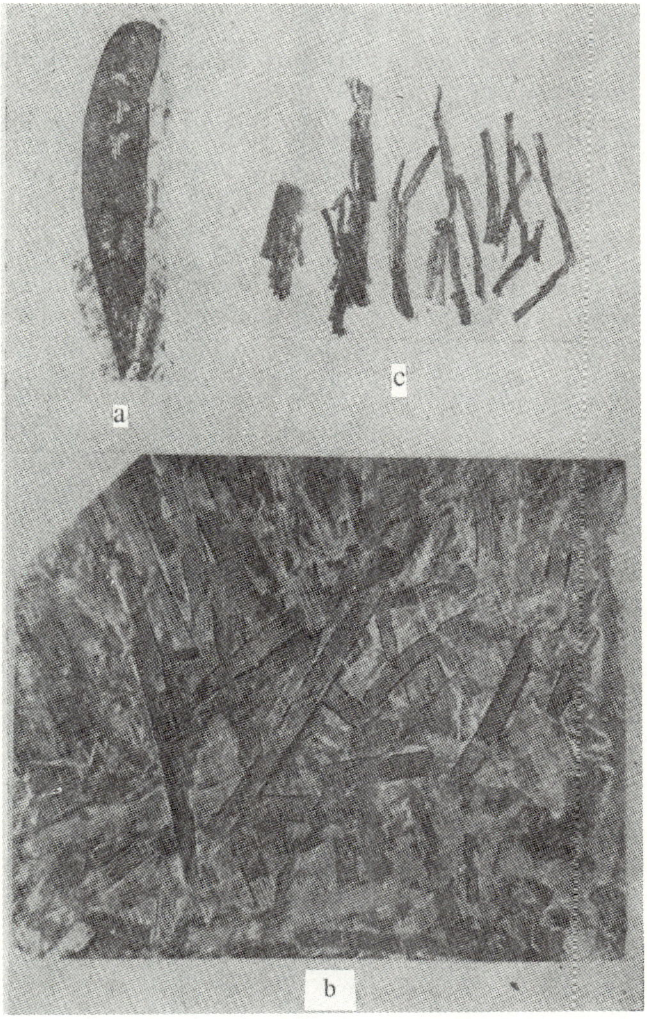

Figure 9. Reduction in leaf size: *Pseudotorellia angustifolia* Dolud. from consecutive horizons (a, b, c) of the Upper Jurassic of the Bureya basin (collection of the author)

Figure 10. Mangrove cordaite, reconstructed by Cridland (1964)

A reduction in the size of leaves often indicates a water deficit, although it is manifested under various conditions: in ericetal plants of upland moors (in connection with nitrogen deficiency), in the crowns of tall trees in a tropical forest, in the tundra, etc. A good correlation is often observed between leaf size and dryness of climate. For instance, the area of a myrtle leaf in Adler is 110 mm^2, while in the vicinity of Rome it is 40 mm^2; in Italian buckthorn, leaf area is 400 mm^2 in Italy and 270 mm^2 in France; in feijoa, leaf area is 300 mm^2 in Tomsk, 100 mm^2 in Yalta, but 35 mm^2 in Dushanbe (after Zadde, 1970). The correlation between size and climate is much less evident in leaves of magnolia, laurel, oak, etc.

Small-leafiness is particularly characteristic for sclerophyllous formations that develop under conditions of a wet-winter climate of the Mediterranean type. The similarity of life forms in the sclerophyllous formations of the Mediterranean, California, Chile, South Africa, and Australia is very striking, although the forms belong to systematically remote species and have a different evolutionary history (Mooney and Dunn, 1970). The convergence is due to a peculiar combination of several limiting factors: insufficient moisture in the warm period, poor soils, frequent fires, etc.

It must be borne in mind that small-leafiness does not mean reduction of the blade – it may be primary (e. g. in the Paleozoic Lycopsida). In order to speak of reduction, we must compare sizes of leaves from consecutive stratigraphic horizons. Axelrod (1958) points out that leaves of plants belonging to the Madro-Tertiary geoflora are only a half or a third of the size of leaves in the Arcto-Tertiary geoflora and a third or a quarter of the size of those in the Neotropico-Tertiary geoflora. We have as yet few examples of changes of leaf size in plants of the same species in successive stratigraphic horizons. In the Upper Jurassic taphocenoses of the Bureya basin there is an exceptional wealth of leaves of the gingkoalean *Pseudotorellia angustifolia* (Figure 9). In a number of beds they form "foliar roofs," that is, there is a continuous cover on the surface of the layer. In such cases hundreds of specimens can be counted and measured on one lump of coal. It was found that in the lower part of the coal-bearing series directly bordering on the marine strata about 80% of all specimens comprise leaves 6–7 mm in width with eight veins. In higher horizons (Azanov section) there is a predominance of narrow linear leaves 3–4 mm in width with four to six veins (about 70%). The narrow and the broad leaves were formerly referred to different species, but they are linked together by forms transitional in size in all the leaf "populations" and have the same epidermal structure. In the upper horizons, where small leaves predominate, the abundance and diversity of Bennettitales sharply increase; in Jurassic floras this is an indication of warming and, possibly, of a certain dryness of the climate.

It should be noted that a reduction in leaf size may be to some extent related to wind-pollination, since large leaves hinder the transport of pollen (Faegri and Pijl, 1966).

d) **Leaf morphology.** Bailey and Sinnot (1916) were the first to show that the ratio of simple and compound leaves, of entire and dissected leaves, of leaves with smooth and serrate margin, and with pinnate and palmate venation is related to climate. Subsequently many investigators adopted this criterion. For instance, Endo (1931) presents the following figures for forests in Japan (Table 3; see also Tanai and Onoé, 1961):

Table 3. Structure of the leaf margin in different climates (after Endo, 1931)

Climate	Leaf margin	
	smooth	dissected
Warm-temperate	56%	44%
Temperate	19%	81%
Frigid	8%	92%

A comparison of the *Woodwardia* flora of Japan with recent forests (Endo, 1968) is presented in Table 4.

Table 4. Changes in the percentage of leaves with different margin structure as a factor of climate (after Endo, 1968)

Area	Leaf margin	
	smooth, %	bicrenate, %
Micronesia and Philippines	90	0
Japan:		
Kagoshima	46	2
South of the Kimamoto Prefecture	33	12
Sendai	21	14
Sapporo, Hokkaido	12	27
Tokyo	15	38
Paleogene flora	35	14

The structure of the leaf margin has been widely used in paleoclimatological analyses of both the North American (MacGinitie, 1941; see also Table 5; Wolfe and Barghoorn, 1960; Dorf, 1969; Wolfe, 1971) and European (Schwarzbach, 1968) Tertiary floras. Axelrod and Bailey (1969) assume that for conclusions to be drawn relating to climate, data on the morphology of the margin must be supplemented by an analysis of leaf size. Small, serrate leaves predominate in the cold zone of the northern hemisphere, large, smooth-edged leaves in that of the southern.

Table 5. Changes in leaf morphology as a factor of climate (after MacGinitie, 1941)

Flora	Margin		Length		Structure		Venation	
	smooth	dissected	more than 10 cm	less than 10 cm	simple	compound	pinnate	palmate
Chalk Bluffs, Eocene	46%	54%	59%	41%	79%	21%	76%	24%
La Porte, Eocene	71	29	35	65	85	15	65	35
Goshen, Oligocene	61	39	53	47	88	12	82	18
Weaverville, Oligocene	47	53	60	40	81	19	60	40
Medicine Bow	67	33	45	55	83	17	60	40
Laramie, end of Cretaceous	71	29	45	55	87	13	64	36
Panama, Recent	88	12	50	44	85	15	83	17

The leaves of mangroves from three families: Combretaceae, Rhizophoraceae, and Avicenniaceae, show a substantial convergent similarity (Stace, 1966). They are fairly large, flat, and simple (rarely dimorphous), with pinnate venation and a smooth or slightly serrate margin. These characters are of little value in aiding us to recognize Mesozoic and Paleozoic mangroves, although they do not seem to contradict my hypothesis that certain Glossopteridae belong to mangrove plants.

Palmate venation is scarcely encountered in the leaves of tropical plants (except in lianas). Characteristic for them are droplike cusps and very small cells in the network of veins, without blind endings (Dorf, 1969).

Xeromorphous plants are characterized by hard, leathery blades and a whorled leaf arrangement. Sclerophyllous shrubs of various families display striking convergence in the morphology of the small, mechanically strengthened leaves. Xeromorphous characters, just like size reduction, are manifested in alpine, arctic, and marsh plants. On the other hand, deciduous leaves of desert plants sometimes do not have xeromorphous characters. In mosses and ferns adaptation to a water deficit consists mainly in resistance of the protoplasm to dehydration. Nevertheless, Gleicheniaceae, *Matonia,* and other ferns living in clearings and on the edges of forests have xeromorphous (heliomorphous), leathery leaves. In sphagnums a correlation is observed between the length of the branches, the shape of the leaves, and water supply (Green, 1968).

Among the Mesozoic ferns, xeromorphous characters are distinctly manifested in *Weichselia.* This fern has dimorphous fronds that are up to 1.2 m long and are arranged in a rosette at the apex of the shoot. Daber (1968) believes that they were periodically shed (this accounts for the exceptional abundance of leaves found in beds). The opposite-row pinnules are set at an angle to each other, forming a characteristic "butterfly" pinnate structure. They were leathery, hypostomous,with a papillose cuticle unusually thick for ferns. The fertile leaves were arranged in a vertical cluster inside the rosette formed by the sterile leaves. They were longitudinally folded, so that the distance between the pinnae of opposite rows was only 1.5–3 cm. This space was also occupied by the "butterfly"-type folded fertile pinnules, with reduced blade. For a long time *Weichselia* was considered a xeromorphous plant inhabiting sand dunes. Daber clearly demonstrated the wrongness of this interpretation. Apparently, *Weichselia* dwelt in coastal marshes or treeless bogs. My data on the *Weichselia* beds in the Suifun basin fully bear out this conclusion. Here is found yet another fern with a "butterfly" pinna structure – *Alsophilites nipponensis* (Krasilov, 1967).

Among the Mesozoic gymnosperms longitudinal folding or whorling of long leathery leaves is observed in the bog bennettite *Nilssonopteris* (Krasilov, 1967).

Polymorphism of leaves, a common phenomenon in recent plants, has been very poorly studied on fossil material, since different forms of blades are more often than not described as different species. One of the few reliable instances of polymorphism is seen in the Cretaceous conifer *Sequoia reichenbachii* (Gein.) Heer, which, like *S. sempervirens,* has leaves of two types: short, awl-shaped, and long, linear. It is interesting that the proportion of these two types of leaves was determined by the habitat conditions, as in the recent sequoia. Squamiform leaves predominate in sequoia growing in river valleys subject to flooding. Analogously, shoots of *S. reichenbachii* from coal-bearing facies have short, awl-like leaves, whereas in coal-free facies specimens with long leaves predominate (Figure 11).

e) **Anatomical characters.** The most common adaptations to a moisture deficit are a reduction in cell size accompanied by a decreased surface to

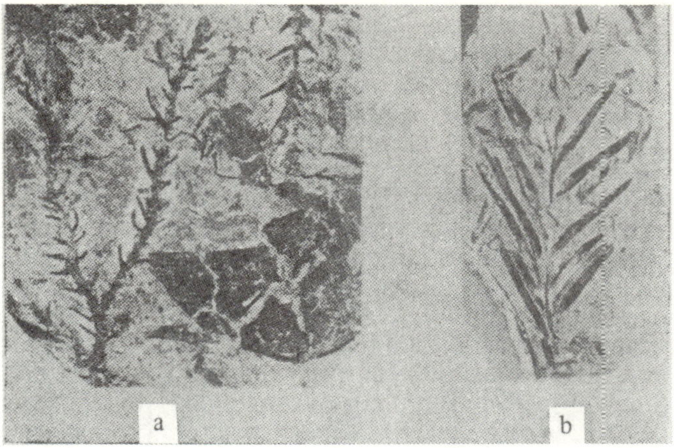

Figure 11. Variability of the leaves in Cretaceous *Sequoia reichenbachii*: a) bog facies; b) river channel facies; Upper Cretaceous of Sakhalin (collection of the author).

volume ratio and the development of mechanical tissue in sclero-xerophytes and of water-storage tissue in succulents. The ratio of the volume of cortex to wood is higher in xeromorphous than in mesomorphous plants (Fahn, 1964). Let us remember in this connection the mighty bark of the arborescent Lycopsida. The primary cortex was shed from the old part of the trunk, which was covered only with phelloderm. The density of stomata on the cortex of *Lepidodendron* was exceptionally great: up to 450 per mm^2. The cortex was apparently photosynthesizing. In the opinion of Thomas (1966), this explains the contradiction between the large size of the plant body and the small number of leaves, which were preserved only on the last-order shoots. The total amount of photosynthesizing tissue was increased on account of the cortex. Proceeding from these data and from the abundance of cortex remains in beds, I suggested (Krasilov, 1969a) that cortex shedding in lepidophytes has a physiological significance, as in some recent xerophytes with photosynthesizing cortex (in the given case it may be related to oligotrophy).

Xeromorphous plants typically have a dense network of veins, lignified cells of the fibrovascular bundles (in sclerophyllous forms: Zadde, 1970), small conducting elements, an increased number of tracheids and sieve tubes in the phloem strands (in grasses: Nikolaevskii and Nikolaevskaya, 1968), strong development of the mechanical tissues and sclerenchymous coverings of the conducting bundles, thickened cell walls, reduced intercellular spaces (and accordingly a reduced inner surface: Fahn, 1964), strongly developed palisade parenchyma, and relatively weakly developed spongy tissue (Vasilevskaya, 1965a, b; Grossgeim, 1950; Nikolaevskii, 1970; Evenari, 1938; Schields, 1950).

An important indicator of climatic conditions is the number of stomata per unit area of the leaf. It has been shown experimentally (Salisbury, 1928) that an increase in illumination without a change in moisture does not lead to an increase in the number of stomata, whereas a humidity change considerably influences their number. This means that the number of stomata per unit area does not depend on the rate of photosynthesis. Illumination here has an indirect rather than direct influence, altering the moisture. The minute study performed by Salisbury (1928) showed that the number of stomata per unit area depends:

1) on the position on the blade. The number of stomata as a rule increases from base to apex and from the midrib to the margin (i. e. the farther away from the main source of water supply). However, the number actually decreases right near the apex. In the long leaves of monocots (*Carex, Juncus, Iris*) the number of stomata is maximal in the middle of the leaf, while near the base and apex it is smaller. In the given case this phenomenon is related to intercalary growth of the leaf: here the base and apex are the oldest parts, that developed while still in the shade of other leaves. Hence, variations in the number of stomata within one leaf depend on the conditions of water supply and illumination;

2) on the size of the leaf. In each pair of leaves of *Mercuriales perennis* the larger leaf has fewer stomata per unit area. In *Ficaria verna* the correlation with leaf size was found to be positive but not fully distinct. In general in humid conditions differences in the number of stomata connected with leaf size are smoothed out;

3) on the position of the leaf. Yapp (1912) demonstrated that the number of stomata is greater on upper leaves. This regularity is also reflected in Zalenskii's "rule" (see Mokeeva, 1948; Darbinyan, 1960). If at the same time the size of the leaves changes as well, the relation between the number of stomata and the position of the leaf is reinforced;

4) on the situation of the plant in a given stratum. Figures for English trees and shrubs are given in Table 6.

Leaves are almost exclusively hypostomous. Isolated stomata on the upper side are noted in *Populus nigra, Carpinus,* and *Acer campestre.* In the herbaceous stratum 20% of shade species and 52.7% of species on the edge of a forest have amphistomous leaves. Thus, here too we observe a distinct correlation with illumination.

Table 6. Average number of stomata per unit leaf area (after Salisbury, 1928)

	Upper side	Underside	Both surfaces
Trees	0	223	224
Shrubs	0	196	199
Grasses	20.3	145	168.8

The correlation between the number of stomata in conifers and soil moisture is illustrated in Table 7.

Table 7. Relation between the number of stomata per mm^2 and soil moisture (after Voroshilova and Batsamyshina, 1969)

Humidification	Age of needles (years)	Yeddo spruce	Khingan fir	Korean pine
Control conditions	1	45.4	25.0	29.2
	2	35.4	28.0	33.2
	3	38.3	22.0	21.1
Aridifier	1	59.3	32.0	29.0
	2	56.1	25.5	34.5
	3	47.7	34.4	25.0
Waterlogging	1	38.5	37.0	42.5
	2	50.5	32.5	34.3
	3	37.6	34.0	28.8

Here we see that the number of stomata in needles of different age varies differently depending on the moisture conditions.

A distinct correlation between the moisture conditions and number of stomata was found on comparing evergreen xeromorphous plants (in relation to nitrogen deficiency in boggy soil) of Carolina and their more mesomorphous relative living in the Appalachians (Philpott, 1956). All xeromorphous species have small, leathery leaves with a smooth or slightly serrate margin and a thick cuticle. In Table 8, borrowed from Philpott, in each pair of species the one mentioned first is mesomorphous, growing in the Appalachians, and the second xeromorphous. The last column shows the type of minor veins: I – without bundle sheath extension, veins embedded in the mesophyll, with a large number of free endings; II – like I, but the tertiary veins have a sclerenchymous sheath and the polygons are large; III – minor veins with bundle sheath extensions, once or twice dichotomizing branchings penetrate the polygons; IV – veins with bundle sheath extensions, polygons small, intrusive branchings usually not dichotomizing; V – small polygons, no intrusive branchings.

Table 8. Comparison of stomatal density and the nature of minor veins in paired xeromorphous and mesomorphous species (after Philpott, 1956)

Species	Leaf area, μ^2	Stomatal frequency per mm^2	Pattern of venation (explanation in the text)
Clethra acuminata	1186	188	V
Clethra alnifolia	921	432	III
Ilex laevigata	1583	489	II
Ilex glabra	518	489	I
Ilex montana	1813	338	II
Ilex coriacea	411	517	I
Magnolia acuminata	2709	226	III
Magnolia virginiana	783	1053	IV
Pyrus melanocarpa	1223	291	II
Pyrus arbutifolia	812	414	I
Smilax rotundifolia	1604	320	I
Smilax laurifolia	3178	216	I

In general, an increased stomatal frequency in xeromorphous plants as compared with mesomorphs has now been definitely proven both for woody plants and for grasses (Nikolaevskii, 1970; Poplavskaya, 1966); moreover, this feature is apparent not only in plants of arid zones but also in marsh, arctic (Vikhreva-Vasil'kova, 1966), and alpine forms (Salisbury, 1928). In all cases an increase in the number of stomata goes along with a reduction in leaf size. We may assume that an increased stomatal frequency is not an adaptation to a water deficit but a side effect of reduced blade size. In fact, reduction in leaf size occurs mainly on account of a diminished size of the cells. Small leaves therefore have more cells per unit area. If we characterize the density of stomata not in absolute terms but by Salisbury's stomatal index: $I = \frac{100\,S}{E + S}$, where S is the number of stomata per unit area and E the number of cells per unit area, then the difference between mesomorphous and xeromorphous leaves is greatly reduced (Salisbury showed that shade and sun leaves with a different number of stomata per unit area have the same stomatal index).

What actually happens is that a reduction in the leaf surface as an adaptation to reducing transpiration leads to an increased number of stomata, in other words, to enhanced transpiration. In order to explain this seeming contradiction, we must take the following into account:

1) Reduction of the blade is connected mainly with the heat regime of the leaf, not with transpiration. The total leaf surface is larger in some small-leaved trees (e.g. in conifers: Fahn, 1969; Whittaker and Woodwell, 1968) or ericetal plants of upland moors than in large-leaved forms. At the same time, it has been shown that in the desert, leaves with an area exceeding 1 cm^2 suffer from overheating (Gates et al., 1968).

2) The stomata of xeromorphous leaves are generally smaller, so that a marked increase in their number results in a relatively small increase in the area of the apertures. Apart from this, some of the stomata do not function here as the apertures are blocked with wax (Fahn, 1964). And finally, a hair cover or whorling substantially cuts down on transpiration, counterbalancing the increased number of stomata.

3) Increased transpiration does in fact take place in some xeromorphous plants. Hemixerophytes transpire more than mesophytes (Grigor'ev, 1968). This accounts for the increase not only in the number of stomata, but also (albeit less considerable) in the stomatal index in xerophytes. Intensified transpiration may be related to insurance against overheating (Keller, 1933) or to other conditions of photosynthesis (small leaves do not form mosaics and receive more nonfiltered light: Thoday, 1931). It may be that the amphistomousness of xeromorphous leaves has the same significance, although

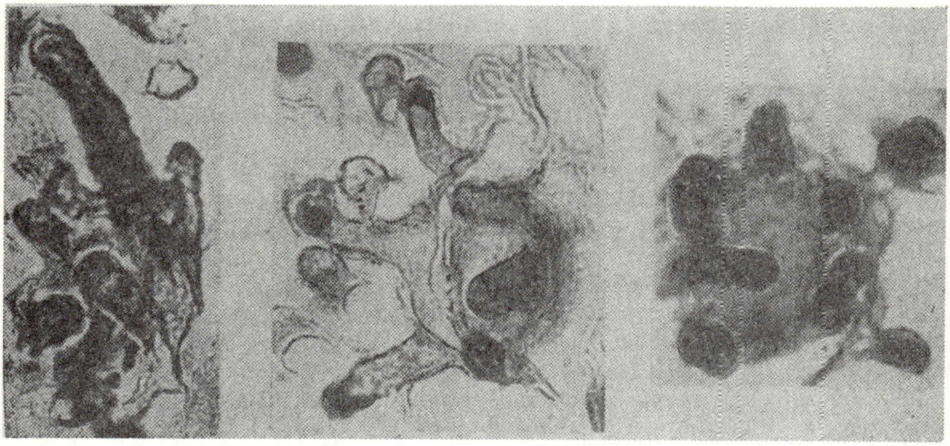

Figure 12. Stomata of Early Cretaceous *Gingkoites*, protected by hairs, X 474 (collection of the author)

it is more likely that it is an expression of the general tendency toward isolaterality that is a property of xeromorphous leaves.

It must be borne in mind that an increased size and decreased density of the stomata may be related to polyploidy. Working with fossil plants, it is hard to decide whether these changes have an ecological or a genetic cause. I have come up against this difficulty when comparing two closely related species, *Czekanowskia acifolia* Krassil. and *Hartzia angusta* Krassil. (the genus *Czekanowskia* differs from *Hartzia* only in the forked leaves – for the rest the two are very similar) that have essentially the same epidermal structure. The former derives from Upper Jurassic and the second from Lower Cretaceous deposits of the Bureya basin and they show a very different size and number of stomata:

Species	Width of leaf, mm	Number of rows of stomata	Average number of stomata per mm²	Length of stomatal apparatus, μ (with accessory cells)
Czekanowskia acifolia	1–1.3	6	20	150–180
Hartzia angusta	0.6–1	8–9	36	100–110

Since the sea was closer in the Upper Jurassic, we may assume that the differences are associated with ecological factors.

Other characters of xeromorphous leaves that may be recognized in the fossil state: a thick cuticle, running along the anticlinal walls (although *Rosa persica*, for example, has a thin cuticle: Keller, 1933), embedded stomata (though again not without exceptions: Keller, 1933), usually straight anticlinal walls of cells (sinuosity of the cell walls depends on the ratio of protraction of the cells of the mesophyll and epidermis). Xeromorphous leaves are often pubescent and covered with both living and dead trichomes that form a protective layer. Living trichomes in some cases have a water-storing function (Mokeeva, 1948). Among the ancient gymnosperms the Paleozoic conifers *Lebachia* and *Ernestiodendron* had quite strongly pubescent leaves, differentiating them from all the later conifers. The leaves of the Mesozoic Gingkoales were often more pubescent than adult leaves of *Gingko biloba*. Among the Siberian representatives of this group we find species with embedded stomata protected by a ring of hairs (Figure 12). In their degree of pubescence they resemble the juvenile leaves of *Gingko biloba*. This similarity can hardly have anything to do with recapitulation. The reason is rather that the adult leaves of the Mesozoic Gingkoales were more similar in terms of moisture regime and conditions of illumination to the juvenile leaves of gingko growing in botanical gardens of the middle latitudes. The most complicated and abundant trichomes among the Mesozoic gymnosperms were possessed by the leaves of *Nilssonopteris*, which is apparently a marsh plant. This genus also shows leaf whorling and other xeromorphous characters.

Mangroves can to some extent be recognized according to epidermal signs. Stace (1966) mentions that in the families Rhizophoraceae, Avicenniaceae, and Combretaceae the mangrove species are distinguished by their thick cuticle, straight epidermal walls, strongly cutinized guard cells, and a tendency toward the transformation of the anomocytic stomatal apparatuses into cyclocytic (the stomata are diacytic in *Avicennia*). The stomata are embedded and almost always protected by hairs. In mangroves of the family Combretaceae the number of stomata is the same on both leaf surfaces, or else there are more on the upper side. Another typical feature is that the system of venation is not reflected in the topography of the epidermal cells.

The Triassic pteridosperm *Lepidopteris ottonis* has amphistomous leaves, the stomata sometimes being more numerous on the upper side and sometimes more numerous on the underside. When combined with a thick cuticle, this feature may indicate that the plant is a mangrove (Townrow, 1966).

f) **Shedding of assimilative organs.** The morphology of leaves provides some clues as to a plant's adaptation to seasonal leaf fall. Thus, deciduous plants more often have leaves with a rigid, entire or shallowly dissected blade that offers resistance to wind (as in aspen) and a long, flexible petiole with lever action (Prokhanov, 1965). With such a structure, defoliation occurs rapidly. Fully analogous adaptations of this type are shown by *Populus tremula,* the Cretaceous *Trochodendroides,* and *Gingko.* Fossil Gingkoales with an almost entire blade and long petiole appeared way back in the Triassic, but it was only toward the end of the Cretaceous that they became widely distributed as components of conifer-broadleaf forests. Such ancient gingkos as *Sphenobaiera* had large, estipulate leaves deeply dissected into long, linear lobes; they were probably not shed. They are often found attached to a spur. What probably happened was that the brachyblasts were shed with the leaves, but there is no proof that this was a seasonal phenomenon.

On the other hand, the Cordaitales, *Glossopteris* and *Pseudotorellia,* which, judging from taphonomic signs, were deciduous, had rigid, estipulate leaves with a short line of detachment. The theory held by Delevoryas (1968) that the cycadeoid Bennettitales were deciduous is borne out by the analogy with the deciduous xerophilous cycads (*Encephalartos*) as well as by the distribution of these plants on the periphery of the Mesozoic arid zones and by taphonomic features. Daber (1968), as already mentioned, suggests that the xeromorphous fern *Weichselia* shed its leaves.

Finds of brachyblasts with attached leaf clusters are known in the Cordaitales, Glossopteridae (Plumstead, 1958, not fully authenticated), *Pseudotorellia* (Krassilov, 1969; here shoots without leaves are found much more often, and we have reason to assume that the leaves were shed independently of the brachyblasts), *Sphenobaiera* (Kräusel, 1943), *Nilssonia* (see Krasilov, 1969a), *Phoenicopsis* and *Czekanowskia,* for which such a form of preservation is especially characteristic, and conifers with *Pityophyllum* leaves. The seasonal shedding of brachyblasts, that form huge accumulations in beds, can be considered as proven only for the Czekanowskiales.

As mentioned above, the Paleozoic lepidophytes were probably cortex-shedding plants.

Deciduous cataphylls (scales) are known in the Cordaitales (*Lepeophyllum*), Gingkoales, the bennettites *Pterophyllum* and *Pseudocycas* (*Cycadolepis*), and the Cycadaceae (*Deltolepis*). The connection between these cataphylls and other organs (only a few have been found in organic association) is confirmed by morphological and epidermal characters. Some of them were probably bud scales.

g) **Taphonomic characteristics**. Theories on the adaptive value of certain characters are based mainly on comparisons with analogous features in recent plants that serve as models in reconstructions. However, the same features may have had a different adaptive significance in ancient plants (a good example of this is cauliflory; see below under C.). In addition, many characters (e. g. xeromorphous ones) do not provide unequivocal indications of habitat conditions. Therefore, as far as possible all hypotheses should be checked by means of taphonomic observations. The most important of these are the following:

1) In order to interpret adaptations to moisture regime, we can divide plants according to the facies distribution of the remains into anthracophilic ("flöznahe") and anthracophobic ("flözfremde": Meien, 1970b). The first to make such a subdivision of floristic elements were Davies in the Carboniferous of England and later Gothan and Gimm (1933) in Permian deposits of Europe. Subsequently, anthracophiles and anthracophobes were established in the Late Paleozoic of the Donets Basin, North America, and other regions (see Fisunenko, 1965, 1967; Remy, 1952, 1958; Cridland and Morris, 1963; and many other publications). Devonian and Mesozoic beds have been studied less in this respect. In the Devonian Nemejc (1962) isolates two groups: a bog group – *Thallomia, Sciadophyton, Rhynia, Hicklingia, Cooksonia,* and a mesophytic group – *Taeniocrada, Zosterophyllum, Gosslingia, Bucheria, Hedeia, Yarravia, Psilophyton, Pseudosporochnus, Asteroxylon.* I have tried to apply this classification into anthracophiles and anthracophobes to the most important genera of Mesozoic plants (Table 9).

None of these genera can be said to occur only in coal-bearing or only in coal-free facies. It is the predominant distribution that is to be kept in mind here. Besides, for some genera the facies confinement changes depending on

Table 9. Anthracophilic and anthracophobic plants in the Mesozoic

Group	Anthracophiles	Anthracophobes
Filicales		
Marattiaceae	*Marattiopsis*	*Danaeopsis*
Osmundaceae	*Osmundopsis*	*Todites*
	Cladophlebis	
	ex gr. *denticulata*	
Schizaeaceae	*Klukia*	*Stachypteris*
	Ruffordia	
Gleicheniaceae	*Gleichenites*	
Matoniaceae	*Matonidium*	
	Phlebopteris	
Camptopteriaceae	*Clathropteris*	*Dictyophyllum*
		Thaumatopteris
Weichseliaceae	*Weichselia*	
Dixoniaceae	*Coniopteris*	
Pteridiaceae	*Onychiopsis*	
Arthrophyta	*Equisetites*	
	Neocalamites	
Lycopodiaceae	*Isoetites*	*Pleuromeia*
	Nathorstiana	*Lycopodites*
Cycadophyta	*Caytonia*	
	Nilssonia	
	Ctenis	
	Heilungia	
Bennettitales	*Anomozamites*	*Anomozamites*
	Pterophyllum	*Pterophyllum*
	Nilssoniopteris	*Ptilophyllum*
	Dictyozamites	*Zamites, Otozamites, Pseudocycas*
Gingkoales	*Pseudotorellia*	*Gingkoites*
	Sphenobaiera	*Sphenobaiera*
	Erethmophyllum	*Glossophyllum*
Czekanowskiales	*Czekanowskia*	*Hartzia*
	Phoenicopsis	
Coniferales	*Araucarites*	
	Elatides	
	Sphenolepis	
	Podozamites	
	Pityophyllum	

the geographical latitude. For example, in Lower Cretaceous deposits of the Maritime Territory conifers with *Pityophyllum* leaves occur primarily in coal-free facies, while in the north, in the Bureya and Lena basins, they are found in coal-bearing facies. Of course, anthracophily and anthracophobia are not equivalent in value to hygrophily and mesophily, but as a supplement to other characters they may help in the reconstruction of adaptations.

2) In order to identify mangrove plants we must observe what forms are confined to coastal-marine deposits and associations with remains of marine phytoplankton and marine fauna. According to these signs (in combination with some morphological features) the Triassic lycopod *Pleuromeia,* the pteridosperms *Lyginopteris* (Teichmüller, 1952), *Pachypteris* and *Lepidopteris,* the bennettite *Ptilophyllum,* and some other plants are supposed to belong to mangroves (see Part II, section 12B).

3) In order to show the presence of seasonal leaf fall we must take note of lithological and paleontological signs indicating the season in which the layer is formed. Strictly speaking, an investigation of this kind is possible only in sections with a seasonal stratification. In this connection we must remember that a bed is not always formed directly after the shedding or dispersal of organs. For instance, the pollen of plants flowering in spring falls into autumn-winter layers after reflotation (Part II, section 5). Heer (1855 and other works) paid great attention to the season of burial, guiding himself by associations with particular species of insects or reproductive organs (e. g. clays with cones and seeds of sequoia are an autumn formation: Heer, 1861), but his observations mostly refer to series in which seasonal stratification is not expressed. Merker (1961) expressed some ideas on the seasonal development of the prothallia of *Rhynia* that are yet to be corroborated.

C. **Adaptations of the reproductive sphere**

a) Adaptations to the dispersal of disseminules

We touched on this aspect in Part II, section 5, and here we shall just go over certain points. The ancient isosporous plants mostly had no adaptations to dispersal other than their small size and light spores. Spores more than 200 μ in size (the arbitrary borderline between micro- and megaspores) appeared at the end of the Early Devonian (Chaloner, 1970). Potonié (1953) mentions that many of the morphological characteristics of Paleozoic megaspores are connected with adaptation to transport by water. First, they differ in their very large size (let us note that the water-dispersed isospores of liverworts are usually also very large). The average size of recent megaspores is 1.5 mm, of Late Cretaceous Senonian megaspores 1 mm, and of Carboniferous megaspores 2 mm. However, frequent among these are megaspores with a diameter of more than 3 mm, and giants up to 11 mm in diameter (*Cystospora gigantea* Zerndt.) are encountered. Many megaspores have hairlike processes that increase their surface considerably. For instance, *Triletes ramosus* Arnold has an equatorial zone of anastomosing hairs that approximately doubles the megaspore diameter. In addition, air is trapped in the hairs, enhancing buoyancy. Second, buoyancy is promoted by equatorial outgrowths: the auricles of *Valvisisporites* Ibr., *Triquitrites* Wilson and Coe, and *Tripartites* Schemel, equatorial frill, and saccus. But the majority of the sculptural elements of megaspores – tubercles, papillae, spines, shagreened surface – may be interpreted as adaptations to anchoring and maintenance of immobility during germination. Some of them had a special "anchor": a beak-shaped process in the region of the tetrad scar in *Lagenicula* megaspores (that probably belong to *Lepidodendron*). I observed such a feature in megaspores of the herbaceous Jurassic *Lycopodites.* Many Paleozoic seeds have a ribbed surface, a common adaptation to transport by runoff. The seeds of *Stephanostoma,* described from Paleozoic deposits of India with a *Glossopteris* flora (Pant and Nautiyal, 1960), have an oval form and measure 1–1.5 × 1–1.1 mm with conical spines about 0.3 mm long.

Potonié thinks that the buoyancy of the Paleozoic seeds *Aetheotesta elliptica* was provided by air chambers in the testa (similar to the aeriferous tissue in the pericarp of *Scirpus, Potamogeton,* etc.). It may be that the interpretation of some morphological and anatomical features of Paleozoic seeds (e.g. the multilayer sarcotesta of *Pachytesta* with large intercellular spaces in the broad parenchymal zone: Taylor, 1965) should be sought in connection with adaptation to water transport.

From Upper Devonian deposits in Western Siberia, Anan'ev (1963) described a formation that he thought to be a winged seed and that actually does remind one of symmetrically winged seeds of the *Samaropsis* type. But this formation is too different from the seed structures of Devonian plants (Pettitt, 1970), and therefore it is doubtful whether his interpretation is correct. Winged seeds appeared in the Cordaitales, in which the greatest diversity of morphological types is observed – seeds with a continuous winglike margin (*Thysanosperma* Zalessky), with two symmetrical wings (as in *Samaropsis* Goeppert), with asymmetrical outgrowths (*Craspedosperma* Zalessky). The symmetrically winged seeds were carried over without any changes to the Lebachiaceae and have been retained in some conifers up to the present. Seeds with a unilateral wing appeared in the Permian (*Sylvella*: Sukhov, 1969), but they did not become widespread until the Mesozoic. Some dominant types of Cretaceous angiosperms also had seeds with a unilateral wing (*Trochodendrospermum*: Krasilov, 1971b), but they were much smaller than in the Pinaceae. The pubescence of the Paleozoic seeds of *Physostoma elegans,* that may have increased their buoyancy, was at the same time a preadaptation to anemochory. Jurassic deposits of Karatau have yielded the peculiar seeds *Problematospermum* (Turutanova-Ketova, 1930) with tufts of hairs. This is the first parachute mechanism in Mesozoic seeds (Figure 13).

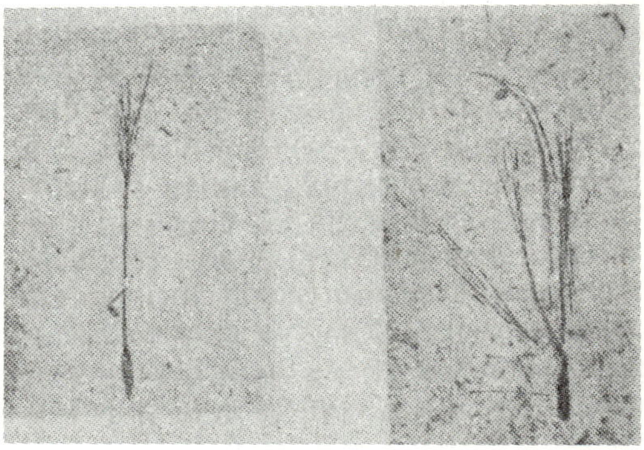

Figure 13. Very ancient seeds with hair tufts from the Upper Jurassic of Karatau (collection of M.P. Doludenko)

Caytonia and *Czekanowskia* (*Leptostrobus*) had very small seeds that could be effectively transported by the wind without any kind of adaptations apart from lightness.

Dispersal of sporophylls apparently took place in the Paleozoic *Lepidocarpon.* A sporophyll with a single megasporangium containing one functioning megaspore was detached whole (Scott, 1900). The spherical sporophylls of *Pleuromeia* that enclosed the megasporangia were probably adapted for transport by currents. This mode of reproduction is in general analogous to the dispersal of cone scales with ovules in conifers. Winged cone scales appeared back in the Permian in this group. They are typical for the Voltziaceae and Araucariaceae.

It is not excluded that the arrangement of ovules on the leaf surface promoted the dispersal of seeds enclosed within a convolute leaf (Ridley, 1930). *Spermopteris* from Permian deposits of North America had taeniopteroid leaves the margins of which bore ovules in two rows (Cridland and Morris, 1960). Paleoecological observations show that *Spermopteris* is a plant of sod habitats. Its reproductive organs have entered the fossil record precisely because they were successfully transported by water. In the Glossopteridae dispersal probably took place originally with the aid of a cupule. Reproductive organs of the *Scutum* type (Plumstead, 1958) had a rounded cupule with a multitude of seeds. The pedicel of the cupule was partly intergrown with the rhachis of the sterile leaf. A further stage of intergrowth with the leaf is illustrated by *Cistella*; here the cupules are sessile and do not have wings. In *Lanceolatus* the fertile organ is embedded in the tissue of the unchanged leaf of the *Palaeovittaria* type, to which the function of dispersal probably passes.

We may thus assume that the evolution of the reproductive organs of the Glossopteridae and of the Pteridospermae as a whole was related to adaptation to dispersal. It is to be noted that transformation of the supporting bracts and perianth into a flight apparatus is observed not only in flowering plants but also in some species of *Ephedra.*

Finds of strobili of *Ullmannia* in a rhynchocephalian stomach (Weigelt, 1930) make us ponder the possibility of zoochory. Seeds (or cupules?) with distinct processes that probably served for attachment have been described from the Triassic of Arizona (*Carpolithus chinlea* Daugherty, 1941). Our hypothesis on the manner in which these seeds were dispersed is correct only if it is true that the Triassic reptiles did in fact have a hair cover (Dehm, 1965).

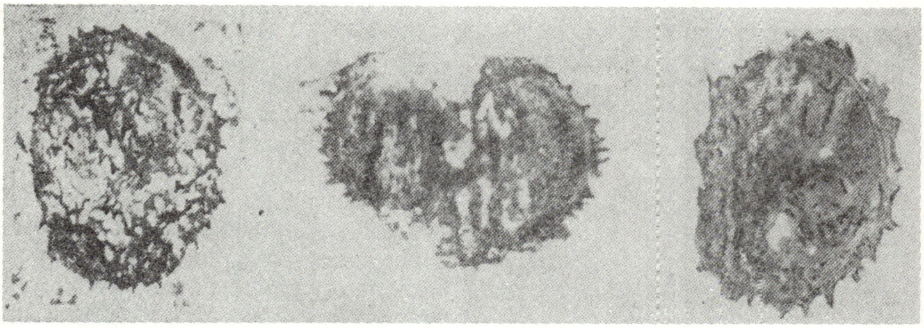

Figure 14. Capsules with barbs, X 7, Upper Jurassic, Tyrma River (collection of the author)

From Upper Jurassic–Lower Cretaceous deposits on the Tyrma River I described bivalvate capsules whose outer surface is covered with conical processes (Figure 14). The apexes of these processes are filled up with a dark-colored substance and apparently had a secretory function. This find may give some clues concerning adaptation to zoochory in the ancient angiosperms. Among the plants accompanying the giant herbivorous dinosaurs, only "*Withamia*" (Seward, 1895) was equipped with stout spines.

b) Adaptations to pollination

The productivity of the sporangia was considerable in the progymnosperms. For instance, the fertile branch of *Tetraxylopteris* consisted of four main segments that were formed as the result of two consecutive dichotomies. Each of these in turn dichotomized three times. The number of microsporangia in one such fertile complex was from 2800 to 4704 (Bonamo and Banks, 1967). However, it is practically impossible to decide which of the progymnosperms "still" produced spores and which "already" produced pollen. Even such a character as the presence of a saccus cannot serve as a reliable criterion in this respect (Chaloner, 1968).

The Paleozoic Pteridospermae had variously developed pollen-trapping devices: lagenostoma and salpinx. Neither in them nor in the Cordaitales or Mesozoic gymnosperms has a haustorial pollen tube been discovered,* although pollen is often found inside the micropyle of practically all seeds studied in the corresponding fashion except in *Amphorispermum*. The constant absence of pollen in the pollen chambers of *Amphorispermum* suggests that the pollen-trapping surface was here situated outside the ovule. The first plants to have a fully developed stigma were the Mesozoic Czekanowskiales (*Leptostrobus*: Krasilov, 1968; Figure 15).

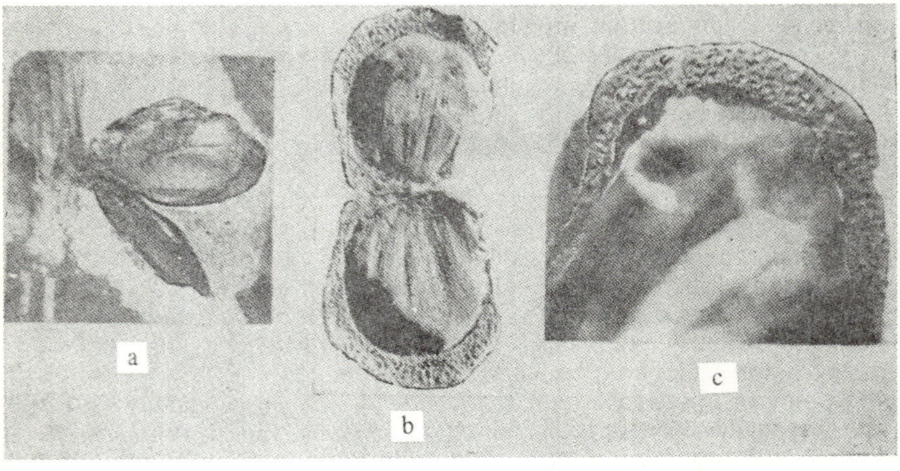

Figure 15. Bivalvate capsules (a, b, × 3.5) and stigma (c, × 7) of *Leptostrobus*, Jurassic of the Bureya River (collection of the author)

* Rothwell (1972, Science, Vol. 175, pp. 772–774) recently described the germination of prepollen and the formation of a branching tube in Paleozoic ovules.

The secretion of a pollination drop has been demonstrated on the Carboniferous ovules *Callospermarion* (Stidd and Hall, 1970), where pollen grains (*Vesicaspora*) entrapped in the drop were found on the micropyle.

Saccate pollen grains were possessed by many Paleozoic pteridosperms and Mesozoic Caytoniales and conifers. Harris (1956) described coprolites of a small animal that abounded in pollen of *Caytonia*. Zoidophily may have operated here. The sacless pollen grains of the Mesozoic gymnosperms are usually small and have a smooth exine (the larger pollen grains of some Bennettitales may be related to polyploidy), which in general fits the description of anemophilous sacless pollen with a typical size of 20–30 μ (Faegri and Pijl, 1966). An exception is the widely distributed pollen *Classopollis*, which apparently belongs to conifers. Frequent finds of *Classopollis* tetrads and of conglutinate grains imply adaptation to entomophily. Interestingly, these characters combine with the development of a distal pore (pseudopore). The pollen of the Cretaceous angiosperms mostly (except for the diverse, often very small grains of the *Tricolpopollenites* type) falls into the size bracket of entomophilous pollen and has a complicated structure.

Bisexual flowers are considered by many to be an adaptation to entomophily. They were present in the Mesozoic Bennettitales. According to Sharma (1968), about 70 species of *Williamsonia* flowers are known, 16 of which have only microsporophylls, 4 seem to be bisexual, and the rest are female. He assumes that the *Williamsoniae* were in most cases dioecious, and male plants were rarer than female plants (as in the Cycadaceae). The male flowers are large, open, and without a perianth (*Weltrichia*). The female flowers are usually found closed, with bracts closing above the gynoecia. It is not clear if they opened at all or were cleistogamous. In the latter case we should have to assume the development of apomixis in the Bennettitales. This would explain the reduction of the androecium, the rarity of male flowers (total reduction of the male sphere does not usually take place in apomicts, since there is an episodic sexual process in the majority of them; moreover, pollen promotes the development of apomictic seeds), and the small (in comparison with macroscopic remains) amount of pollen found in beds. In the Jurassic of the Ukraine I found a *Williamsoniella* flower (Figure 16) with fully developed microsporophylls and immature gynoecium. Protandry probably occurred here, or else the gynoecium was aborted and the flowers functioned as male ones. Delevoryas (1965) showed that the majority of flowers of the Cycadeoidea were female, with or without remains of a reduced androecium. In bixesual flowers the microsporophylls were not pinnate, as formerly thought, but looked like segments of an orange and were intergrown between themselves and with the axis. A flower of this kind did not open, and it seems that self-pollination was the sole possibility. In many

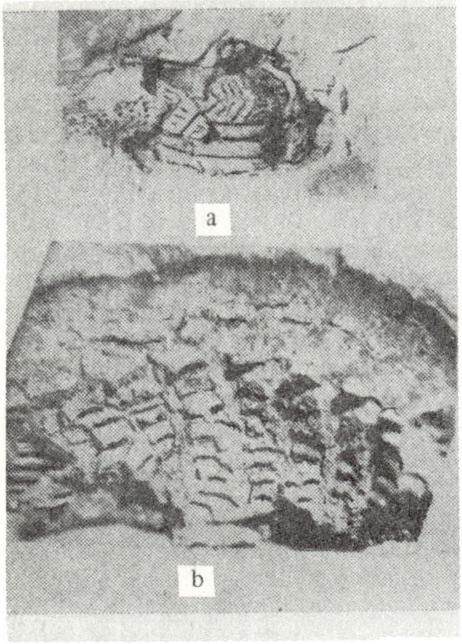

Figure 16. Jurassic flower of a bennettite:

a) arrangement of microsporophylls, ×3; b) wall of microsporophyll with synangia, ×7; Jurassic of Kamenka (collection of the author). The gynoecium, unlike the androecium, is immature (protandry).

cases, however, the gynoecium was aborted and the flowers functioned as male ones. Upon ripening, the microsporophylls were shed or split open, and the pollen was borne away by wind or insects. Many flowers bear injuries, indicating that they were visited by insects. On the other hand, signs of androecial reduction and the relative rarity of male flowers in the Bennettitales may attest to apomixis. The fact that dispersed pollen is relatively rare in these plants may be attributed both to entomophily and to apomixis. According to Chiguryaeva (1970), an increase in the amount of abnormal pollen in transitional epochs (Rhaetic, Lower Miocene, etc.) may be related to the development of apomixis. This interesting hypothesis has not yet been sufficiently bolstered by factual material.

c) Cauliflory

This feature is usually mentioned as one of the most important characteristics of trees in a tropical rain forest. In fact, about 1000 species with cauliflory have been established in tropical forests, although the ecological significance of this is not quite clear. A distinct correlation is noted between cauliflory and pollination by bats (chiropterophily) or dispersal by these animals (chiropterochory, sometimes passing into ornithochory). Thus, we may assume that cauliflory in flowering plants is a specific adaptation to pollination and is confined to tropical forests for the reason that it is here that the appropriate pollinators are present in sufficient numbers (Faegri and Pijl, 1966). These considerations, and also the taxonomy of cauliflorous species (they are present only in advanced families) contradict Corner's assumption that cauliflory is an original feature (at least for flowering plants). As for pre-Cenozoic plants, here cauliflory was naturally related to other factors. Potonié (1953) believes that in Paleozoic plants it was an adaptation for protection of the reproductive organs against heavy rains.

4. EVOLUTION OF THE LIFE FORM

The ancestors of the terrestrial plants are not known to us (Schopf, 1970). The theory of an isomorphous alternation of generations in the first land plants seems to be borne out by new data on Devonian rhyniophytes. At first, Merker (1959 and other works) came to the conclusion that the "rhizome" of *Rhynia* is its gametophyte, since it is not vascularized (the stele of the vertical shoot ends blindly in a polyhedral group of cells), dies before the vertical shoot, has a more variable form than the shoot, and is covered with unicellular rhizoids. Later, Pant (1962) put forward a still bolder hypothesis, according to which *Rhynia* had a vascularized gametophyte. And, finally, the isomorphous alternation of generations was confirmed by the discovery of archegonia (Lemoigne, 1969 and other publications).

Many authors went along with Leclerq (1954) in refusing to see in the rhyniophytes the original group of terrestrial plants, since these were supposedly specialized marsh plants (or xerophytes, which later became specialized as halophytes: Merker, 1959). Zimmermann (1963) at present thinks that *Rhynia* was a specialized descendant of the primitive rhyniophytes. The huge autochthonous bed of *Rhynia* discovered in a 40-cm layer of a fossil peat bog scarcely leaves room for doubt that these were plants which lived in marshes and were evidently well adapted to do so. This, however, does not contradict the notion of the archaicness of the rhyniophytes, for it is natural to assume that even the very first land plants were specialized according to their need. Investigators who believe the rhyniophytes were relatively advanced doubt altogether that the paleontological record can accurately date the appearance of terrestrial flora. In their view, the land plants emerged and underwent a long process of development before their first remains were buried. Without dwelling on this point, let us just note the conclusion reached by Chaloner (1970, p. 368), based on an enormous amount of palynological material: diversification of the spores of land plants essentially began in the Early Devonian, and there is nothing to confirm Axelrod's assumption of prolonged pre-Silurian evolution ("there is no foundation for Axelrod's apprehension that there had been a major phase of pre-Silurian land-plant evolution, virtually unrepresented in the fossil record"). The undifferentiated terminal branches of *Rhynia* are in fact Zimmermann's telomes, in which such elementary processes as polarization, formation of the apical meristem, etc. had already been completed (Zimmermann, 1952). Further evolution of the gametophyte, as is known, proceeded under the badge of reduction, whereas the fate of the sporophyte was more complex.

a) Growth

Plants differ from animals (except for corals) in that their growth is confined to specific points of the body. Growth (and accordingly the meristem) may be apical, intercalary, marginal, or basal. Apical growth, which is the main factor in increasing the size of the plant, occurs by division of a single apical cell, which is clearly expressed in the Bryophyta and Equisetales and less distinctly in the other pteridosperms, or by division of the apical meristem, that consists in angiosperms of two more or less distinct zones – tunica and corpus – and in the majority of gymnosperms of radially arranged cells. Ontogenetic investigations on fossil plants are reported here according to the survey made by Delevoryas (1964). Kidston and Lang described the shoot apex of *Asteroxylon,* that is protected by a closely adjoining ring of scales. Melchior and Hall (1961) studied *Calamites* with a pyramidal apical cell, as in *Equisetes.* Morgan (1959) showed that in terms of the nature of growth the arborescent fern *Psaronius* in general differs little from the recent representatives of this group. Increase in size occurs exclusively on account of growth of the primary body, without secondary thickening, and is accompanied by complication of stele structure, the addition of new orthostichies and the concomitant replacement of old ones, and a switch from spiral to verticillate phyllotaxis. On the whole, the stem thickens upward, the cover of adventitious roots downward.

One of the results of Morgan's minute investigations was that several "species" of *Psaronius* that had been isolated on the basis of the stem diameter and number of orthostichies were united. They are actually different ontogenetic stages. Therefore, the structure of a stem fragment of *Psaronius* depends on its age. At first it was believed that the same applied to the arborescent Lycopsida, but Walton (1935) and later Eggert (1961) showed the relationship between the structure of a fragment and its position. In the early stage of development the stem has a siphonostele with a primary xylem of large tracheids, is covered with a thick cortex, and bears rows of large leaves. Secondary xylem and periderm then develop. Contrary to old reconstructions, the lower part of the stem, covered with periderm, did not bear pulvini, while the upper part of the stigmariae showed no traces of roots. In the primary branches the stele has a smaller diameter, fewer rows of leaves, and less secondary tissue. In the secondary branches all these elements are further diminished, and so on. This, according to Eggert, is how growth is restricted according to the principle of **apoxogenesis**. In *Lyginopteris,* according to Andrews, size limitation is also achieved by a reduction in the potential capacity for growth at the transition from the main stem to branches of higher and higher orders, but here substantial differences are observed in the size of the primary body. *Medullosa,* according to Delevoryas, has a polystelic stem, and inner periderm surrounds each stele. The primary cortex is also shed here. A special feature is that the primary tissues apparently retain the capacity for growth when secondary tissue is already beginning to form. Despite the thickening, the steles become increasingly discrete as growth proceeds. I would note that, according to certain observations, in the arborescent Lycopsida the strongly developed protoxylem also preserved the capacity for growth after the development of metaxylem (Lemoigne, 1966).

b) Size

Quite popular among zoologists is Cope's hypothesis ("law") of increase in size in the course of phylogenesis (for an analysis of this see Rensch, 1954). Kozo-Polyanskii (1922) earlier asserted that progress "goes hand in hand" with size increase, but up to a certain point, beyond which shrinking and drying out ensue. If we take *Rhynia* as the point of departure, we see that a very considerable increase in size took place in the general evolution of the plant world, though the matter proved more complicated in certain phyla. For the angiosperms some authors (Tsvelev, 1969) postulate that development proceeded from herbaceous to arborescent forms, but most investigators favor the scheme of Sinnot and Bailey: from trees to grasses. Takhtadzhyan (1943, etc.) links this process with neoteny, against which Serebryakov (1962) pits his view that the period of development up to the first flowering may be longer in grasses than in trees. He considers that along with descending there were ascending lines in the evolution of size, both lines possibly being connected with xeromorphogenesis (Serebryakov and Serebryakova, 1969). Stebbins (1965) postulates for the first flowering plants the habit of a low tree or shrub (Hallier advanced a similar hypothesis back in 1912), while Meeuse (1967) opts for the polyphyletic hypothesis and considers that the very first Mesozoic Angiospermae were represented by all life forms except therophytes. It is clear even from this very sketchy survey that we are still far from definitive conclusions in this matter.

An increase in size demands first of all enhanced mechanical durability. This is achieved by increased hydrostatic pressure, silica infiltration of the cellulose membranes (in horsetails and grasses), and other means, the chief one of which is lignification. Lignin apparently appeared independently in the vascular plants and giant mosses of the genus *Dawsonia* in connection with increased size (Siegel, 1969).

Bower (1930) points out that an increase in size leads to a smaller ratio of body surface to volume. The result is physiologically inefficient. A reduced surface/volume ratio is compensated by morphological complication, primarily an increase in the surface of conducting tissues. The stelar apparatus becomes more complicated due to crenulations, medullations, the development of metaxylem, or the segregation of the conducting bundles. Bower presents a table illustrating the increase in the diameter of the stele and in its complication in connection with size increase in the series *Rhynia–Asteroxylon* (Table 10).

Table 10. Correlation between stele size and structure in the Psilophytales (after Bower, 1930)

Species	Diameter of stele, mm	Form of stele
Rhynia gwynnevaughanii	0.156	cylindrical
Hornea lignieri	0.300	cylindrical
Asteroxylon mackiei	1.800	stellate, with 4 rays
Asteroxylon elberfeldense	2.200	stellate, with pith and 10 rays

In the ontogenesis of many pteridophytes we see a recapitulation of phylogenetic complication of the stelar anatomy related to size increase. The preservation of a large protostele in some ferns (*Gleichenia montaguei*) is regarded by Bower as fetalization. Still, the main pathway leading to truly large dimensions is the development of secondary wood.

The discovery of *Callixylon* stems with secondary wood in association with fernlike fronds of *Archaeopteris* was the first step in the study of the phylogenetically extremely important group of progymnosperms. The *Callixylon* stems grew to a diameter of 5 feet and a height of 28 feet (Beck, 1970). The structure of the metaxylem is in general as in the conifers. The anatomy of the plants united in this group attests to the unity of origin of the eustele of the cycadophytes (*Lyginopteris*) and coniferophytes by way of division of the primitive protostele in the Calamopityaceae (*Stenomyelon, Calamopitys*) and its transformation into a system of cylindrically arranged sympodia.

The height of *Callixylon* (to 10 m) may indicate the size from which the evolution of the various groups of seed plants began (Figure 17). Florin (1954, etc.) conjectures that the primary conifers were low trees with whorled branches.

Figure 17. *Archaeopteris*, reconstructed by Beck (1970)

According to my data (Krasilov, 1969a), the original Gingkoales with large, estipulate, deeply dissected (*Sphenobaiera*) or entire (*Glossophyllum*) leaves and small seeds were small trees forming part of the underbrush of conifer forests. The increase in size that took place in the Mesozoic went along with the appearance of larger seeds, the development of an entire leaf blade with a long petiole, deciduousness, and promotion to the role of dominants of the plant cover. Only an ascending line seems to have been expressed here.

All we know for sure about the Nilssoniales is that they had leafy shoots of two types: long shoots and brachyblasts, with a cluster of leaves (both types are represented in my collection). But even this is enough to refute Meeuse's motion (Meeuse, 1967) of the Nilssoniales as plants with a short underground stem and a leaf rosette. Judging from the morphology of the leaves and of the loose, pendent megastrobili of *Beania* with a mobile axis, the Nilssoniales were, rather, small trees. Similar assumptions may be made concerning the habit of *Czekanowskia* with leaf bundles on the brachyblasts and very long, pendent male and female strobili. On the whole, the *Czekanowskia* foliage resembles that of *Pinus.*

Neither can we agree with Meeuse regarding the reconstruction of the Caytoniales and Corystospermales (underground stem with a leaf rosette). Harris (1964) described branching fragments of a stem (or large branch?) of *Caytonia.* Mesozoic pteridosperms of the *Lepidopteris* type have been compared to semisubmerged herbaceous hydrophytes on the basis of the fact that in such species as *L. stormbergensis* there are more stomata on the upper side of the leaf (Harris, 1959). However, according to Stace (1966; see Part III, section 3), such a distribution of the stomata in combination with a thick cuticle is characteristic for woody mangroves of the family Combretaceae. The rather massive "cupular disks" of *Lepidopteris* also speak against a herbaceous habit.

Thus, all the Mesozoic groups of seed plants that can claim to be the ancestors of the Angiospermae were primarily woody in habit. This contradicts both Tsvelev's hypothesis (Tsvelev, 1969) on the primariness of the herbaceous flowering plants and Meeüse's notion of the original diversity of life forms in these plants, although there is, of course, no basis for supposing that the ancestral group was represented by only one life form. The transition from trees to grasses has been comparatively well studied in many groups of angiosperms and is seen to be realized in different ways. The intermediate stage may be the prostrate shrub, the orthotropic or creeping subshrub, etc. (Serebryakov and Serebryakova, 1969; Kolishchuk, 1968).

We should not doubt that in the flowering plants there was, together with a descending also an ascending line of size evolution.

c) Branching

The following types of branching are observed in the very ancient land plants: 1) dichotomous; 2) dichotomous with a "bud" in the bifurcation (this "bud" may be an artefact related to flattening of tissue upon fossilization, but all the same, it is characteristic for some *Hostinella, Protopteridum, Psilophyton,* and *Gosslingia*: Höeg, 1942); 3) pseudomonopodial: *Psilophyton* (Höeg, 1942), *Timerophyton, Krithodendron,* and others (Siegert, 1965); 4) H-shaped (*Zosterophyllum*: Walton, 1964; *Kaulangiophyton*: Gensel et al., 1969; *Crenaticaulis*: Banks and Davis, 1969, etc.); 5) of the epiphyllous branch type: *Enigmophyton* (Höeg, 1942).

The variations of branching within one genus (*Psilophyton*, for example) point to the relative ease of transition from one type to another and contradict the taxonomic theory of Greguss (1961), who made this character the basis of his classification and excluded the group *Sigillaria–Isoetes* from the Lycopodiales, relating it instead to the cycadophytes. It is to be noted that also the arrangement of the sporangia is a highly variable sign in Devonian plants: *Psilophyton* had terminal and lateral sporangia (Anan'ev and Stepanov, 1968). Described for *Asteroxylon* are terminal and lateral sporangia and sporangia forming a fertile region on the main axis (Lyon, 1964). The relation between the type of branching and climate is manifested in the fact that tropical plants usually show monopodial branching, which under conditions of a cold or arid climate is often transformed into sympodial branching (Takhtadzhyan, 1964).

H-type branching is specific to Devonian plants, while epiphyllous branches are present in *Botryopteris antiqua* (Galtier, 1969), are observed in some ferns (Troop and Mickel, 1968), and apparently played a certain role in the evolution of the reproductive organs (the gonophyll theory: Melville, 1960).

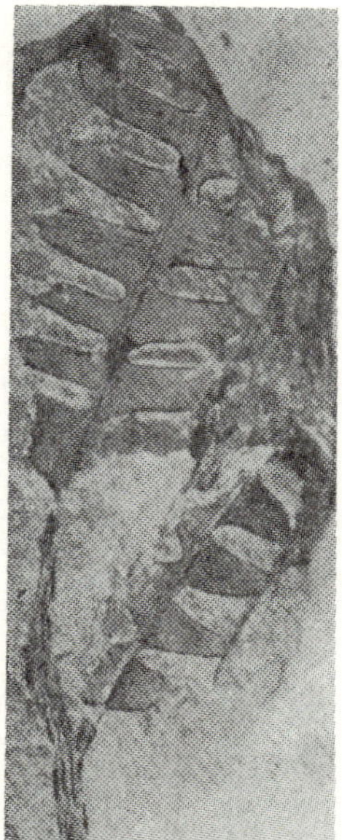

Figure 18. Shoot of *Nilssonia*, showing that in growth habit this Mesozoic plant had little in common with the recent cycads (collection of the author)

Reduction of branching is traced in all lines of the Gymnospermae beginning from the progymnosperms (see, for example, Wilde's work on the Podocarpaceae: Wilde, 1944). If we consider the Mesozoic plants of the group *Nilssonia–Ctenis* (Figure 18) as the ancestors of the recent cycads (which is not absolutely certain), we may say that this group evolved in the direction of reduced branching and pachycauly (we have already mentioned more than once the brachyblasts and other features of branching in *Nilssonia*). This line is more convincingly traced in the Bennettitales, where thin, furcate stems preceded nonramified short, thick stems of the type in the Cycadeoidea (Delevoryas, 1968).

The type of branching of the first flowering plants may be indirectly judged from the morphology of the leaves and reproductive shoots. They were all more or less small-leaved (simple leaves prevailing) and had small flowers clustered into paniculate or spicate inflorescences (*Trochodendrocarpus*: Krasilov, 1971b). These characters definitely contradict Corner's "pachycauline" theory. Besides, all the most likely predecessors of the angiosperms (see above) were leptocauline plants. Pachycauline trees appear at the end of the phylogenetic series in both the gymnosperms and angiosperms (Parkin, 1953). Corner's "law" (Corner, 1967), according to which a pachycauline "starter" is present at the base of the phylogenetic tree of angiosperms and all its branches, is in fact groundless, like the whole Durian theory. Corner (1954) vividly describes the fruits of *Durio*, to which the combination of red and black lends something demonic. But if it is demonic, then it is probably also archaic. There are no other proofs that the durian is primary or even ancient. Since Goethe's time, this seems to be the only case where a theory based on a single, merely esthetic impression has gained such wide popularity.

d) Leaf

The relationship between leaf and stem has attracted keen interest among phytomorphologists since Goethe's time (for a historical survey see Stewart, 1964; Maekawa, 1965). Maekawa mentions that with the discovery of the Psilophytales, Potonié's pericaulome theory became modified into Zimmermann's telome theory, with which Agnes Arber's "partial-shoot" theory (Arber, 1950) and Maekawa's own theory of "leaf classes" are in some way or other connected.

The apical stem of pteridophytes is totipotent, that is, it gives rise to all the tissues. In seed plants sometimes only the tunica of the apical meristem takes part in the formation of the leaf primordium, while in other cases both the tunica and the corpus participate in this (Clowes, 1961). In the initial stages the stem also grows due to division of the cells of the leaf primordium. Thus, it follows that starting from the ontogenesis there is no absolute boundary between leaf and stem. On the other hand, stem increment on account of growth of the leaf primordium must not be interpreted in the sense that the leaf is the true apex of the shoot, while the stem is formed by the leaves (phytons). It is better to think of both of them as integral parts of the shoot (Philipson, 1949).

Some pteridophytes show dichotomous branching by way of division of the apical cell (or group of apical cells) into two equal parts. With monopodial branching the buds are borne below the apex or (adventitious buds) on the old part of the stem. An axillary bud is formed when the age of the leaf equals several plastochrones (periods of growth between the laying down of two consecutive leaves). A branch is thus formed of an older meristem than a leaf. In ferns success has been achieved in converting a leaf primordium into a branch, but such experiments have failed with seed plants (Clowes, 1961).

New paleobotanical discoveries shed light not only on the relation between leaf and branch but also on the origin of large and small leaves. Many authors, following Linnaeus, divided leaves (or more correctly trophophylls) into phylloids (in the Lycopsida) and cauloids, and the latter into microphylls (coniferophytes), macrophylls (cycadophytes), and mesophylls (angiosperms). However, there is no weighty evidence for an essential difference between the trophophylls of the lycopods and pteridophytes. On the other hand, a study of the progymnosperms seems to confirm a different origin of micro- and macrophylls (Beck, 1970). The most primitive progymnosperms, Aneurophytales, still did not have true leaves. The lateral branches ramified abundantly and ended in small, forked terminal twigs. The branches were distributed in various planes, or else the entire system was more or less flattened. The Upper Devonian *Rhacophyton* occupied an intermediate position between the Aneurophytales and the more advanced gymnosperms. The terminal segments of the lateral branches here resemble more a leaf. Within the genus *Archaeopteris* transitions are observed from dichotomously branching terminal segments to pinnules with an almost entire blade. Such, according to Beck's theory, was the origin of the small simple leaf of the coniferophytes. The lateral organs of *Archaeopteris* were formerly described as compound leaves. In fact, these organs are flattened and look like leaves, but their "rhaches" have a radial rather than dorsoventral structure of the conducting system (Carluccio et al., 1966). Apart from this, supporting leaves are present at the base of the lateral organ and its units ("plumes"). This organ is therefore to be considered not a leaf but a branch. Yet it is quite probable that the compound leaves of the pteridosperms with "intermediate" pinnules on the rhachis are a modification of the lateral branching system of *Archaeopteris.* To bolster this idea, some "stem" characters of leaves of Carboniferous pteridosperms have been presented: *Tristichia, Tetrastichia, Chapelia,* and others, as well as leaves of such angiosperms as *Kigelia* and *Quarea rhapalocarpa* (Grushvitskii and Skvortsova, 1970) with a radially symmetrical conducting system of the petiole. Hence, the simple leaves of the coniferophytes derived from the terminal segments of the branching system of the progymnosperms, while the compound leaves of the cycadophytes arose as a result of a flattening of the entire system; this, as Beck points out, does not rule out a secondary genesis of a compound from a simple leaf.

The theory concerning the origin of the leaf blade in the progymnosperms apparently links this process with a change in genetic structure, i. e. with the development of constitutional potentials for leaf formation. However, the Devonian *Gingkophyton* and related forms, some of which also belong to the group of progymnosperms, had fairly broad and almost entire leaves. This gave rise to the assumption that the absence of a leaf blade in the Aneurophytales has an ecological rather than a constitutional significance. The morphological series from furcate terminal twigs to an almost entire leaf blade may reflect an amelioration of the climate at the end of the Devonian.

A transition from simple to compound leaves is observed in some groups of flowering plants (*Acer, Rubus,* Vitaceae: Eames, 1963; Parkin, 1953) or, conversely, from compound to simple by way of reduction or fusion of leaflets (Proteaceae, *Bauhinia,* and others). These data can scarcely help resolve the question of the original type of leaves in the angiosperms. Hypotheses of a large pinnate leaf (Corner) or of an entire magnolium-type leaf take no account of paleobotanical data according to which the leaves of the first angiosperms were for the most part simple, small, with an entire or (more often) laminal blade. The quite large group of Cretaceous angiosperms had a peculiar, later almost extinct type of leaves, with sympodial branching of the midribs and segments (*Dewalquea, Proteophyllum, Fontainea,* and others). The Late Triassic and Jurassic plants *Scoresbya* and *Imania,* among others, had leaves with such a structure (Krasilov and Shorokhova, 1970). This type of leaf may have been ancestral for the angiosperms (or one of the ancestral types).

At the same time there is evidence that ecological factors determined the form of the leaf in the very first flowering plants. The distribution of the Early Cretaceous angiosperms of the Maritime Territory shows an interesting feature (Krasilov, 1967): only compound (*Sapindopsis*) and laminal leaves are found in the Suchan basin and only entire leaves in the Suifun basin. The vegetation was of a different nature in these two areas: in the Suchan basin a redwood-type forest, dominated by Taxodiaceae (*Elatides, Athrotaxopsis,* etc.), and in the Suifun basin savanna or chaparral with a predominance of Bennettitales. The differences in the vegetation were apparently related to an orographic factor – the effect of the mountains separating the two basins on the distribution of atmospheric precipitation.

The homology between leaf and branch is corroborated by an analogous variability under the influence of climate. Zimmermann (1969) showed that the compound leaves of Cenozoic and recent ferns (pteridophylls) show certain fundamental differences from the Paleozoic forms. The former are characterized by an anadromous structure, the latter by a catadromous structure. The catadromous leaf is built on the principle of a mirror image: a pinna departing, let's say, clockwise branches in such a way that its first pinnule will also be directed clockwise. A change of symmetry with the transition to anadromy (in a pinna departing clockwise the basal pinnule is directed counterclockwise) is connected by Zimmermann with a change in the polarization of the axis. I think that the change in branching here, as in many other cases, can be explained from the point of view of allometric growth, specifically, the suppression of apical growth. The first segment of an anadromous branching system should apparently be considered as the primary apex of the main axis, that is displaced laterally by the rapidly growing lateral branch. In this case an anadromous system may be interpreted as a special variant of sympodium – monochasium (one branch grows higher than the main axis).

The Mesozoic pteridophylls are, according to Zimmermann, an intermediate link between the Paleozoic and Cenozoic pteridophylls, demonstrating a transitional catadromous-anadromous structure and palmate leaves of the "Dipteridaceae type." In fact, in the Paleozoic we hardly know of any ferns with palmate fronds, while in the Mesozoic they are widespread in a number of families: Camptopteridae, Matoniaceae, Schizacaceae (*Lygodium*, beginning from the Cretaceous), and others. Considering the pteridophylls as a whole, including the leaves of pteridosperms as well, we see that a tendency toward the formation of a palmate leaf is observed as early as in the Paleozoic (*Syniopteris*), but it was only in the Triassic that pteridosperms with "Dipteridaceae-type" leaves appear (Krasilov and Shorokhova, 1970). The interpretation of the palmate leaf presented in this work is analogous to the interpretation of the anadromous leaf: a palmate leaf is formed as a result of inhibited growth of the main axis (see Hirmer and Hoerchammer, 1936) and its overgrowth in the given case by two sympodially branching lateral pinnae (dichasium).

If we take into account that palmate leaves developed in parallel and more or less simultaneously in different groups of pteridophylls and that this process is essentially analogous to and coincides in time with the transformation of the catadromous into the anadromous system, it is natural to conclude that the evolution of pteridophylls is related to climatic change.

A correlation between changes of a leaf (or leaf arrangement) and climate is observed in fossil neuropterids (retardation: Laveine, 1966), *Sphenophyllum* (Asama, 1966), *Nilssonia* (reduced width of leaves during the Jurassic and Early Cretaceous: Asama, 1968), and conifers. Decussate leaves (reduction of the internodes – mesomes) arose in Triassic conifers under arid climatic conditions (Lemoigne, 1967). Among the Taxodiaceae, *Sequoia* with heteromorphous leaves is intermediate between *Metasequoia* with a distinct seasonal rhythmics and *Sequoiadendron* (without such a rhythmics). At the end of the vegetation period an apical bud is formed on the shoot of *M. glyptostroboides,* while in *S. sempervirens* at the same time the length of the leaves and internodes is reduced (incomplete bud formation). In the next vegetation period an area of shoot is formed with short leaves and long internodes; this part branches and corresponds to the long shoot of *M.glyptostroboides* (that also branches). There follows a nonbranching part with long leaves and long internodes, which corresponds to the short shoot of *M.glyptostroboides,* that bears longer leaves with longer internodes than are present on the long shoots. Thus, both plants show the same rhythmicity of development, the differences being only quantitative (Schwarz and Wide, 1962).

We pointed out in the preceding section that polymorphism of leaves in the Cretaceous *Sequoia reichenbachii* and the recent *S. sempervirens* coincides and is governed by ecological factors in the same way. Prolonged preservation of a definite system of polymorphism of the leaves is also demonstrated by the genus *Cercocarpus,* in which two forms of blade, large and small xeromorphous, are traced from the Miocene (Searcy, 1969).

e) Reproductive organs

1) **Distribution.** Chaloner (1970) showed that the spores of terrestrial plants underwent differentiation during the Devonian and heterospory developed. From the standpoint of the ecology of reproduction, the appearance of megaspores marks **the transition from wind dispersal to water dispersal.** The adaptations of the Paleozoic megaspores to transport by water were described by Potonié (1953; see Part III, section 3).

The purpose of the megasporangium coverings, that developed parallelly in the progymnosperms and the Lycopsida, was to protect the megasporangium against desiccation and from mechanical injury; they also constituted a further adaptation to transport by water (Part III, section 3). It is interesting that the integument of the very ancient seed of *Archaeosperma arnoldii* (Figure 19) was covered with spiky processes (Pettitt, 1970). What was the purpose of these spikes? Could they have afforded protection against the first insects (Hughes and Smart, 1967), that appeared in the Late Devonian?

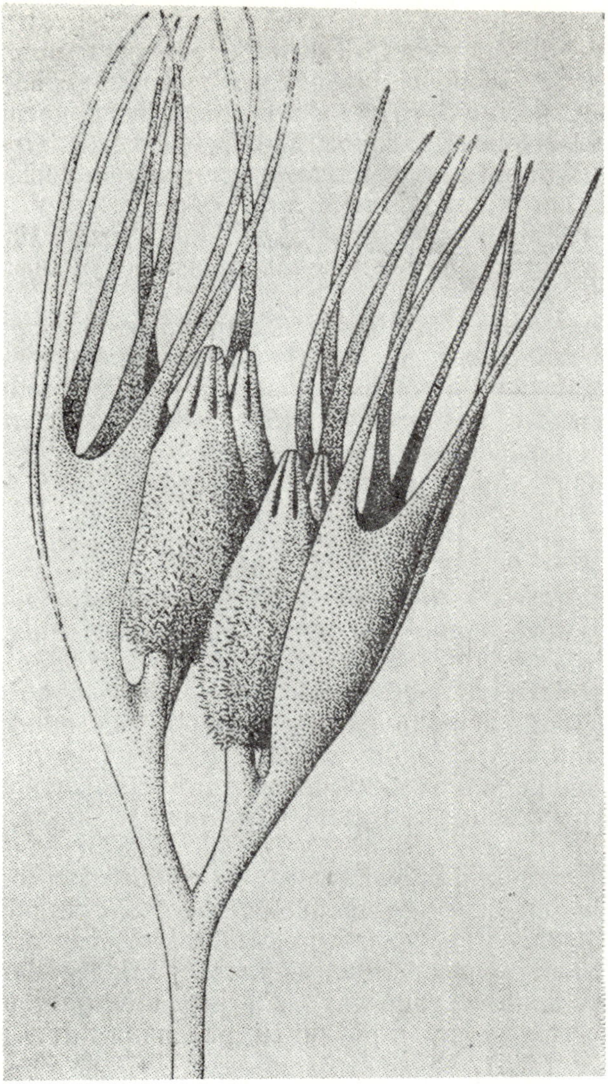

Figure 19. Cupule complex of Devonian *Archaeosperma arnoldii,* reconstructed by Pettitt (1970)

This seems to be a very attractive hypothesis. It may be that the ancient cupules with their long, thin terminal processes served such a function.

In the course of evolution the contradiction was resolved between the various functions of seeds: the function of protecting the ovum apparatus, that demanded massive covers, and the function of distribution, that called for a lessening of weight. Enhanced transportability was achieved by two principal means – the development of processes that increased the surface (winged seeds) of the tuft (Figure 13) and union with another, more transportable organ (the winged cone scales of conifers; the winged cupules of *Glossopteris,* etc.). However, the cardinal solution to the contradiction is apparently zoochory, since with this means of transport massive (edible) covers do not diminish but enhance transportability.

In discussing the reasons why the Angiospermae climbed to the top of the ladder in the plant world, many investigators have drawn attention to the role of the ovary. This, in the opinion of some (Golenkin, 1927; Pervukhina, 1970, and many other works), acts as a moist chamber, safeguarding against desiccation, while others (Faegri and Pijl, 1966, and other works) see its main purpose as constituting protection from insects. I see no point in discussing this problem, as the ovary cannot be considered a new organ. It is merely a modification of the bivalvate polyspermous cupules, the prototype of which can be said to be the cupules of *Leptostrobus* (Krasilov, 1970b). The very ancient cupules of *Archaeosperma* accommodated two ovules, while some of the Late Paleozoic cupules held as many as 60 ovules. A tendency is observed in different phyla of pteridosperms toward the formation of monospermous cupules (similar to the evolution of fruits from follicle to achene) and toward the union of the seed covers. A reduction in the number of ovules is related to anemophily (Robertson, 1904), since with this mode of pollination a large number of the ovules in polyspermous reproductive complexes go unpollinated. We may therefore assume that the preservation of polyspermous cupules was associated with entomophily. On the other hand, the very small (in comparison with other gymnosperms) sizes of the seeds of the most closely related angiosperms of the Mesozoic groups Caytoniales and Czekanowskiales imply a different explanation: unlike the other gymnosperms, the large seeds of which are adapted for dispersal by water or wind on account of their winged processes, here adaptation to wind transport consisted in a marked reduction in the size and weight of the seeds (as in the seeds of the Orchidaceae). A parallelism is displayed here both in the protection of the seeds and in their reduction in size.

Increased size of seeds is correlated with a reduction of their number. At the same time, polyspermous capsules with an edible wall and small, inedible seeds are the chief adaptation to zoochory. Thus, a reduction in the size and an increase in the number of ovules, related to anemochory (and maybe partly to entomophily), served as a preadaptation to zoochory.

In Part III, section 3 we mentioned paleontological finds that led us to assume zoochory in Paleozoic conifers and Jurassic Caytoniales. But until birds and mammals emerged on the scene, zoochory probably had only an episodic significance. Archeopterices have been found in Upper Jurassic deposits, while mammals probably appeared in the middle of the Jurassic period (Simpson, 1965). Correlation of the main datings disproves the fairly popular point of view (going back to Gothan and Potonié) that animals "lagged behind" plants with respect to evolution. On the contrary, it has been discovered that the evolutionary episodes coincided in time in both, or even that plants followed closely on the heels of animals. For instance, in the Wealden of Europe, where angiosperms are virtually absent, remains of mammals with tribosphenic molars have been found (*Aegialodon, Parendotherium*: Clemens, 1968). In the Albian, together with the first signs of adaptive radiation and far-flung spread of flowering plants, more advanced representatives of the same group of mammals emerge. During the Late Cretaceous mammals developed in cenoses dominated by reptiles and became remarkably diversified by the end of this epoch. The same happened with the angiosperms, which in the Cretaceous apparently still did not attain the status of dominants in plant formations (contrary to a widespread conviction; a quantitative analysis of Late Cretaceous beds clearly bears witness to the dominant role of conifers: Krasilov, 1971b). Only when the dinosaurs became extinct and the mammals developed vigorously in the Paleogene did the plant world assume an appearance similar to its present one.

2) **Pollination.** The development of angiospermy (in the broad sense of the word) may be regarded as a negative factor where pollination is concerned, because it makes it difficult for the pollen to reach the nucellus. This contradiction is resolved by several adaptive means: development of a salpinx (Paleozoic ovules), transfer of pollen inside the integument of insects (the pollen of *Macrozamia* is carried by wind, but the micropyles are situated so deep inside the cone that the pollen apparently would not reach them without the aid of insects living inside the cone: Baird, 1938), and germination of the pollen outside the ovule.

The last means is already observed in conifers, the pollen tubes of some forms (*Agathis*) possessing a strong erosion capacity. A haustorial pollen tube is formed also in gingko. Burlingame (1913) considers pollen tubes a primitive character, but no traces of these formations have been discovered in either the Paleozoic or the Mesozoic gymnosperms.* Apparently, the pollen tube mechanism attained importance only with the appearance of the stigma (at first in the Jurassic *Leptostrobus*). Faegri and Pijl (1966) think that there was no true pollination without a stigma and the associated modifications. The stigma as a mechanism of selecting pollen, contributing to the

* See footnote to p. 116.

economy and efficiency of pollination, has played a cardinal role in the adaptive radiation of the Angiospermae.

Although the prickly integuments of *Archaeosperma* served as protection from Devonian insects, it is most unlikely that entomophily was a feature of the progymnosperms. If we consider that such plants as *Tetraxylopteris,* with its 3000–5000 sporangia in one fertile complex, produced pollen instead of spores, then the high productivity and also the morphology of the grains point to anemophily.

Zoidophily is usually considered a progressive adaptation as compared with anemophily. It is more economical and more effective, since here the pollen is transferred directly from the sporangium to the ovule and there are consequently fewer losses due to the pollen not reaching its goal and ovules remaining unpollinated. The peculiarity of zoidophily consists in the mutual adaptation of plant and pollinator, whereas with abiotic pollination it is only the plant that adapts, not the pollinator. Therefore, although many anemophilous plants have extremely refined pollinating mechanisms (see Pervukhina, 1970; I would add that hydrophily in the family Hydrocharitaceae is quite an economical mode of pollination), still their specialization remains one-sided. The drawback of zoidophily (as, moreover, of any refined mechanism) lies in that it requires more stable conditions than anemophily. A seasonal climate is thus more favorable for anemophily than for entomophily. Further, the vegetation of the temperate latitudes differs from the tropical in being less diversified and having larger populations. Here there are relatively few dominant species, and the sizes of the populations are so large that anemophily proves effective enough. Finally, a lack of animal pollinators is also an important circumstance. It is only in the tropics that pollinators are always available in sufficient numbers. In addition, there is hardly any wind beneath the canopy of a tropical forest, and it is hard for the plants of the lower strata to count on wind pollination. All these considerations explain the relationship between the means of pollination and climate and the predominance of anemophilous species among the edificators of forests in the temperate zone. Autogamy is also governed by climate (Stebbins, 1951; Faegri and Pijl, 1966).

Among the recent gymnosperms insect pollination has been found in Cycadaceae (*Encephalartos* and *Zamia* are apparently pollinated by beetles, *Macrozamia* by small bees) and Gnetales. Faegri and Pijl think that the Coniferales and Gingkoales are the only originally anemophilous plants. However, the progymnosperms – the forerunners of all the seed plants – were probably pollinated by the wind (see above).

Unfortunately, fossil material is of little help in a study of the evolution of pollination. The size and morphology of the saccate pollen of many pteridosperms, cordaites, and Paleozoic conifers clearly attest to anemophily, which is thus original not only for the coniferophytes but also for the cycadophytes. The latter as a rule were not as large and grew in the underbrush, this (in the presence of a gentle wind beneath the canopy) being conditioned by a tendency toward insect pollination. Nevertheless, the first pollen with signs of entomophilous adaptation – *Classopollis* – appeared in the Early Mesozoic conifers Cheirolepidiaceae. It is remarkable that here reactivation took the place of the proximal germination apparatus and a distal pseudopore developed. The parallel development of pores in the Cheirolepidiaceae and Angiospermae shows that this character is in some way connected with entomophily or, more exactly, with the coevolution of plants and insects (these pores may have facilitated sucking-out of the protoplast and therefore promoted more regular visits on the part of higher pollinators).

Bisexual flowers (bennettites) may be interpreted as an adaptation to insect pollination. Many authors (see Takhtadzhyan, 1964) consider bisexuality a prerequisite for entomophily. However, the primariness of the unisexual sex organs of the Cycadaceae leaves no room for doubt, and yet some of them are entomophilous. Pollinators visit female and male organs for different reasons: the former because of the pollen, a source of protein, the latter for the sweet pollination drop (present in *Gnetum*) or to lay eggs. The integration of the sex organs in the absence of such adaptations as a stigma, dichogamy, etc. seems inevitably to lead to self-pollination. It is thus not by chance that bisexual strobili are rarely encountered in gymnosperms. The only group for which they are characteristic, the Bennettitales, displays a tendency, on the one hand, toward the formation of a unisexual flower and, on the other, toward self-pollination and cleistogamy (see the preceding section).

Finds of predominantly unisexual flowers in Cretaceous deposits oblige me to support the hypothesis of the primariness of the unisexual flower in the Angiospermae (see Heslop-Harrison, 1958). If we assume that at this early stage in the evolution of the angiosperms bisexual flowers prevailed, it is hard to understand why they did not become fossilized along with the unisexual flowers. Arguments in favor of allogamy in the primitive angiosperms (Faegri and Pijl, 1966) and the combination of insect pollination and wind pollination (Pervukhina, 1970) seem very convincing to me.

5. GENERAL TREND OF EVOLUTION OF LIFE FORMS

The discovery of the gametophytes of the Rhyniaceae confirms the assumption of strong development of the haploid phase in the first land plants.

However, on the whole these gametophytes are similar to the prothallia of ferns and *Psilotum* and give no hints that the Bryophyta originated from the Rhyniaceae by way of reduction of the sporophyte (the reduction hypothesis: Kashyap, 1919). The Devonian liverwort *Hepaticites* (*Pallaviciniites*) *devonicus* (Hueber, 1961) with a cylindrical rhizome and dichotomously branching thallus hardly differs from the recent Pallaviciniaceae. If the liverworts reached such a high level of organization as early as in the Devonian, it is logical to assume that they are even more ancient than the rhyniophytes. The theory of an ancestral group with characters of liverworts and *Rhynia* (Anthorhyniaceae: Mehra and Handoo, 1953) and predominance of the haploid phase finds confirmation in these paleobotanical discoveries.

Dawsonia superba probably demonstrates the size that is maximal for a vertically growing gametophyte – 50 cm. All the while the sporophyte parasitized on the persisting gametophyte its growth was limited by the size of the latter, since a parasite may not be larger than its host. Subsequent increase in size of the sporophyte apparently exhausts the gametophyte, as occurs in *Anthoceros* (Proskauer, 1960). The sporophyte takes root in the soil. How far this scheme corresponds to reality only the future will show.

It is often said that the progressive development of terrestrial plants consists in increasingly successful adaptation to life on land and in the overcoming of the need for water for the reproductive process. This notion has captured the minds of botanists for more than a century now. Many consider xeromorphogenesis to be a general trend in the evolution of plants as a whole and of flowering plants in particular (see Serebryakov, 1962). A hypothesis that has gained wide popularity in the USSR is that of Golenkin (1927), which attributes the victory of the angiosperms over the gymnosperms to their more successful adaptation to the dryness of air and solar radiation. Golenkin suggests that a marked change of meteorological conditions occurred in the first half of the Cretaceous period, and for the first time in the history of the Earth heavy cloud cover was succeeded by prolonged sunshine. The result was that the gymnosperms, ill-adapted to conditions of long-term illumination, became extinct, leaving the angiosperms to step into the foreground (mainly owing to their having an ovary, a moist chamber that guarded the ovules against desiccation). Senyaninova-Korchagina (1959), Pervukhina and Vakhrameev (1947), Sinitsyn (1967), and many other investigators found a weighty set of evidence in support of Golenkin's hypothesis. In recent years this hypothesis has been linked up with that of the expanding Earth (Tsvelev, 1969).

Even though there is a certain logic and appeal in both Golenkin's hypothesis and the whole set of views on which it hinges (independence from water, conquering of the land, etc.), there are a number of weak points which should not be passed over. There is no scope in this chapter to discuss the paleoclimatological principles of Golenkin's theory, which are by no means faultless (judging from the spread of red beds and salt-bearing series in the history of the Earth and its plant world, there were periods of a much more intensive aridization than that in the Early Cretaceous; increased solar emittance should produce greater, not lesser cloudiness: see Brooks, 1949). Here I shall just mention some ecological aspects of the problem:

1) Mechanisms permitting fertilization to become independent of water are present in the lower plants (the various explosive mechanisms in fungi, adaptation to the transfer of gametes by insects in the Bryophyta, etc.). Would it have been necessary to complete a long path of evolution in order to obtain what was achieved right at the beginning?

2) Ecologically, the appearance of megaspores meant a transition from wind dispersal to water dispersal, since the Paleozoic megaspores were too heavy to be effectively carried by the wind and, furthermore, they possessed adaptations enhancing buoyancy (see Part III, section 3). The first ovules were not adapted either to anemophily or to zoochory. In this case transport by water seems the only possibility. Hence, the transition from spore-bearing to seed-bearing plants is associated with adaptation to dispersal by water. From this point of view the hypothesis on an increasing independence from water as a main evolutionary trend hardly appears plausible.

3) Germination of the pollen tube outside the ovule makes fertilization more dependent on moisture than direct penetration of the pollen into a pollen chamber.

4) The transformation of several ovaries into a moist chamber (Pervukhina, 1970) only bears out the idea of the angiosperms' being more dependent on moisture for fertilization than the gymnosperms, which have no need of a moist chamber. The examples Pervukhina presents of the influence of the moisture regime on the development of the female reproductive sphere in a number of flowering plants is yet another confirmation.

5) The ovary of the angiosperms is not to be considered an innovation in either the morphological or the functional sense. The cupules of the Paleozoic pteridosperms and the capsules of the Czekanowskiales are homologous and functionally analogous to the ovary.

6) Xerophilous and photophilous species are frequent among the Bryophyta, Sellaginiales, and ferns and common among the Cycadaceae. *Encephalartos* is admirably adapted to life on dry savanna; *Welwitschia* dominates in the Namib desert.

7) It was the xerophilous bennettites that first began to die out in the middle of the Cretaceous period. The moisture-loving Nilssoniales were still numerous in the Late Cretaceous, while *Gingko* retained its status as

a subdominant in the Arcto-Tertiary flora. The Caytoniales and Czekanowskiales died out at the end of the Cretaceous, despite their having ovarylike chambers. The conifers were less affected than the other groups by extinction during the Cretaceous.

Raunkiaer (1934, 1937) held that his proposed series of life forms from evergreen phanerophytes with unprotected scales to therophytes reflects the historical sequence of their appearance in connection with a progressively worsening climate from the Carboniferous to the Holocene. In so doing he grossly simplified the history of climates.

Reduction series from trees to perennial and then to annual grasses are traced in different adaptive lines, among which most importance attaches to cryomorphogenesis and hydromorphogenesis (see Takhtadzhyan, 1943; Grossgeim, 1950).

According to Serebryakova (1971), mesophyllous tropical border-area grasses like the bamboos gave rise to hygrophilous, xerophilous, cryophilous, and halophilous lines. She stresses the significance of photophilization (see also Kleopov, 1941) and (for grasses) trampling by animals as morphogenetic factors. In cases where reduction is connected with an arid climate it cannot always be considered as xeromorphogenesis, since short-lived plants do not adapt to a dry season and "avoid" this (Grossgeim, 1950). The geography of speciation centers gives no evidence in favor of a substantial evolutionary role of xeromorphogenesis.

Among the extinct plants, xeromorphogenesis is more or less clearly expressed only in the Bennettitales, that evolved from leptocauline branching forms to pachycauline forms with a short stem of the Cycadeoidea type that did not branch and was enclosed in a sheath of leaf scales (Delevoryas, 1968).

Evolution under the badge of adaptation to the aquatic mode of life is traced in the extinct Lycopsida and ferns. Judging from the distribution of stigmariae and palynocenoses with *Lycospora* (Habib et al., 1966), the arborescent lepidophytes lived near the inner margin of mangroves. The Triassic *Pleuromeia* with a short (1.5 m high) nonbranching stem was a mangrove halophyte, while the Cretaceous *Nathorstiana* with a strongly reduced stem dwelt on coastal marshes (Daber, 1968). This reduction line leads to *Isoetes,* a semiaquatic or aquatic plant (the series *Sigillaria–Pleuromeia–Nathorstiana–Stylites–Isoetes* is not orthogenetic or in general phylogenetic, but purely morphological, since *Isoetes* appeared in the Triassic simultaneously with *Pleuromeia*).

On the whole, both xeromorphogenesis and hydromorphogenesis limit the evolutionary potentials of a phylum. Evolutionary progress, on the contrary, is related to the acquisition of nonrestrictive (Huxley) adaptations of overall significance.

Another set of hypotheses associates the general trend of evolution of life forms with an increase of variety. Corner (1966), like many other authors, sees the main advantage of the angiosperms as lying in the exceptional "versatility of leaf, stem, and cambium by which they built the majestic forests of the world." But is the plasticity, or versatility, of the vegetative organs (and cambium) of the Angiospermae in fact something exceptional? Among the recent Cycadaceae there are trees, forms with an underground stem, with a coralloid mass of roots, with an inner cavity in the stem that is penetrated by adventitious roots, and, finally, epiphytes (one species of *Zamia*). The life forms of the genus *Gnetum* include trees, lianas, and parasites (Maheshwari and Vasil, 1961). Moreover, these gymnosperms have not yet exhausted their possibilities! As for the extinct forms, naturally not all the life forms have been preserved and by no means all of them have been recognized. Still, there is reason to assume that the Psilophyta were highly versatile. The character of branching varied within one genus from dichotomous to monopodial, leaf arrangement was from spiral to verticillate (in *Drepanophycus*: Kräusel and Weyland, 1933), and the arrangement of the sporangia from terminal to axial. At first glance such forms as *Drepanophycus* and *Taeniocrada* seem to be totally different, but the specimens described by Kräusel and Weyland (1933), Stockmans (1940), and others lead us to suspect that they are one and the same plant.

Woody forms evidently predominated in the genus *Lepidodendron*. The thin axes are usually considered as small branches of the crown. However, Baxter (1965) described more than 200 specimens of *L. serratum* in which the diameter does not exceed 10 cm, the stele is often arranged asymmetrically, and all the leaf scars are on one side. He concluded that *L. serratum* is a liana and that the Lepidophytes included not only trees and lianas but also shrubs, lowbushes, etc.

Among the Gondwana *Glossopteris* species there were forms with a fine stem bearing verticils of leaves (liana?), with leaf bundles on the apex of the nonbranching stem, deciduous trees with leaves on the brachyblasts, etc. (Dolianiti, 1954; Maheshwari, 1966; Pant, 1967; Plumstead, 1958).

We do not know the full spectrum of life forms of *Drepanophycus, Lepidodendron, Glossopteris, Zamia,* and *Gnetum*. But even that part of the spectrum which has been preserved in the paleontological record or in a few recent species shows that the plasticity of the vegetative sphere of these genera was not inferior to the plasticity of such Proteuses of the plant world as *Ficus*. In terms of diversity of life forms the angiosperms contrast sharply only with the conifers, the mediocre versatility of which is apparently related to the uniformity of the reproductive mechanisms and the stability of the karyotype (polyploidy is an extremely rare phenomenon in conifers).

All the same, how are we to explain the "victory" of the angiosperms? Finds of ancient fruits with adaptations for zoochory and of ancient flowers with nectaries (Part III, section 3) show that the early stage in the evolution of the flowering plants was marked by zoochory and zoidophily. These adaptations, apparently quite ancient, became of prime importance only with the spread of progressive types of animals – mammals, birds, and the higher insects – that directly preceded and developed in parallel with the spread of the first flowering plants. In the process of coevolution the angiosperms attained a complexity and perfection of organization that bespoke a new stage in the development of terrestrial ecosystems. There is thus a basis for saying that there is a relationship between the general trend of evolution of life forms and the ecosystems in which they are included.

The question of primary life forms is connected with the problem of centers of origin. With regard to the angiosperms hypotheses have been advanced of a monoboreal (Heer, Engler, Seward, Hirmer, and others), bipolar (Campbell, Hutchinson, Florin), tropical (Axelrod), and West Pacific (Takhtadzhyan) origin. They are often founded on the assumption that the center of settlement is situated in the place where the largest numbers of primitive flowering plants occur at present. Matthew, on the other hand, believed that the primitive representatives of the phylum shifted toward the periphery. This is partly borne out by the rule, established by Vavilov, that recessive characters are concentrated on the periphery of the range. Simpson suggests that searches for centers of origin are simply a quite harmless pastime. It was he who gave rise to the idea that paleontological finds are more important than any speculations as to where and when a particular group arose (Simpson, 1965). With respect to the flowering plants these finds do not tip the scales in favor of any of the above hypotheses (the monoboreal hypothesis lost appeal after a review of the age of the Greenland floras, while the recent discovery of angiosperms in the Lower Cretaceous of Yakutia contradicts Axelrod's hypothesis) but attest rather to a polytopic origin of the group. An analysis of the life forms of the most ancient dicotyledons will be found in the work by Bews (1927), while the relationship between their evolution and changes in physicogeographical conditions was shown by Stebbins (1947), who specified four stages of development. The first, the Mesozoic stage, saw an accelerated evolution of the Hymenoptera and other pollinators and consisted in a diversification of the reproductive sphere in the presence of relatively stable growth habits. The main families were by then established. Cenozoic orogenesis and differentiation of climatic conditions, plus the expansion of the arid and alpine biotopes, led to the emergence of new forms of growth and to a greater diversification of the vegetative sphere. In the Pleistocene the main evolutionary process was hybridization of populations migrating in the wake of the glaciation and the settlement of allopolyploids.

Part IV

VEGETATION

Around the grotto trees grew luxuriantly,
alder, poplar and sweet-scented cypress...

The Odyssey

The reconstruction of vegetation and a study of the patterns of change it undergoes in time and space constitute the prime purpose of paleoecology, which here comes in contact with plant geography, historical floristics, and stratigraphy.

So far, little has been done to establish a procedure for a synecological analysis of the vegetation of the past. The use of quantitative methods is restricted mainly to Quaternary vegetation. Such descriptions of vegetation as "a conifer-gingko forest with ferns," that are commonly presented in works on paleobotany, seem too primitive and nonspecific. For a reconstruction of vegetation to have any value it must be performed in a way that will facilitate correlation with other reconstructions and further study. The situation is complicated by the fact that the paleobotanist often has to deal with form genera, and designations such as "association with *Aquilapollenites*," "association with a predominance of species producing *Aquilapollenites* pollen," or "*Aquilapollenitetum* ass." are considered ambiguous by many. Nevertheless, a nomenclature must be worked out irrespectively of the reliability of reconstructions.

Even more complicated is the mapping of ancient vegetation. Usually, sampling points or beds of macrofossils are plotted on a map and the content of an individual component is indicated or the composition of each taphocenosis is marked on the map in the form of a complex plane locus (Faegri and Iversen, 1950). The task is reduced to outlining areas that are relatively homogeneous in terms of a particular index. It must not be forgotten, however, that with this method all that is obtained is a map of taphocenoses, not of vegetation.

1. RECONSTRUCTION OF VEGETATION

In works published around the middle of the last century, Heer (1855, 1859, and others) already displayed keen interest in the facies limitations of taphocenoses. According to his data, the lower coal-bearing molasse of Switzerland contains remains of *Taxodium* and formations of species accompanying this genus, plus a rich fauna of insects and vertebrates. The paleolandscape is reconstructed on the basis of a correlation with the *Taxodium* bogs of Virginia. In the midst of the bogs were stretches of open water – lakes with *Nymphaea*. The banks of streams showed a riparian vegetation with willows, poplars, alder, water elm, bottle tree, and fan palms. The marine molasse with coastal crags contains a totally different complex, consisting of three components: algae, eel-grass, and hillslope plants (elms, beeches, oaks, acacias, etc.). *Daphnogene polymorpha*, the foliage of which forms a continuous cover on the surface of the plant-bearing layer, probably dominated in the coastal thickets. In the upper freshwater molasse there is a recurrence of the *Taxodium* coalifying complex with sweet gum, *Nyssa*, and sumac, but no longer with palms or Proteaceae, and with more plentiful remains of poplar (these characteristics are imputed to a cooling of the climate).

Heer determined 920 species, which, according to his calculations, accounts for one third of all species in the Tertiary flora of Switzerland. He based this estimate on the fact that the recent flora of the same region includes 2131 species, but 25 families that are common to the Tertiary flora now contain 152 genera, while in the Miocene they had 253 genera. Heer analyzes the spectrum of life forms. Phanerophytes make up 6/7 of the entire complex. Many insects also confirm the forest nature of the vegetation, although it must be taken into account that herbaceous plants have fewer chances of entering the fossil record (it is mainly fruits that are buried). Certain insects and mammals bear witness to the development of herbaceous vegetation.

Heer attempted to interpret the rhythmic alternation of various rock types – marls and sandstones in the Miocene of Switzerland, quartz sandstones, clays, and lignites in lacustrine sediments of Beauvais (Heer, 1861) – as seasonal stratification. He established seasons according to combinations of different organs and their common burial with certain species of insects.

According to Heer, the Beauvais lignites in England have an allochthonous origin and are formed mainly of trunks of sequoia that grew on slopes surrounding a lake. On the other hand, plant remains in the Atanekerdluk ironstones (Heer, 1866) are in an autochthonous bed. This is attested to by the good degree of preservation of the leaves and their association with fruits and seeds of the same plants and also by the remains of insects that lived on them (during transit they came to be buried separately due to their different buoyancy). The enclosing rocks consist 75% of iron oxides. Heer assumed that this indicates a forest type of vegetation, since the deposition of iron in such a form required a huge amount of plant material in the sediments.

Similar inferences are to be found in the works of Göppert, that appeared during the same period. Analyzing the Schossnitz Tertiary flora, Göppert concluded that the burial of plant remains took place in July (fruits of elm and hornbeam, willow branches with catkins, male strobili of Taxodiaceae, absence of twisted autumn leaves) as a result of sudden (catastrophic) submergence beneath a layer of sand and mud. The bed is not autochthonous (there are no roots), but near thickets of bald cypress, oak, zelkova, and willow, which are represented both by branches and by individual leaves. In some places *Taxodium* dominates, in others, plants from the oak formation.

No less interesting are Weber's paleoecological constructions (Weber, 1852), in which beds are distinguished according to the mode of transport of the plant material (wind or water) and its restriction to coal-bearing or coal-free facies.

Thus, from the mid-nineteenth century the main principles were laid down for reconstructing vegetation:

1) taphocenoses have a structure that may be determined from the quantitative participation of species and that is repeated from bed to bed;

2) a comparison with the recent vegetation confirms the presence of a link between the structure of taphocenoses and the original vegetation;

3) the vegetation of the past is imprecisely reflected in taphocenoses, there being significant losses and distortion of information;

4) sedimentological signs and remains of heterotrophic organisms provide additional information on the structure of the vegetation;

5) the structure of a taphocenosis reflects both direct and indirect facies control. The former is expressed in selective burial, a diverse combination of allochthonous and autochthonous elements, and in the form of preservation, while the latter is manifested in the reduction of some plant groups and the enhanced role of others (Heer, as we have seen, used this mechanism to explain the differences between beds in marine and continental facies: many years later palynologists confirmed that a primary burial of plants from hillsides in marine sediments is related to a reduction of coastal vegetation during transgressions: Brinkmann, 1934; Overbeck and Schmitz, 1931; Chaloner and Muir, 1968; Bond, 1968; Vorman, 1952).

A. Discreteness and continuum

The continuum concept, promoted by Curtis and his adherents (Curtis, 1959; Curtis and McIntosh, 1951; McIntosh, 1967; an exhaustive bibliography is given in the last of these works), stimulated a lively polemics that is still being kept up. In the course of a discussion in the periodical "Botanical Review" (1968, Vol. 34, No. 3) it is mentioned that Curtis at first tried to debunk the idea of a plant community as an organism and show that a typical selection of samples overstates the discreteness of vegetation, whereas random sampling and ordination give a more correct picture of its structure.

Curtis asserted that rigid cenotic links are not displayed in the distribution of vegetation; a species grows in a place where it finds suitable conditions. This conviction is in harmony with Gleason's individualistic theory (Gleason, 1926).

The concept of continuum helped in the elaboration of various methods of ordinating the plant cover and in general improved procedures in geobotanical investigations. On the theoretical plane the contradiction between discreteness and continuum and between the association and individualism of species is apparently related to the different degrees of organization of ecosystems. This had been observed by Tansley (1935), who wrote of the "selection" of ecosystems and their evolution toward maximum stability. Although the quasi-organism concept as a whole is founded on too direct analogies between organism and community, the homeostasis of highly organized communities and their equifinal properties, plus their capacity for regeneration justify the organismal approach (Fosberg, 1967). On the other hand, in most cases their completeness is not so high as to totally subordinate the distribution of a species to a cenotic environment. Some paleoecological data have a direct bearing on this topical problem.

Proceeding from the individualistic concept, Mason (1947) denies that plant associations are stable in time, holding that they are never repeated exactly. However, geofloras on the whole preserve their structure for many millions of years. In the part of a geological section that corresponds to the time of existence of a geoflora a recurrence of the same combinations of species is constantly observed, indicating that the community preserved its structure. The *Weichselia–Matonidium* association is repeated several times in the Lower Cretaceous Quedlinburg section, that embraces at least two geological stages: Hauterivian and Barremian (Daber, 1968). Josten (1961) described several long-lived groups in the Carboniferous of the Ruhr basin.

Many recent species groupings have existed at least since the Miocene (the *Taxodium–Nyssa* community is one of the best known examples). It seems that some communities cross the boundaries between geofloras, but this is not certain. For instance, if we consider the Cretaceous *Trochodendroides* to be the equivalent of the recent *Cercidiphyllum* (many support this point of view), then the emergence of the *Metasequoia–Cercidiphyllum* community should be dated to the Cretaceous period (Chaney, 1952). The combination

Agathis–Lygodium, found in conifer forests of New Zealand (Churchill, 1969), is probably still more ancient, since remains of Araucariaceae (including possibly *Agathis*) associate with spores of the *Lygodium* type in Jurassic and Lower Cretaceous deposits. On the other hand, there are a good number of examples of a transition from certain plant groups to others. *Gingko* or its closest relatives, that composed the bulk of the Mesozoic forests, became assimilated in the broadleaf forests of the Arcto-Tertiary flora. *Nilssonia,* a typical representative of Mesozoic communities dominated by gingkos and conifers, coexists with angiosperms at the end of the Cretaceous epoch. The same liverwort species, *Hepaticites yabei,* occurs in the still typically Mesozoic Lower Cretaceous taphocenoses of the Maritime Territory and in Upper Cretaceous taphocenoses of Sakhalin together with angiosperms. These examples are not monotypic and illustrate different variants of transition. *Hepaticites* forms massive accumulations, completely covering the surface of the layer. These accumulations apparently result from accelerated growth of the thalluses on exposed areas of the bottom of the water body, as is observed in recent aquatic liverworts. The switch of forest dominants evidently did not affect the *Hepaticites* biotope in any way. Isolated specimens of Late Cretaceous *Nilssonia* are found together with remains of angiosperms, but in very large beds this genus associates only with ferns (also of Mesozoic habit) and conifers. In this case it is probably better to speak of a refuge rather than of a transition.

The situation is different with regard to the dominants of the overstory, which, like *Gingko,* pass from one forest type to another. The structure of a tropical rain forest in New Guinea (Paijmans, 1970) gives clues to explaining this phenomenon. The closed canopy, formed by various Lauraceae and other dicots, has an average height of 41 m. Above this in the form of an emergent rises *Araucaria hunsteinii,* that forms a very sparse but absolutely pure layer at a height of 47–67 m, which no other species attains. Here we have "a forest within a forest" – an ancient araucarian forest preserved despite the changed composition of the lower strata. Since the Gingkoales were high-stemmed trees as early as in the Mesozoic, and low trees and shrubs apparently predominated among the ancient angiosperms (see Part III), we may assume that the formation of gingko–broadleaf forests began from the "forest within a forest" stage, with an upper stratum composed of one species of *Gingko.* The history of development of the West Siberian vegetation abounds in examples of transition from one type of community to another (Grichuk et al., 1970). For instance, fir, initially an element of "dark"-conifer* taiga, became a component of mixed conifer forests of the southern taiga type. Communities of oligotrophic bogs of a dark-taiga forest

* [Spruce, fir.]

(sphagnums, Ericaceae) became adapted to life on bogs of open birch woodland, and so on (see also Giterman et al., 1968).

The long survival of Pleistocene species also has a bearing on the problem of individualism and continuum. A continuous geological record of *Betula nana* from the Late Glacial to the present has been studied in Northern Europe (Overbeck and Schmitz, 1931; Hutchinson, 1966). This is a much shorter time interval than those mentioned above, but it far surpasses them in terms of the rate at which the vegetation changed. In the Late Glacial period, when *B. nana* was widely distributed, treeless landscapes prevailed. When the climate improved later and forests appeared, the incidence of this species dropped sharply. It is thought to have survived a forestation peak in isolated refuges on slopes covered with sparse forest and to have penetrated again into boggy lowlands only very recently (Hutchinson, 1966).

Communities of the cold-temperate zone of the southern hemisphere also took in genera with a long pre-Quaternary history (e.g. *Podocarpus*: Wardle, 1968).

The characteristics of Quaternary communities are sometimes explained from the viewpoint of the individualistic theory (Cushing, 1967). Unusual combinations of species arose in connection with climatic features or varying rates of migration (Turner and West, 1968). In the opinion of Livingstone (1968), the ratio between the rates of climatic change and migration is such that a climatic climax did not manage to be formed, and in Quaternary deposits we have a succession of quasiclimaxes.

Goff and Zedler (1968) think that the recent communities of the sod forest in the Great Lakes region have an extremely complex structure in which groups linked by long coevolution associate with more or less casual unions of species with a different evolutionary history, that emerged as a result of preadaptation to similar conditions. If for example, *Acer saccharum* and *Fagus grandifolia,* that have analogues in the mesophytic forests of the Appalachians, seem to form a stable association which has existed from the Pleistocene and migrated en bloc, other species derive from different regions that were disjunct during the Glacial. From this point of view Goff and Zedler support the individualistic theory.

The individualism of species is manifested not merely in their passing from one community to another and in the comparative ease with which they assimilate into new groupings but also in selective elimination and drifting apart: deciduous and evergreen oaks (of the type *Quercus pseudolirata* and *Q. virginiana*), that were associated in the Miocene Mascall flora (Chaney, 1959), now grow in different areas.

It is believed that ordination according to various similarity indexes fully meets the requirements of paleoecology, but the possibilities of this method

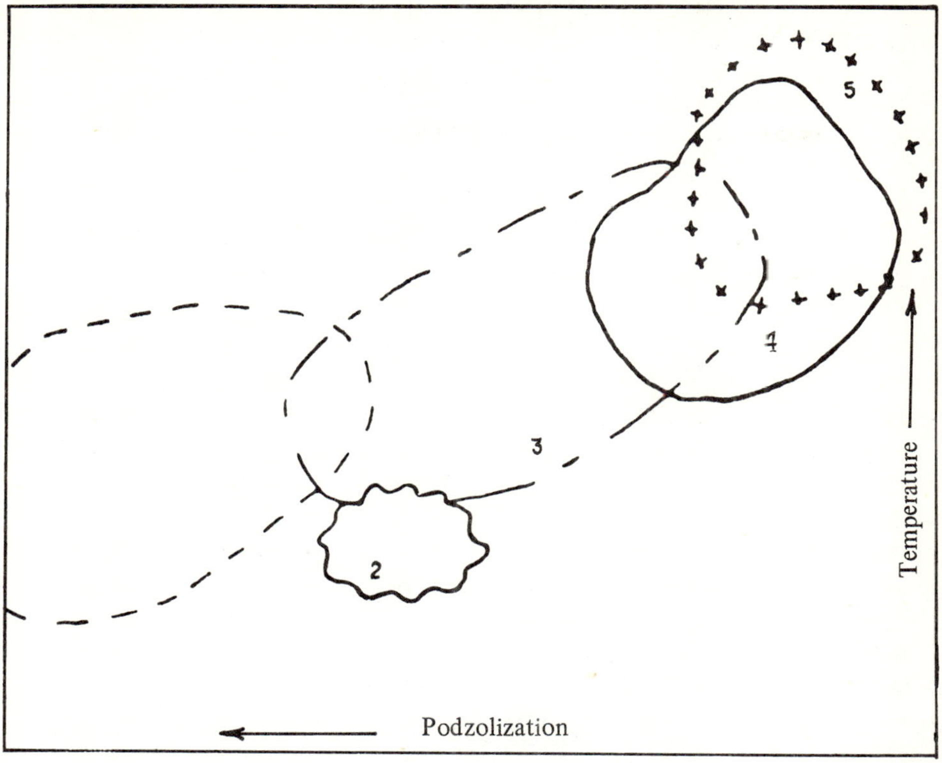

Figure 20. Ordination of Holocene spectra of Michigan according to the quantitative share of genera: *Pinus* (1), *Tsuga* (2), *Acer* (3), *Quercus* (4), and *Carya* (5), after Schaffer and Wilke, 1965

have still hardly been explored. Schaffer and Wilke (1965) used bimeric ordination in studying the structure of mollusk taphocenoses and palynocenoses. Individual samples were compared with Jaccard's coefficient. The difference index was determined as the number bringing the similarity index up to 100. The maximum values of the difference index established the length of the x-axis, on which the other samples were plotted by drawing an arc with a radius corresponding to the difference index between a given sample and the samples plotted at the ends of the axis. The ordinate of each sample was determined by a line drawn perpendicular to the x-axis to the intersection with the arc. The difference index of two samples from the middle part of the x-axis served to plot the ordinate (Figure 20). Although with this method of ordination the axes do not correspond to any particular ecological factor affecting the distribution, the presumed ecological significance is shown on the graph. Palynological data were taken from 21 postglacial profiles in Michigan and Ohio. The distribution of the samples in the system of coordinates is apparently related to podzolization and the temperature gradient. Schaffer and Wilke favor the individualistic concept. Yet another example of ordination of palynocenoses is presented in the work by Maher (1963).

B. The bed as a sample of vegetation

Every bed reflects a certain part of the vegetation, in other words, it can be regarded as a vegetation sample, or specimen. In order to make a rough estimate of the size of the area from which a sample is taken, we must take into account the ratio between the allochthonous and autochthonous elements in the bed. Even though, as we have seen, much attention has been paid to this aspect since Heer's time, there are still no fully reliable methods of defining autochthonous and allochthonous beds or autochthonous and allochthonous elements in one bed. The terms auto- and allochthonous were first proposed by Gümbel for coals. Potonié (1958) calls euautochthonous a bed that is formed in situ while for remains that have not undergone any appreciable transport but are even so not buried in the place where they grew, he suggests the term hypoautochthonous. In most publications, however, both these types of bed are called autochthonous. The most valuable indexes for distinguishing between allochthonous and autochthonous elements are the following:

a) **Sedimentological indexes.** The structural and textural characters of the rock indicating the dynamics of the sedimentation environment are of definite importance in an analysis of an allochthonous bed. Several classifications of cross-stratification have been proposed that to some degree indicate the connection between this and the current regime (Botvinkina, 1962, 1970; Vassoevich, 1950; Kutyrev, 1968; Allen, 1962; McKee, 1964; and other works). Special hydraulic experiments (Menard, 1950; Jopling, 1963) show that the nature of the bed configuration is related to the flow intensity, expressed by the equation using the Froude number:

$$F = \frac{V}{\sqrt{gD}},$$

where V is the mean flow velocity, g the gravity constant, and D the flow depth.

The value $F = 1$ establishes the boundary between tranquil (lower) and shooting (upper) flow regimes. Under conditions of an upper flow regime

(F ⩾ 1) lamination is in the form of flat plates 2–3 mm thick with a system of parallel ribs on the surface (lineation). With a reduction of intensity to the upper limits of the tranquil regime, large ripple marks, bent in plane, appear on the surface of the sand, this being expressed in the formation of large cross-stratified series from several centimeters to several meters thick, alternating with layers without cross-stratification or with a different type of cross-stratification. A further lowering of flow intensity leads to the appearance of small, asymmetrical ripples and a correspondingly continuous sequence of small cross-stratified series up to several centimeters in thickness, with a gradual transition or erosion boundary between them (Allen, 1963, 1964; Figure 21).

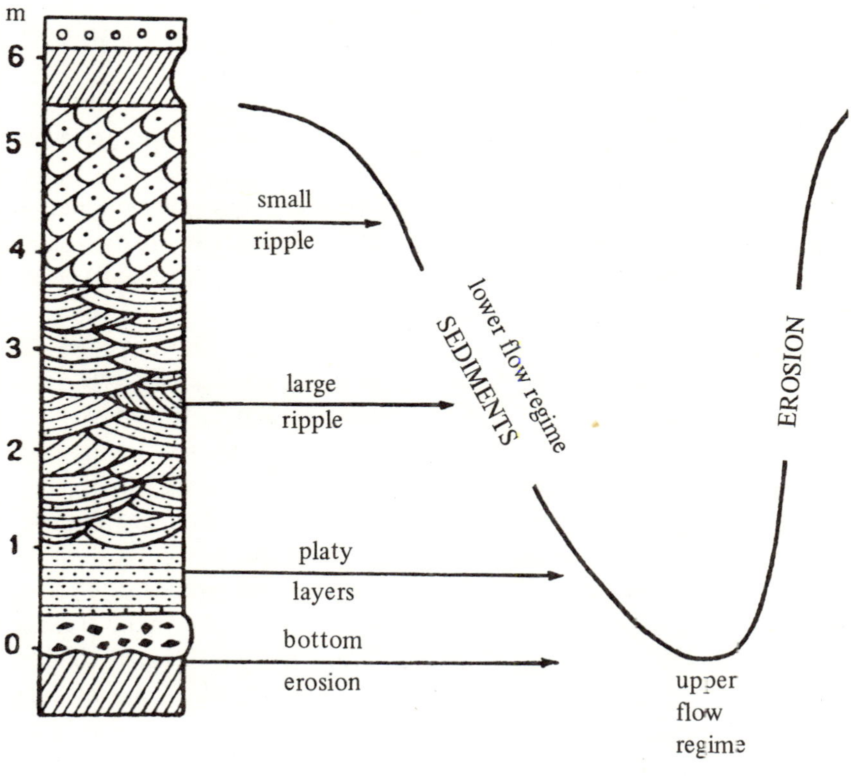

Figure 21. Cyclothem interpretation after Allen (1963). Notation as in Figure 7 (explanation in the text).

A definite orientation of large plant remains, elevation of the anterior end of the stem, and traction furrows indicate the direction of currents. Examples of an analysis of orientation with the construction of rose diagrams are to be found in many works (see, for example, Schenke, 1969).

b) Differentiation. When transported, plant remains behave like sedimentation particles and are sorted by the stream. A concentration of large pollen is observed in alluvium (Permyakov, 1969; Grichuk, 1970, etc.). When spores and pollen are borne by marine currents, grains with an exine that is nonresistant to corrosion are eliminated from palynocenoses. A considerable impoverishment of marine palynocenoses in comparison with continental ones of the same age is therefore an indication of current transport (and vice versa, a rich diversity of palynormorphs attest to the proximity of the shoreline: Habib, 1968).

Clear signs of differentiation of large remains in terms of transportability (e.g. the formation of "seed" and "leaf" beds in the Paleogene of England: Reid and Chandler, 1926) point to an allochthonous burial. In the case of an autochthonous burial organs of the same plant that have a different weight and volume are commonly found in the same place (Heer had observed this).

In lacustrine beds (e.g. in the Jurassic of Lake Karatau: Turutanova-Ketova, 1950) a concentration of fruits and seeds with a flight apparatus is often observed, these representing the allochthonous element here. Chaney (1959) notes that a high percentage ratio of fruits to leaves is evidence of transport (in the Mascall flora a high ratio of the content of fruits to leaves is characteristic for *Acer bolanderi* and a low ratio for *A. scottiae*; the former probably grew on slopes and the latter in the valley).

c) Degree of preservation. It is assumed that remains that have been subjected to long transport must be more deformed than autochthonous remains. However, such factors as the state of the material prior to transport and burial (leaves may be traumatized during life and become twisted upon drying out; desiccated spores and pollen are less liable to be well preserved; seeds that have not lost their germinating capacity are better preserved than seeds that have lost their viability, etc.), diagenesis and postdiagenetic changes, and injuries caused by mechanical or chemical preparation affect the degree of preservation to no lesser extent. On the other hand, remains transported in the suspended state are hardly apt to be deformed. Traces of rounding appear when they are dragged or roll along the bottom, but they do not always indicate prolonged transportation, and therefore the nature of preservation is not of much help when we want to identify allochthonous elements.

d) Ecological discrepancies. Many have thought it unlikely that the palm *Nipa* grew near the coasts of England in the Eocene, and to account for finds of its fruits in the London Clays the hypothesis of long-distance transport by currents was advanced (van Steenis, 1961). It is now obvious that such a hypothesis is without substance. In Quaternary palynology common burial of pollen from plants of different climatic zones is often interpreted from the standpoint of long-distance transport, but we should apparently have recourse to this explanation only as a pis aller.

Finds of fruits of the hydrophytes *Trapa, Brasenia,* and others that lived in calm waters in beds with sedimentological signs of a high flow intensity point to an allochthonous burial (Sukachev et al., 1968). In describing the history of Holocene bog communities, Rybníček and Rybničkcvá (1968) mention that such genera as *Filipendula, Galium,* and *Comarum* are found throughout the section, allochthonousness being indicated when they are combined with the oligotrophic *Eriophorum* and *Sphagnum magellanicum.* These are just a few examples when a nonconformity of ecology enables us to recognize allochthonous elements. If a species is encountered in various facies and in combination with different ecological groupings, we may suspect an allochthonous burial, though our suspicions should be bolstered by data from other sources. For instance, the conifer *Farndalea* is consistently found in Jurassic beds of Yorkshire and Scotland that belong to different facies, but as yet not a single shoot has come to light – only individual small leaves.

e) Quantitative share. Even though many factors affect the content of the remains of a particular species in a bed, the quantitative share is still regarded as the main criterion in identifying allochthonous elements (also in taphocenoses of terrestrial vertebrates: Shotwell, 1964). On the basis of a count of 8922 specimens from 42 surface beds, Chaney (1924, 1959) established a certain relation to the distance from the burial site (Table 11).

Table 11. Quantitative share as a factor of the distance from the burial site (after Chaney, 1959)

Average distance from bed, feet		Content in bed
5.25	4 species	91.26%
25.25	4 species	8.34%
50.00	11 species	less than 0.5%

Although the size of the plant has a marked influence on the amount of material and the conditions of transport (according to Chaney, the probability of *Sequoia* being buried is several thousand times higher than in the case of *Rosa* or *Rubus*), the leaves of the largest trees – *Quercus, Chrysolepis, Pseudotsuga, Arbutus menziessii,* that grow to a height of about 220 feet – are rare in beds. The fragility of leaves (e.g. in *Corylus rostrata* and *Rhododendron*) is of secondary importance. Sparse representation of a species may be due to its being indeed rare or to allochthonous burial. Correlation with related recent species, transportability, the presence of special adaptations to transport, and facies limitation can all help solve the question (see also MacGinitie, 1941). Instances are known where large remains have been borne 50 km and more by rivers (Axelrod, 1938; Odzaka, 1969).

For spores and pollen, as we have mentioned several times, long-distance wind transport is of significance only in treeless areas. Firbas (1949) notes that so much oak pollen is brought to Helgoland, which lies 50 km from the continent, that the trees growing there could be fully pollinated by it. In the north of Finland a 2–6% content of *Picea* pollen is found in samples at a distance of 20–30 km from a spruce forest. A content of *Tilia* pollen amounting to about 0.008% was discovered some 600 km from the northern boundary of linden, and so on. Transport of pollen belonging to *Alnus, Betula, Pinus,* and others over distances of more than 1000 km in treeless areas is not a rare phenomenon. Near the forest edge it is the pollen of trees in the boundary zone that is usually found, but with increasing distance from the skirts the proportion of debris from more outlying zones increases ("Je weiter ein Punkt von der Waldgrenze entfernt ist... so mehr muss daher auch der relative Anteil des aus entfernteren Waldgebieten zugeführten Pollens ansteigen": Firbas, 1949, p. 25). Thus, the pollen of the marginal zone of birch in the tundra is masked by pollen brought from more remote but larger conifer and even broadleaf forests. The effect of distance from the forest is approximately the same in the tundra as in marine sediments (Chaloner, 1968): the regional pollen is featured first and foremost and, as seen from the above examples, the plant zones may be reflected in reverse order.

Many investigators think that a content of a species' pollen above 1% in most cases indicates the presence of this species in the given region (Szafer, 1935, presents different minimum values for a number of species); however, confirmation by finds of macrofossils are at least desirable. According to Watts (1967), "seed rain" is in many ways analogous to pollen rain. The quantitative share of *Picea* and *Larix* macrofossils in a number of cases corresponds to the content of their pollen.

Back in 1916, von Post (1967) spoke of local and regional components of the pollen spectrum. Firbas establishes five groups: local, regional (from a distance of about 0.5 km), transported from the vicinity (0.5–10 km), long-distance-transported (10–100 km), and very-long-distance-transported ("Fernflug," more than 100 km).

In a forest, as noted in Part II, section 5, horizontal transport between tree trunks prevails, and the bulk of the pollen comes from a distance of a few hundred meters (Tauber, 1967; Manten, 1967) or a still smaller distance, more or less equivalent to the height of the trees (Andersen, 1967). Regional pollen, borne by the wind over the forest canopy, is carried in in small amounts, mainly by raindrops.

Groot (1966) mentions that the content of *Tsuga* pollen carried 200 km by water to the mouth of the Delaware River is up to 7%. However, pollen of *Fagus,* which grows together with *Tsuga,* is absent here.

Summing up the above, we can approximately determine the area embraced by an autochthonous sample to be 700 m^2 for leaves (a 50-foot radius according to Chaney) and 30,000 m^2 for pollen. The area indicated in an allochthonous sample averages 10,000 km^2 for pollen and more transportable large organs. Clearly, an allochthonous sample does not reflect the composition of the vegetation but gives only an extremely rough idea of the chief dominants.

An autochthonous sample as a rule shows all the main tree species in the area in question.

Beds have the form of layers or lenses, the extent of which is governed by paleogeographic factors. The distance between adjacent samples from one layer depends on the degree of exposure, the distribution of the boreholes, and other circumstances. In describing the Lower Cretaceous flora of the Suifun basin, I had at my disposal 20 large beds from one stratigraphic unit (the Galenkov suite) on an area of about 100 km^2, i. e. an average of one bed per 5 km^2. In areas that are better exposed (or in which there are more boreholes) the number of sampling points can be greatly increased, but I think that for an area of 100 km^2 it rarely exceeds 1/1 km^2.

As many as 400 species of trees occur on an area of 1 km^2 in a tropical forest (Poore, 1968), while a 100 m^2 sector yields on average from 50 to 80 species (Pires et al., 1953; Fedorov, 1966). It follows that the probability of any of the 400 species' entering a sample in an area of 100 m^2 is about 1/8 in the tropics; on the other hand, the composition of the less rich temperate vegetation will be reflected more precisely. But a bed must not be considered as a random sample, since certain ecological groupings – riparian, bog, mangrove, etc. – have priority over all others. This aspect is discussed in works by Tolmachev (1961, etc.).

C. Information losses

Heer (see above) estimated losses amounting to 2/3 in the case of the burial of the Miocene vegetation of Switzerland. His evaluation method is by no means accurate. Still, it is noteworthy that according to the data presented by Lawrence (1968), information losses in an Oligocene taphocenosis with *Crassostrea* come to 75%.

Information losses with the transition to the fossil state were examined in Part II, sections 3, 5, 7, and 9 and are taken into account in classifying beds (part II, section 13) and taphosuccessions. If we designate the area occupied by parent plants (that yielded material for burial) as PD (after Fagerlind, 1952), the coverage of a given species as U, and its pollen (the category of organs here is in general indifferent) productivity per unit area as Pr, the share of a species in the primary sum of pollen is UPr, after transport, UPrDe/100 (De% is the transport factor), after fossilization, UPrDeCo/10,000 (Co% is the conservation factor), and after extraction, UPrDeCoMa/1,000,000 (Ma% is the manipulation factor symbolizing the losses upon extraction). Superimposed on these losses are reworking, washout, and weathering. Distortion of information occurs due to differing productivity, sedimentation rate, transportability, mechanical durability, chemical stability, and recognizability of the remains.

The following are of prime importance, according to various observations:

1) for pollen – the frequency of flowering, sporangium productivity, and the mode of pollination. The varying degree of resistance of the membranes and reworking on the whole play a secondary role (although under certain conditions their importance increases: Firbas, 1949; Faegri and Iversen, 1964);

2) for seeds and fruits – size and adaptations to transport;

3) for leaves – size of the plant and seasonal leaf fall (Tolmachev, 1961).

D. Types of vegetation

After taking account of allochthonous burial and the extent of information loss, the next logical step is to ask oneself whether we can distinguish the main types of vegetation according to taphocenoses. The answer to this question is provided by a comparison of subfossil beds with the recent vegetation of a given region.

Many investigations of this nature have been performed in the case of palynocenoses. The transport of pollen by air streams was studied in the nineteenth century, at the beginning of which traps with glycerine were used (Hesselman, 1919), and the first surface samples were correlated with the recent vegetation in Northern Europe.

By now numerous analyses are available for almost all vegetation zones and embrace both marine and continental sediments (Leopold, 1964).

The nature of vegetation may be evaluated by the following indexes:

a) Ratio of arboreal and nonarboreal pollen. Analyses of surface samples taken in forested areas show that the ratio of nonarboreal pollen (NAP) to arboreal pollen (AP) is in the large majority of cases less than 100% here, whereas in treeless areas it often exceeds 1000% (Firbas, 1949). For example, in the Urals the transition from forest to steppe is expressed in pollen spectra by a drop in AP from 95% to 24–40% and a corresponding rise in NAP from 1.7–23.3% to 59–73% (Anan'ev and Karevskaya, 1970). In the Carpathians, Caucasus (Klopotovskaya, 1968), and other mountain massifs (Tsukada, 1958; King, 1967; Richard, 1968) the vegetation of open spaces can be recognized just from a few pollen grains of herbaceous plants. Taiga spectra on the whole yield an analogous ratio (Permyakov, 1964, 1965; Prokhorova, 1965; Shakhova and Kolpakov, 1966; Grichuk, 1970; and others). Only occasionally does a high pollen productivity of *Calluna* and some other herbaceous plants distort the ratio in a few samples, the more so that the pollen of these plants is not as broadly scattered by the wind (Firbas, 1949). In a study of surface samples in northern Finland, Aario (1932) came to the conclusion that even if the share of NAP is high in forest spectra, this pollen shows a much smaller diversity than in treeless areas. An approximate equivalence is often observed between locally produced NAP and AP brought in from the vicinity both in forestless (about 50%) and in forested (about 5%) areas (Faegri and Iversen, 1950).

In open spaces the proportion of AP increases on account of long-distance transport. According to Aario's data, in a zone of pine forests pine pollen constitutes up to 80%, while in a birch zone it accounts for only 20%, and in the tundra again 80%. Owing to the drifting in of AP, the spectras of tundra and forest-tundra are almost identical (Davis, 1967).

The distortions with increasing distance from the forest in the tundra and from the shore in the sea are in many ways analogous. Faegri and Iversen believe that marine samples are altogether unsuitable for reconstructing vegetation. Nevertheless, in a number of cases they adequately reflect the dominant formations (Shatilova, 1968; Cross et al., 1966; Rossignol, 1969; Caratini, 1968; etc.). Pollen of polynias* and maris** predominate in the bottom spectra of the Caspian Sea, except in its southeastern part, where there is more pollen of broad-leaved trees of the Caucasus and northern Iran (Mal'gina and Maev, 1966). In the silts of the Aral Sea, where the concentration of pollen grains is relatively high due to the wide distribution of pelitic sediments, there is a predominance of pollen belonging to drought-resistant semishrubs that dominate in the desert and solonchak vegetation of the Turan plain. Only near the mouths of the Amu-Darya and Syr-Darya is a high content noted of pollen belonging to rushes and aquatic plants (Vronskii, 1970).

* [Open stretches of water amidst ice.]

** [Shallow, often hummocky bogs.]

b) Absolute pollen frequency (APF) is measured as the number of grains per unit volume of sediment. This index depends not only on the pollen productivity of the vegetation but also on the rate of sedimentation (and the rate of decomposition of peat, which is also related to the rate of growth of a peat bog). All the same, a change in APF may be a sign of forestation or, on the contrary, of forest degradation (Aario, 1932; Tsukada, 1967). To eliminate the effect of the sedimentation rate on APF, another index is introduced, the pollen deposition rate, determined either as the number of grains per unit area per unit time or by multiplying APF by the rate of sedimentation (in cm/year). The latter is determined by the radiocarbon method (Davis, 1969).

c) Composition of tree species. Tsukada (1967) mentions that a high content of pollen belonging to trees of pioneer groups attests to degradation of a climax forest. For macrotaphocenoses this criterion has especial significance, since remains of herbaceous plants are poorly preserved, and conclusions on a nonforest nature of a vegetation are often based on the composition of the trees. Thus, the presence of seeds of herbaceous plants is a good sign of the development of steppes or prairies (Elias, 1942), but seeds become buried only in the absence of a filter in the form of riparian thickets (Watts and Bright, 1968). If such a filter is present, the remains of riparian plants alone – *Salix, Platanus, Populus,* and others – may point to prairie development (Chaney and Elias, 1936). Krishtofovich (1955) determined from Neogene deposits in the vicinity of Odessa remains of alder, poplar, and willow, that is, exclusively riparian plants. Since species of a sod formation are not represented here, he assumed that the "plateau of that time was covered with steppe vegetation" (see also Dorofeev, 1968).

d) Sedimentological signs. The influence of vegetation on the formation of relief is well known. Seward (1933) thought that only with the development of land vegetation did the hydrographic web assume stability and conditions appeared favoring burial (see also Meien, 1970a). If this is so, then in the Devonian at the latest the patterns of formation of river valleys were in general the same as they are today (Allen, 1970). Paleoecological works occasionally mention changes in geomorphological characteristics – in the form of river valleys (Paepe and Vanhoorne, 1967), the steepness of slopes (Rahn, 1969), in the formation of terraces (Brunner, 1970) in connection with forestation or, on the contrary, forest degradation and the effect of these processes on sedimentation. Podzolization of soils or the formation of a laterite profile, a change in the peat content of bogs and the concomitant modifications in the bog communities, an altered nature of weathering of igneous rocks (the disappearance of such minerals as epidote and hornblende during the weathering of granites on the territory of a tropical rain forest:

Zonneveld, 1968) – all these signs may be used for reconstructing a vegetation type. In this connection special attention is paid to red beds. Many authors consider them a sign of treeless landscapes (see, for example, Gol'bert et al., 1968). However, paleoecological investigations (already discussed in Part II, section 11) have convincingly shown the incorrectness of this assumption (Daugherty, 1941; Clapham, 1970; Brunner, 1970).

In the opinion of Erhart (1956), vegetation determines the type of sedimentation: the distribution of forests promotes chemogenic sedimentation (limestones, dolomites, siliceous rocks), whereas in treeless areas terrigenous sediments and weathering crusts are formed. This was also mentioned by Lichkov (1945) and other authors.

E. Plant formations

In the ecological-physiological classification of vegetation going back to Humboldt and Grisebach the formation was the main unit. Clements also isolated formations on the basis of dominant life forms. Later the term formation became discredited somewhat by its being used for units of different rank (see Aleksandrova, 1969), and some authors either did not single out formations at all or considered them as auxiliary units (Mirkin, 1968). Meanwhile, in many classifications, particularly of forest vegetation, the formation retained its original meaning as an ecological-physiognomic unit, established according to dominants (Ellenberg, 1968; Dokhman, 1960; Sochava, 1969; etc.). In the classification of the vegetation of the past the formation will apparently not be quick to lose its significance as a main unit, since the identification of groupings according to dominant forms is here the only method that can be employed in practice.

The concept "dominant" is not fully determined either. By dominant what is usually meant is simply the most numerous species or (sometimes) the species of the tallest trees towering above the canopy.

Another definition of dominance exists, according to the share of the species in the yield (McNaughton, 1967). In animal ecology and (rarely) in plant ecology the most competitive species is called the dominant (Naughton and Wolf, 1970).

However, since the concept of the dominant grew up in connection with physiognomic classification of vegetation, it is obviously more correct to designate thus a species whose life form determines the physiognomy of the community. In this sense frequency of occurrence, as a characteristic related at the same time to abundance and coverage, is more important for identifying the dominant than mere numbers. In order to stress the difference between the physiognomic importance of a species and its effect on other species or its role in the formation of the phytocenotic environment, dominants and edificators are distinguished (edificator is Braun-Blanquet's term, which was introduced into Soviet geobotany by V.N. Sukachev). Edificator is a cenological concept, applicable only to highly organized "associated" (Norin, 1968) groupings. Various classifications of dominants exist in geobotany (Shelyag-Sosonko, 1968).

The possibility of identifying formations in the vegetation of the past depends on whether we can correctly establish the dominants according to quantitative relationships in taphocenoses.

From the analysis of surface beds of large organs in the Redwood Creek sequoia forest (Chaney, 1924) it follows that the chief dominants *Sequoia, Alnus, Quercus,* and *Umbellularia* constitute more than 85% both in the vegetation and in the taphocenoses. With the assistance of three colleagues, I performed a rougher estimate of the percentage composition of 11 main tree species and their leaves in the alluvium of a small stream on the eastern shore of Amur Bay in the area of Vladivostok. The rectangular sampling area of 1000 m^2 embraced part of the floodplain and the northern slope. Trees with a breast-height diameter of more than 10 cm and leaves in 1 m^3 of alluvium were counted.

Table 12. Relation between the abundance of leaves in stream alluvium and the percentage of trees of the same species in the surrounding forest on the east shore of Amur Bay, Maritime Territory

Species	Content of leaves, %	Content of trees, %	
		on the slope	in the floodplain
Quercus mongolica Fisch. ex Turcz.	34.01	31.00	8.34
Tilia amurensis Rupr.	19.33	3.03	6.16
Acer pseudosieboldiana (Pax) Kom.	14.65	22.34	22.80
Fraxinus mandschurica Rupr.	12.34	10.07	9.75
Carpinus cordata Blume	12.32	19.72	16.65
Acer barbinerve Maxim.	2.38	7.15	5.25
Alnus japonica Sieb. et Zuc.	2.11	6.12	17.51
Phellodendron amurense Rupr.	1.39	0.38	1.04
Ulmus propinqua Koidz.	0.54	1.14	2.45
Juglans manschurica Maxim.	0.22	0.15	–
Corylus heterophylla Fisch. ex Bess.	0.21	–	10.05

The content of the main dominants, oak, linden, and maple, is reflected quite well in the bed (Table 12). The vegetation of the slope is represented more accurately than that of the floodplain. The high content of *Alnus japonica* and *Corylus heterophylla* in the floodplain (the latter forms thickets right near the bed) was not expressed in the quantitative share of the leaves. The amount of *Fraxinus, Phellodendron,* and *Juglans* was high, probably due to the compound leaves, and so was that of *Tilia,* on account of the large size of the leaves. Distortion occurs also owing to the difference in the durability of the leaves (oak is well represented, but there is a low content of the fine-leaved chosen maple and hornbeam). I can find no explanation for the low content of leaves of *Ulmus* and *Acer barbinerve.*

Palynological analyses of surface samples in most cases confirm the theoretical possibility of recognizing dominant genera of vegetation in palynocenoses (Permyakov, 1964, 1966; Kotseruba and Rybakova, 1965; Janssen, 1966; and many other works).

Having studied the content of pollen in suspension in the mouth of the Delaware, Groot (1966) pointed out that the main genera of woody vegetation in northern Delaware and eastern Pennsylvania, especially *Quercus,* which is dominant in the most widely distributed type of forest here, are well represented in it (an exception is the entomophilous maple). Even pollen of *Tsuga,* which dominates in the forests of northern Pennsylvania 150–200 km from the sampling area, makes up 7% of the spectrum. The vegetation of salt marshes is represented by pollen of Chenopodiaceae. There is more pollen of herbaceous plants and fewer spores in suspension than in bottom samples, this being due to the different resistance of the grains and also to the change that has taken place in the vegetation (reduction of the forest) in recent years.

At the same time, sharp differences in the productivity and stability of pollen grains cause marked distortions in a number of cases. The pollen of *Populus* is poorly preserved, and in places where poplar is a forest-forming species (in Canada, for example) palynocenoses do not give a correct picture of the dominance relations (Firbas, 1949). Vronskii (1970) mentions that the riparian woodland forests dominated by poplar are scarcely reflected in the Aral Sea spectra.

The role played by larch has been constantly underestimated in palynocenoses of Northern Europe (Firbas, 1949). Larch is missing from surface palynocenoses in the larch taiga of northern Yakutia (Shakhova and Kolpakov, 1966). In marine palynocenoses near the coasts of New Zealand such important forest dominants of this country as *Agathis* and *Libocedrus* are absent (Koreneva, 1964). In palynocenoses from alluvium of the African savanna there is hardly any pollen of *Acacia,* whose species are dominant here (Bonnefille, 1969). Pollen of *Prosopsis,* a dominant of the shrub vegetation in the south of North America, is not represented in surface samples (Jelinek, 1966). Laurel forests with *Castanopsis* in the Pleistocene of Taiwan are recognized only from the pollen of the latter, since laurel pollen has not been preserved (Tsukada, 1967c). Such cases appear fatal for a palynological analysis, but they are relatively rare.

The role of *Acer* is also generally underestimated due to the low productivity and poor resistance of the pollen during diagenesis (Sangster and Dale, 1964), and the same applies to *Carpinus* and *Fraxinus* mostly on account of the instability of the pollen. Grichuk (1970) notes that the birch-aspen forests of the West Siberian Plain are ill-reflected in surface samples. On the other hand, *Pinus* pollen is a real trouble-maker in palynology. Its abundance, exceptionally high transportability, and stability make it literally ubiquitous, and it often dominates in the palynocenoses of localities where there is very little or no pine (Traverse and Ginsburg, 1967).

The problem of distortion of information was raised by Hesselman immediately after the publication of von Post's method in 1916. Iversen (Faegri and Iversen, 1964) divided trees into groups on the basis of their pollen productivity. To eliminate distortion, the percentage of alder, birch, hazel, and pine pollen (group A) has to be divided by four. This method was approved for the Scandinavian countries, but Jonker (1951) considers it too subjective. Differences in the pollen productivity of trees, according to Faegri and Iversen, amount to 1 : 10, 1 : 20, and in extreme cases to 1 : 40. Fagerlind (1952) writes that this applies only to anemophilous trees. If entomophilous plants are also taken into account, the product factor ratio (PFR) ranges from infinity to nil.

Curtis (1959) classified Wisconsin trees into seven generic groups with a relative productivity of 35 : 15 : 7.5 : 5 : 2.5 : 1.6 : 1. Davis (1963) proposed a method of interpreting pollen spectra by means of the index R, calculated from the formula

$$\text{Ra} = \frac{\text{percentage of pollen of species a in the spectrum}}{\text{percentage of species a in the vegetation}}$$

or in abbreviated form:

$$\text{Ra} = \frac{\text{Pa}}{\text{Va}}.$$

The meaning of R is shown by the following example (after Davis, 1963, with a slight adjustment): if species a is uniformly represented in vegetation samples A and B, while accompanying species b and c are represented to different extents in these samples, then the percentage of pollen a will prove different, but the ratio Ra : Rb : Rc is constant:

	A			B		
	V	P	R	V	P	R
a	50%	10%	0.2	50%	12.5%	0.25
b	10	10	1.0	30	37.5	1.25
c	40	80	2.0	20	50.0	2.5

The example below shows how the spectrum is interpreted with the aid of R:

	P	R	$\frac{P}{R}$	V
a	64	10	6.4	32
b	28	5	5.6	28
c	8	1	8	40
			20	

The correctness of the interpretation, according to Davis, depends on:

1) the accuracy of the pollen determination and the effect of selective fossilization;

2) the difference in R; in the case of small differences a corrective may introduce additional distortion;

3) the homogeneity of the vegetation; Fagerlind (1952) and Livingstone (1968) also emphasize the importance of this factor. Fagerlind thinks that with considerable heterogeneity an investigation using pollen diagrams loses all meaning;

4) the sizes of forest clearings;

5) periodic changes of pollen productivity.

An interpretation, using R, of a spectrum from lake sediments with dominance of *Pinus* pollen showed that the percentage of *Pinus* was greatly overstated due to the large number of species with a low R. In fact, the role of *Pinus* was relatively unimportant in the vegetation, and there is no basis for speaking of a dry and warm "pine period."

Davis determined the importance of a species in the vegetation according to the basal area. Comanor (1968) also used the coverage. His aim was to verify whether R stays the same from sample to sample (Table 13). The results proved negative, which, in my opinion, does not contradict Davis' theory that provides for fluctuations of R. Still, some of Comanor's conclusions and the methodological aspect of his work are of much interest. He compared two forest sectors in New Jersey that we will call A and B for short. Oak dominates in A and pine and oak in B. For the main species a calculation was made of the percentage content in the valley, where sampling was performed, and on the surrounding area within a radius of 2.5 km, according to both basal area and coverage. These indexes were determined at first on a quadrat and then for the whole territory, the area occupied by a given forest type being measured by means of aerial photographs.

Table 13. Comparison of R values for two sectors (after Comanor, 1968)

Genus	Area A						Area B	
	for V_1			for V_2			for V_2	for V_1
	L	R	Av.	L	R	Av.		
Quercus	0.5	0.5	0.5	0.7	0.6	0.6	0.5	0.3
Acer	0.2	0.3	0.2	0.1	0.1	0.1	1.8	0.6
Alnus	—	—	—	1.9	1.7	1.8		
Betula	4.2	7.1	5.6	3.0	7.1	5.1	—	103.0
Carya	1.8	4.2	3.0	1.5	3.9	2.8	—	—
Castanea	—	—	—	4.3	—	—	—	—
Fagus	9.0	35.0	21.9	9.0	35.0	21.9	—	—
Fraxinus	0.4	1.7	1.0	0.5	2.4	1.5	—	—
Nyssa				0.1	0.5	0.3	10.0	1.7
Ostrya								
Carpinus	3.5	—	—	0.9	3.5	2.2	—	—
Tilia	1.0	—	—	1.0	—	—		
Pinus	1.0	0.8	0.9	2.2	1.8	2.0	0.6	0.9
Tsuga	11.7	41.0	26.4	13.7	41.0	22.3	—	—
Chamaecyparis	—	—	—	—	—	—	4.6	6.1

Note. V_1 stands for the content of the species determined by the basal area; L, local vegetation; R, regional vegetation (in a 2.5-km radius); Av., the average for L and R.

On the whole, excluding the entomophilous genus *Nyssa,* each of the chief tree genera is represented in the spectrum by more than 1% of pollen. The pollen content of the main dominants (oak in A, pine and oak in B) is rather low. This is especially remarkable because outside the limits of an oak or pine forest their content is as a rule overstated. The content of oak is the same in A and B, but Comanor thinks this is a chance coincidence, as in B the percentage of oak is low due to the high content of *Chamaecyparis* and *Betula* pollen. Let us remember that although birch has a high pollen productivity, its dominance is reflected only directly in birch forest spectra, not beyond the forest borders (Firbas, 1949). The pollen of Cupressaceae is not distinguished by high transportability either. Thus, in local spectra the importance of dominants (even with such a high productivity as in pine) **proves to be understated due to the exaggerated picture presented by locally abundant species (genera) with nontransportable pollen.**

Livingstone (1968) proposed a method for standardizing diagrams according to the deviation of R from the average value in all samples. The formula drawn up by Andersen (1967) allows for determination of the difference between the pollen content of a species in the area where its share in the vegetation is equal to a and in the area where it is equal to nil. The index is calculated as the ratio $\frac{\text{Pa}-\text{Po}}{\text{a}}$.

In general, as many authors acknowledge, the methods of statistical interpretation of spectra are far from perfect. Moreover, they can only be used for Quaternary palynocenoses. No such methods are available for more ancient palynocenoses or for macrotaphocenoses, although certain general assumptions hold good here too. In analyzing the Eocene Wilcox taphoflora, Berry (1930) had come to the conclusion that a wide distribution in taphocenoses is more important for determination of the dominant form than local abundance. Chaney (1959), in touching on this aspect, wrote that only a combination of broad distribution and high percentage content in separate beds is a reliable indication of dominance. In the Mascall taphoflora *Taxodium* meets this requirement, whereas *Gingko,* also found in all four beds, is represented everywhere by a small number of specimens. If a species is usually represented in several beds, it is, according to Chaney, a dominant of lowland vegetation. A species that is present in the majority of beds but is abundant only in one or a few beds probably dominated in the slope vegetation, descending in places into the valley (for instance, *Carya bendieri* is represented in one bed by 658 specimens but in another by just a few individuals).

Chaney classifies all species into abundant (more than 6.66%), common (0.45–6.66%), casual (0.09–0.44%), and rare (less than 0.09%). These quantitative symbols have a different significance for different beds. While a 0.45% content in Mascall characterizes common species, in other floras it will be too low even for casual ones. Apart from this, in the determination of dominants frequency of occurrence probably has a certain advantage over quantitative presence. It is worthwhile comparing frequency in different beds (layers) and for a taphoflora as a whole. The frequency of species (or genus) a in a bed is found from the formula

$$\text{Fa} = \frac{\text{Na } 100}{\text{N}},$$

where Na is the number of samples with species a (some of them may bear several impressions) and N is the total number of samples.

Frequency values for the main genera of the Jurassic flora of Kamenka in the Donets Basin (for a more detailed analysis see Part III) in calculations for 100 lumps of coal from each of layers I, III, IV, and V are shown in Table 14.

Table 14. Comparison of the frequency of occurrence of the main genera in four beds of Jurassic flora

Genus	Fa				Fax	Deviation Famax–Fax
	I	III	IV	V		
Todites (williamsonii)	0	2	87	80	42.2	44.8
Cladophlebis	0	82	10	12	26.0	56.0
Sagenopteris	0	5	15	0	5.0	10.0
Nilssonia (orientalis)	1	10	0	0	2.7	7.3
Ptilophyllum (pecten)	0	1	8	8	4.2	3.8
Nilssoniopteris	0	80	0	0	20.0	40.0
Czekanowskia	90	0	0	0	22.2	67.8
Podozamites	77	1	0	0	19.5	57.5
Elatides	2	50	75	12	34.7	40.3
Pityophyllum	0	100	0	0	25.8	74.2

Note. Frequency calculated on the basis of Table 16; Fa stands for frequency in given layer; Fax, average frequency in layers I, III, IV, and V; Famax, highest frequency for the given genus.

Comparing the frequency of genera in layer I, we see that *Czekanowskia* and *Podozamites* seem to be the most likely dominants. The highest frequency is shown by *Pityophyllum* (100%), after which come *Czekanowskia* (90%) and *Todites* (87%); however, going according to the pattern established above, it is *Todites* and *Elatides* that are to be considered the most probable dominants for the entire taphoflora, since they have the highest average frequency and the deviation of the maximum from the average values is smaller than in *Pityophyllum* and *Czekanowskia*. Our 10 genera are thus distributed among Raunkiaer's frequency classes (Raunkiaer, 1934):

A	0–20%	5 genera
B	21–40%	4
C	41–60%	1
D	61–80%	0
E	81–100%	0

For classes A, B, and C these figures correspond to Raunkiaer's law:

$$A > B > C \gtrless D < E.$$

Zero values for classes D and E apparently corroborate the above-noted law that the frequency of the dominant species in a taphocenosis is lower than the actual frequency.

Miki (1941) identifies the dominants of the Pliocene vegetation of Japan on the basis of an analysis of frequency in combination with the most widely distributed species *Pinus trifolia.*

In concluding this section, we shall briefly discuss the accuracy of reconstruction according to palynocenoses and macrotaphocenoses. Weber (1893) had already pointed out that palynocenoses give a better picture of vegetation than beds of seeds and other large organs. For the Quaternary period this is undoubtedly true. Chaloner (1968) claims that neither type of taphocenosis has an advantage, for they merely supplement each other ("neither the macroflora nor the microflora is 'more complete' or 'more correct.' Each gives an incomplete picture of the contemporaneous flora, the two pictures have large areas of overlap but each complements the omissions of the other"). This is only partially correct. Quaternary palynology is far superior to all other paleobotanical methods, which here merely play an auxiliary role, although finds of macrosize organs sometimes introduce substantial correctives into an interpretation of vegetation. For example, poplar pollen is poorly preserved and is omitted from palynocenoses, while the leaves are excellently preserved and can fill in the blank. A many-faceted study of pollen and macroremains (see, for instance, Watts and Bright, 1968; Rybníček and Rybníčková, 1968) gives admirable results. But still the role of macroremains is subordinate. The situation is different in investigations devoted to pre-Cenozoic vegetation. The fact is that in many cases we cannot determine pollen affinities or classify pollen according to formal indexes. The impossibility of reliably identifying the pollen of such important groups of plants as Cycadales, Bennettitales, Gingkoales, and Czekanowskiales, that play a completely different role in the structure of plant formations in former epochs, places pre-Cenozoic palynology in a difficult position. There were probably many more entomophilous and apogamous plants in many of the ancient floras than in the recent flora of Northern Europe. In view of these circumstances, macrotaphocenoses emerge into the spotlight in pre-Cenozoic paleoecology.

F. Consortia and other groupings

Some authors say that from taphocenoses (especially palynocenoses) one can reconstruct only very large units of the plant cover – formations or groups of formations (Prokhorova, 1965). Harris (1952) wrote that an analysis of association in the Yorkshire taphoflora is possible only with the aid of a computer. At present computer group analysis is practiced in animal paleoecology (Fox, 1968) and plant paleoecology (Dale, 1970; see also Oltz, 1971, Micropalaeontology, 17, No. 2).

As shown above, an autochthonous bed corresponds to a small part of a vegetation (part of an association or a patch of mosaic) and permits us to reconstruct some fragment of the phytocenotic structure. The concept of structure includes: the spatial distribution of the components (association, mosaic pattern); vertical structure (stratal or synusial); the population makeup; the physiological structure determined by consortive relations.

A consortium is an aggregate of organisms of different trophic levels that are linked with a certain autotrophic central species (Mazing, 1966). The following finds make it possible to restore fragments of consortive relations in phytocenoses of the past:

1) Fungi. Roots of cordaites with fungal hyphae were known in Linnaeus' time. He wrote several works on the subject, and the conviction grew that these remains represent a mycorhiza. However, Cridland (1962), who made a critical revision of all the material, was eventually persuaded that the fungi in roots of cordaites were in fact parasites or saprophytes, rather than symbionts.

Berry (1925) notes that the leaves of Paleozoic and Mesozoic plants were attacked by fungi to a lesser extent than those of angiosperms. This curious phenomenon has still not been explained, although it should be pointed out that remains of ascomycetes are found quite often on the leaf cuticles of Mesozoic ferns, cycadophytes, and conifers (see Krasilov, 1967). Well-preserved fruiting bodies on leaves have been described from the Eocene of Geiseltal (Rock, 1939), from Tertiary sediments of Australia, and other localities.

2) Galls. As far as I know, the most ancient galls were described by Florin (1938) on the shoots of Paleozoic *Lebachia piniformis*. These are kidneylike formations measuring 8–12 by 5–7 mm; they somewhat resemble the male strobili of the same plant but are situated laterally instead of terminally. A. I. Turutanova-Ketova showed me formations resembling galls on leaves of *Yuccites* from the Lower Triassic (data not published). Vishnu-Mittre (1956) gave a brief description of gall-like structures on shoots of Jurassic conifers. Harris described as the new genus *Wonnacottia* leaves of the *Anomozamites* type from the Jurassic of England; they are covered with tubercles, which at first he took to be sporangia but later decided that they were galls (Harris, 1969). Probably of the same nature are the seedlike formations described by G. Tomas (1911) from the Jurassic of Kamenka. My collection

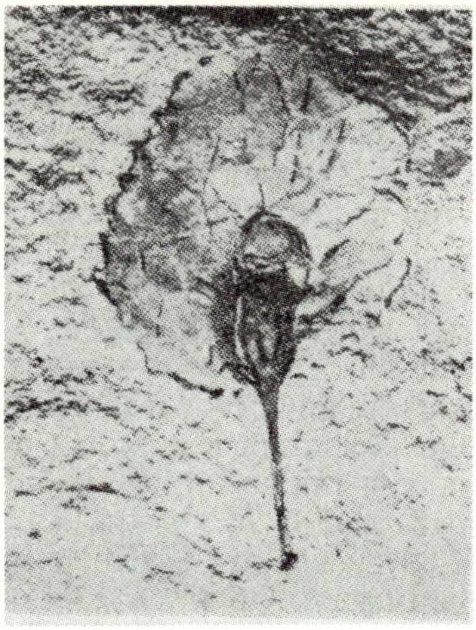

Figure 22. Gall on a *Trochodendroides* leaf from Upper Cretaceous deposits of Bureya Tsagayan (collection of the author)

from the Upper Cretaceous of Bureya Tsagayan contains remains of galls on *Trochodendroides* leaves (Figure 22). The causative agents of galls in the Paleozoic and Cenozoic were apparently not gall midges (Cecidomyiidae), which appeared later, but other insects. There have been finds of gall midges in Baltic amber and in cones of Tertiary sequoia (Mohn, 1960; Mamaev, 1968). Galls from the Paleogene of England were described by Chandler (1963). Leaves injured by insects are found in the Paleozoic of Eurasia and Gondwana, as well as in younger sediments. Of particular interest are the injuries that Delevoryas (1968) reported on the flowers of Mesozoic Cycadeoidea.

3) Plant remains in the stomachs or excrement of extinct animals. In the preceding sections we have more than once mentioned finds of *Ullmania* strobili in a rhynchocephalian stomach (Weigelt, 1930) and of *Caytonanthus* pollen in coprolites of a small Jurassic animal (Harris, 1956). We may judge the diet of the Siberian mammoth from the pollen preserved in its stomach. Pollen of herbaceous plants, that probably passed through the stomach of cave bears, has been described from phosphate horizons of caves (Klaus, 1967). Myospores from coprolites of herbivorous Mesozoic reptiles have also been studied (Waldman and Hopkins, 1970).

Marks of teeth on *Calamites* stems, as Seward writes, conjure up before the mind's eye a living picture of life in the Paleozoic forest. Chandler (1964) described round hollows in the endocarp of *Rutaspermum ornatum* which were probably the work of a small rodent.

G. Recent analogues

In reconstructing the vegetation of the past, we always have to use some or other present-day model. Certain contemporary plant groups are chosen for this purpose more often than others, and of these we may mention in particular the so-called relict groups, that is, those in which the largest number of species has been preserved that are common to the Tertiary or even more ancient floras. They are sometimes considered as practically unchanged parts of Tertiary vegetation. In fact, it would probably be more correct to say that individual fragments of the structure of past phytocenoses have been preserved, i.e. assimilated by later ecosystems. Apart from floristic similarity, an analogy is also seen in the physiognomic characteristics of certain recent communities (for example, the evergreen forests of Shol in India as a relict of the Neogene vegetation: Vishnu-Mittre and Gupta, 1968).

Analogies of the paludal forest in the Carboniferous period were mentioned in Part II, section 12. *Sequoia* dominated in the vegetation of coal-bearing landscapes of the Cretaceous period, but in terms of the composition of the lower strata these communities had nothing in common with redwood. As pointed out by Koreneva (1964), they can be more readily compared to the recent vegetation of New Zealand. The overstory here is dominated by *Agathis* (corresponding to *Araucarites* of the Cretaceous forests), *Podocarpus, Phyllocladus* (possibly corresponding to the Cretaceous *Protophyllocladus*), *Libocedrus* (the so-called "*Thuja*" *cretacea* of Cretaceous forests), and many Lauraceae and Proteaceae. The second layer features various Araliaceae, the leaves of which resemble those of some Cretaceous forms, Icacinaceae (with this family are linked the very ancient fruits of *Onoana* and a whole series of later finds), Winteraceae, and others. A characteristic feature is the abundance of ferns; some of them (*Gleichenia microphylla*) differ little from the Cretaceous ferns.

In New Zealand *Agathis* is encountered together with *Lygodium* (Churchill, 1969); analogous associations (*Araucarites* and spores of Schizaeaceae) probably existed as early as in the Jurassic.

The vegetation of Florida is commonly thought to be an analogue of the "lignite" landscapes of Europe. Here are distributed dense thickets of sclerophyllous shrubs (oaks, *Carya, Sabal, Vaccinium*) and, rising above them, tropical forest outliers – hummocks. This vegetation remains stable in the presence of frequent fires. Bald cypress (*Taxodium*) in river valleys, bogs with red maple (*Acer rubrum*), where sweet gum, magnolium, Lauraceae, and bayberry also grow, mixed forest with *Magnolia, Carya,* and *Aralia* on slopes (Monk, 1968; Watts, 1969) – all these are fragments of Tertiary forest formations that have survived intact to this day. The vast treeless bogs Everglades were also apparently a typical feature of lignite landscapes (Thomson, 1949).

The Appalachians, the Yangtse valley in China, and Central Japan are considered the main refuges of the Arcto-Tertiary flora (Li, 1952). Indeed, such ancient angiosperms as *Trochodendron, Cercidiphyllum, Euptelea,* and others are preserved here. In river valley forests of the Appalachians *Betula lenta* associates with *Magnolia fraseri* and *Liriodendron* (Mowbray and Oosting, 1968). Kolakovskii (1961) believes the plant world of Colchis to be the refuge of the Late Pleistocene vegetation (Miocene species such as *Tectocarya lusatica* are preserved in the Pliocene flora of this area: Kolakovskii, 1969). The complex of summer-green deciduous forests of the Colchis type includes alders, oaks, zelkovas, hornbeams, chestnuts, and other varieties.

The vegetation of river valleys and mountain passes of Colchis is very diverse. Dominant here are chestnut, beech, and hornbeam forests with an admixture of linden, maple, and elm. There are many evergreen shrubs: *Rhododendron,* cherry laurel, and *Philadelphus*; on calcareous soils are box thickets with a thick moss cover of tangles of moss. Rocky substrates bear both the hygrophilous ferns *Hymenophyllum* and *Woodsia* and the xerophilous *Arbutus*. The aquatic vegetation of Colchis is exceptionally rich. Here are many large lakes with *Salvinia, Myriophyllum, Potamogeton, Hydrocharis,* water chestnut, and others. *Hibiscus ponticus* occurs in bog thickets. The sphagnum bogs with *Molinia, Rhynchospora,* and *Carex* are similar to the northern bogs of the Atlantic type, but here grow Pontic rhododendron, azalea, sarsaparilla, and fern royal.

Despite some remarkable coincidences, appreciable differences between the Tertiary and recent communities are always revealed upon minute analysis.

While development of the plant cover followed a trend of impoverishment and elimination of certain ecological groups, in the ancient communities we find associations of species that are not in evidence today. In the Neogene Mascall flora (Chaney, 1959) there are oaks with lobed leaves of the *Quercus pseudolyrata* type and evergreen oaks *Q. virginiana*. These are no longer found together due to the increased differentiation of the plant cover. A change of tolerance is one of the possible reasons for the lack of correspondence. Thus, *Cornus mas, Thelycrania sanguinea, Coninus,* and some other species of mesophytic forests became assimilated in sclerophyllous formations of the Mediterranean (Negru, 1969). Condit (1938) mentions that the pre-Pliocene oaks, which are common in intracontinental floras of the Mascall, Latah, Bridge Creek types, etc., appeared on maritime lowlands in large numbers during the Pliocene. He attributes this to a change in tolerance, although such factors as a general change of climate could apparently have played a decisive role.

It may be that under new climatic conditions a biotope change did not in fact mean a change of tolerance. For instance, with aridization of the climate the Neogene plane tree passed from the rich mesophyllous forests into the riparian communities of the semiarid vegetation, and later was eliminated altogether (Chaney and Elias, 1938).

Conclusions on tolerance that are based on the distribution of recent analogues are to be made with great caution. It would be a mistake to think that the Tertiary forms of *Metasequoia* did not descend below 1200 m above sea level like the recent *M. glyptostroboides*. The metasequoias may simply have been felled below this mark.

H. Animals

Finds of insects or vertebrates adapted to a particular type of vegetation also have a certain importance for reconstructing vegetation, as they help fill in gaps of information. This device was used widely by Heer. New data on the ecology of extinct animals often upset deeply rooted notions on vegetation. For example, it was believed for a long time that the brontosaurs inhabited bogs. However, their limbs are adapted to solid ground and their teeth to hard plant food (Bakker, 1971). It is more likely that they lived in sod forests and, like the elephants, trampled the underbrush, maintaining an open, park appearance of the vegetation. Thanks to their long neck, they could strip off leaves from a height of 14 m, which tells us the minimum canopy height of the Jurassic forest.

2. PALEOSUCCESSIONS

I call paleosuccessions the sequence of changes of taphocenoses in a geological section (Krasilov, 1969b). This sequence may be related to the conditions of burial (taphogenic paleosuccession: see Part II) or else to a change of the actual phytocenoses (cenogenic paleosuccession). The majority of Soviet geobotanists accept the classification of successions proposed by Sukachev. The following types are distinguished: **florogenesis** (auto- and allogenic) – changes as a result of speciation; **selectogenesis** – regrouping and selection of phytocenoses; **syngenesis** (or succession proper) – development from pioneer to climax communities (in this work we understand syngenesis to mean the evolution of communities in the widest sense); **endoecogenesis** – transformations as a result of a change of biotope with interaction between vegetation and environment; **hologenesis** – changes related to the evolution of the Earth's surface (corresponding to Gams' secular successions); **geitogenesis** – changes in the presence of external influences of local significance. I suggest distinguishing mutational paleosuccessions (florogenesis, selectogenesis) and migrational paleosuccessions, where shifts of the boundaries of communities are the principal factor. Actually, migrations in some way or other alter the genofunds of populations and in the final analysis lead to a mutational transformation (see Mason, 1947).

A change of taphocenoses usually indicates shifts of vegetation zones in a drainage basin that are serially linked up into a unified system, or what is known as a **catena**. Against a background of fluctuating catena elements an overall trend of development is manifested, and catenas are reorganized and replaced. A sequence of phytocenoses that replace each other within a single adaptive zone is called here a **cenophylum**. A discussion of catenas, cenophyla, and the factors involved in their development is presented below.

Owing to the superposition of taphonomic factors, the changes of taphocenoses in consecutive layers of a section do not adequately show the change that has taken place in the vegetation. There are a host of ways of measuring similarities, some of which are rather complicated and hardly applicable in paleoecology. The procedure most often adopted is that using the Jaccard coefficient of commonness in different versions. The Jaccard coefficient is determined as the number of species common for two taphocenoses, expressed as a percentage of the total number of species.

If we compare not two but n taphocenoses with $S_1, S_2 \ldots S_n$ species in each one, we may use the Koch coefficient $\frac{100\,(t-S)}{(n-1)\,S}$, where $t = S_1 + S_2 + \ldots S_n$, and S is the number of common species.

Simpson (1943, 1947) proposed several coefficients that are also related to the Jaccard index, e.g. $\frac{C}{N_1} \times 100$, where C is the number of common species in two compared taphocenoses and N_1 is the number of species in the smaller of them. Some of Simpson's indexes have found application in paleobotany (Oltz, 1969).

Another index is Chekanovskii's (Greig-Smith, 1957): $\frac{a+b}{2c} \times 100$, where a and b are the numbers of species in two samples and c is the average number of species in a sample.

In order to clarify whether any change at all has occurred, let us assume that both taphocenoses being compared illustrate the same community. The Jaccard coefficient should then equal 100%, but in reality, even without taphonomic distortions, it will be different from 100% due to the uneven distribution of species in the area (Greig-Smith, 1957). It is assumed that the relation between the number of species and the number of individuals may be logarithmic (Williams, 1947) or logarithmically normal (Preston, 1948). In the first case it is expressed by the equation $S = \alpha \ln (1 + \frac{N}{\alpha})$, where

S is the number of species, N is the number of individuals in the sample, and α is the index of diversity. If N is large in comparison to α, the "1" may be neglected (Greig-Smith, 1957). Then

$$S = \alpha \ln \frac{N}{\alpha}.$$

Using this formula, we can determine the expected number of species in both samples, provided that they are taken from one association, and compare it with the number observed

For a logarithmically normal relationship Preston proposed the formula $N_{a+b}^{\frac{1}{z}} = N_a^{\frac{1}{z}} + N_b^{\frac{1}{z}}$, where N_a and N_b are the numbers of species in samples taken from one association, N_{a+b} is the expected total number of species, and z is a constant, equal approximately to 0.27. An example of calculating the total number of species by Preston's formula is given in the next section.

Fagerlind (1952) showed that in the presence of a certain amount of taphonomic distortion a straight line on a palynological diagram does not correspond to a status quo of vegetation. If an association has two groups with respective contents R and 100−R in horizon A and r and 100−r in horizon B, a change in the association can be expressed by the equation

$$C_r = \frac{R:(100-R)}{r:(100-r)} = \frac{R(100-r)}{r(100-R)}.$$

Due to the distortions on the diagram, the corresponding change is

$$Cd = \frac{D(100-d)}{d(100-D)},$$

where D and d are the diagram equivalents of R and r.

Fagerlind's index is $K = \frac{Cd}{Cr}$.

If we take the maximum values of K and construct diagrams according to $\sqrt{Kmax}$ and $\frac{1}{\sqrt{Kmax}}$, these diagrams denote the limits of deviation of the diagram values from the real values in relation to fluctuations of K. According to Fagerlind's theory, we can speak of a real change of R only in a case where the upper limit for point A proves to be below the lower limit for point B.

A. Catenas

As we discussed in Part IV, section 1B, an autochthonous bed may be considered as a sample from an area of about 700 m^2. Let us assume that this area is part of a floodplain near a river channel. Having only one bed (1), we do not obtain any information either on the vegetation of the central floodplain or on the marginal parts of the river valley (unless allochthonous elements tell us something). However, as was shown in Part II, section 12, there is a buildup of alluvial deposits in the section on account of lateral migration of the stream and the successive change of the facies. Thus, the next bed (2), which is situated above the first within the cyclothem, will give us information on the vegetation of another part of the river valley. Higher up in the section, in the next cyclothem, the facies sequence is repeated, and we may again come across a bed (3) that corresponds to the channel part of the floodplain. If the vegetation did not undergo any appreciable change during the time that the cyclothem was formed, beds 1 and 3 may be considered samples from one channel zone. In the upper part of the second cyclothem bed 4 will correspond to bed 2, and so on. In the presence of a sufficiently complete set of facies we obtain a more or less true picture of the distribution of vegetation zones in the river valley. Together they make up a complex that characterizes the valley vegetation as a whole. Analogously, in deltaic deposits (Part II, section 12) we have a rhythmical sequence of prodelta, delta platform, intradelta, etc. facies and a sequence of the corresponding taphocenoses, each of which makes it possible to restore a part of the vegetation of some or other morphological element of the delta plain or of the elevations flanking it. Again we obtain a complex the components of which are linked by successional transitions. In the geobotanical literature such groupings are called Keller's ecological series, succession complexes (Kiseleva et al., 1969), exchange-dominant complexes (Mirkin, 1968), or microzonal series (Isachenko, 1969), but I prefer the term **catena**, which was initially used for soil series (see Walter, 1971) and is employed by some authors to denote the sequence of plant groups in a river valley or on a slope (Greenway and Vesey-Fitzgerald, 1969). I hold that the catena is one of the fundamental concepts of plant paleoecology that are of considerable significance for paleogeography and stratigraphy.

I call the components of a catena associations. This concept is often given a completely different meaning, but most authors define the association as a spatial entity of lower rank (Nitsenko, 1966; Kamyshev, 1956). Associations are isolated on the basis of dominants or determinants, this depending largely both on general classification principles and on the type of vegetation. An interpretation of the association that is similar to that adopted in the present work is to be found in Kiseleva et al (1969). Associations are named after the characteristic species or dominants. It is also recommended that diagnoses be published and that the author also indicate the date of the original description of each association (Neuhäusel, 1968). As more and more descriptions appear, this taxonomic procedure will evidently become indispensable in paleoecology.

Table 15. Taphocenoses of the Kamenka flora

No. of layer, from bottom to top	Rock	Boundary with underlying layer	Thickness of layer, cm	Stratification	Facies	Taphocenosis
I	Ferruginous sandstone, rust-brown, conchoidal bog iron ore		20	Indistinct	Coastal marshes	Phyllo-caulocenoses, casts and molds of stems, three-dimensional incrustations of leaves and strobili, voids from acicular leaves and seeds
II	Light-gray, lumpy, gypsiferous clay	erosion	50	Indistinct	Lagoons inside a mangrove zone?	Phyllocenosis, sparse, haphazardly oriented, dark brown incrustations with dispersed organic matter
III	Brown carbonaceous clay with slickensides	sharp	68	Indistinct	Swamped and silting-up water body on the delta plain	Rhizo-phyllocenosis; leaves brown, filling the rock to overflowing, crumpled and torn, incrusted with iron oxides with an admixture of organic matter
IV	White argillite-like clay, greenish sapropelic clay at the base of the intercalation	sharp	10	Fine horizontal, slightly wavy	Lacustrine (gyttja)	Phyllocenosis; rust-brown remains incrusted with iron oxides, forming leaf roofs
V	Siltstone, at the bottom clayey, yellowish white	gradation	170	Continuous sequence of fine cross-bedded series of the asymmetrical migration ripple type	Channel streams or delta arm facies	Phyllocenosis; leaves brownish gray or pale yellow, incrusted with iron oxides, crumpled, bent

Number of specimens		Saturation coefficient, Kc	Deformation coefficient, Kd	Hypoautochthonous species		Allochthonous species
Z	total			arboreal	herbaceous	
341	419	20.95	81.38	*Nilssonia vittaeformis, Czekanowskia rigida, Leptostrobus laxiflorus, Podozamites lanceolatus*	*Klukia exilis, Equisetites columnaris*	*Nilssonia orientalis, Williamsoniella* sp., *Phoenicopsis angustifolia, Elatides setosa*
0	5	0.10	0	*Ptilophyllum pecten*	absent	absent
649	941	13.87	67.90	*Sagenopteris* cf. *phillipsii, Nilssonia orientalis, Nilssoniopteris vittata, Pityophyllum angustifolium, Schizolepis grandis, Carpolithes* sp.	*Cladophlebis denticulata*	*Todites williamsonii, Coniopteris burejensis, Dictyophyllum rugosum, Ctenozamites* sp., *Nilssonia donetziana, Ptilophyllum pecten, Podozamites lanceolatus, Elatides setosa*
266	649	64.90	40.98	*Sagenopteris* cf. *phillipsii* (+ *Caytonia*), *Ptilophyllum pecten, Elatides setosa*	*Todites williamsonii, Dictyophyllum rugosum, Cladophlebis denticulata, Cladophlebis haiburnensis*	*Todites princeps, Otozamites iziumensis, Elatocladus* sp.
140	235	1.38	59.57	*Ptilophyllum pecten, Elatides setosa*	*Todites williamsonii Coniopteris latiloba, Dictyophyllum rugosum, Cladophlebis denticulata, Cladophlebis haiburnensis*	*Nilssonia vittaeformis, Ptilophyllum acutifolium*

Of course, in the scheme presented above the restoration of the catena has been greatly simplified. Vague interpretations of taphocenoses and the superposition of incidental factors blur the picture in many cases. It is therefore essential that the data obtained be compared with material pertaining to other sections. If they yield the same or even a similar sequence of change of taphocenoses, the catena has been restored correctly. Let us illustrate this by looking at a Jurassic deltaic catena in the Donets Basin (Kamenka).

At the base of the plant-bearing section occur glauconitic sandstones with marine fauna, accumulations of fine detritus on the surface of the layers, and a few remains of *Coniopteris, Nilssonia vittaeformis, Gingko, Elatides,* and *Podozamites.* The percentage content was not worked out for this layer, since there are not many remains and the predominance of *Coniopteris, Gingko,* and *Podozamites* is readily determined visually. In the overlying stratum of intradelta sediments without marine fauna that were formed during a marine regression and contain autochthonous equisetum beds in ferruginous rocks, layers (and accordingly also taphocenoses) I–V were isolated (Table 15). For each layer mention is made of the nature of the boundary with the underlying layer, the thickness, stratification, and the facies (hypothetically, proceeding from the lithology, finds of marine fauna, and the type of beds).

All the taphocenoses fall into the category of macrotaphocenoses. They are classified according to the prevalence of certain organs (see Part II, section 13). Separate counts were made of the number of whole and deformed (marked with the symbol Z) specimens for each taphocenosis. Abundance, or the **saturation coefficient** (Kc) is determined as the ratio of the number of specimens to the volume of enclosing rock. The **deformation coefficient** (Kd) is the ratio of the number of deformed specimens to the total number of individuals in a given taphocenosis, expressed as a percentage. Autochthonous and allochthonous elements were identified according to characters reviewed in the preceding section: in the first place according to the quantitative share, account being taken of the transportability, durability, and state of preservation of organs and ecological characteristics.

The quantitative share of each species is shown in Table 16. Layer II is not included in the table because only *Ptilophyllum* was found here. The category of organ is designated by the following symbols:

br – brachyblast
cl (caulis) – stem
fac (folium aciculatum) – acicular leaf
fc (folium compositum) – compound leaf
fl (folium lineare) – linear leaf
fr – frond

pn – pinna
pl – pinnula
rm (ramus) – branch
sm (semen) – seed
sq (squama) – scale
str – strobile

Taphonomic characters are indicated as follows:

– remains lie in the bedding plane
Z remains lie at an angle to the bedding plane, deformed
* remains form a continuous cover on the surface of the layer (foliar roof)

The taphocenosis of glauconitic sandstones (prodelta deposits with marine fauna) is dominated by tough, leathery leaves of *Gingkoites* and *Podozamites*, which are probably allochthonous, as in the "drifted plant beds" of Yorkshire. Remains of *Coniopteris* are numerous but fragmentary. In the above-delta part of the section taphocenosis I (iron ores) is a combination of two synusia: monospecific stands of *Equisetum* in an autochthonous bed and arboreal vegetation with *Czekanowskia* and *Podozamites* as dominants (Podozamito-Czekanowskietum). The allochthonous element here is represented by *Williamsonia* and strobili of *Elatides*. Apparently, these are strobili of the plants (*Ptilophyllum, Elatides*) whose leaves and shoots dominate in taphocenosis IV (white clays). *Phoenicopsis* is also probably in an allochthonous bed.

The monospecific *Ptilophyllum* taphocenosis in the gypsified clays is remarkable in that this bennettite is presumed to belong to the mangroves (see Part III).

In taphocenosis III (brown clays) there is a mingling of remains of two associations: a bog forest with *Pityophyllum* (forming a foliar roof) in combination with the bog bennettite *Nilssoniopteris* (Nilssoniopteridi-Pityophylletum) and a forest with *Elatides* (mainly strobili and small twigs, oriented at an angle to the bedding plane). The high deformation coefficient is probably connected with movement of the sediment before it solidified.

Taphocenoses IV and V are extremely similar in composition and probably reflect the same community; this may be verified by comparing them with the aid of Preston's formula:

$$N_{a+b}^{\frac{1}{z}} = N_a^{\frac{1}{z}} + N_b^{\frac{1}{z}},$$

where N_a and N_b are the numbers of species in two samples from the same population, N_{a+b} is the total number of species in the population, and z is a constant that with a logarithmically normal distribution is equal to 0.27.

Table 16. Quantitative share of species in the Kamenka taphoflora

Species	Layer I			
	organ	number of specimens		
		–	Z	total
Todites williamsonii	0	0	0	0
Todites princeps	0	0	0	0
Klukia exilis	pn	1	15	16
Coniopteris burejensis	0	0	0	0
Coniopteris latiloba	0	0	0	0
Dictyophyllum rugosum	0	0	0	0
Cladophlebis denticulata	0	0	0	0
Cladophlebis haiburnensis	0	0	0	0
Equisetites columnaris	cl	24	81	105*
Caytonia sp. (cf. *C. nathorstii*)	0	0	0	0
Sagenopteris cf. *phillipsii*	0	0	0	0
Ctenozamites sp. (cf. *C. cycadea*)	0	0	0	0
Nilssonia orientalis	fr	1	1	2
Nilssonia donetziana	0	0	0	0
Nilssonia vittaeformis	fr	0	18	18
Williamsoniella sp.	str	0	2	2
Ptilophyllum pecten	0	0	0	0
Ptilophyllum acutilobum	0	0	0	0
Otozamites iziumensis	0	0	0	0
Nilssoniopteris vittata	0	0	0	0
Czekanowskia rigida	fac, br	20	130	150*
Phoenicopsis angustifolia	fl	2	2	4
Leptostrobus cf. *laxiflorus*	sq	0	3	3
Podozamites lanceolatus	fl	20	96	116*
Elatides setosa	str	0	3	3
Elatocladus sp. (cf. *Taxus jurassica*)	0	0	0	0
Pityophyllum angustifolium	0	0	0	0
Pityocladus sp.	0	0	0	0
Schizolepis grandis	0	0	0	0
Carpolithes sp.	0	0	0	0

Layer III				Layer IV				Layer V			
organ	number of specimens			organ	number of specimens			organ	number of specimens		
	–	Z	total		–	Z	total		–	Z	total
pn, pl	0	5	5	fr, pn	188	130	318*	pn	53	68	121
0	0	0	0	0	0	0	7	0	0	0	0
pn	0	6	6	0	0	0	0	0	1	0	0
0	0	0	0	0	0	0	0	0	1	0	0
1	0	1	1	pn	12	0	0	fr	1	0	1
fr,						29	41	pn	0	18	18
pn	32	84	116*	pn	9	14	23	pn	11	23	34
0	0	0	0	pn	44	37	81	pn	7	6	13
0	0	0	0	00	0	0	0	0	1	0	0
0	0	0	0	str	0	1	1	0	0	0	0
ph	12	5	17*	fc, ph	45	8	53	0	0	0	0
pn	0	2	2	0	0	0	0	0	0	0	0
fr	8	12	20	0	0	0	0	0	0	0	0
fr	2	5	7	0	0	0	0	0	0	0	0
0	0	0	0	0	0	0	0	fr	0	1	1
0	0	0	0	0	0	0	0	0	0	0	0
fr	2	0	2	fr	18	2	20	fr	5	9	13
0	0	0	0	0	0						
0	0	0	0	pl	1	0	0	fr	1	5	6
fr	32	80	112*	0	0	0	1	0	0	0	0
0	0	0	0	0	0	0	0	0	0	0	0
0	0	0	1	0	0	0	0	0	0	0	0
0	0	0	0	0	0	0	0	0	0	0	0
fl	1	0	1	0	0	0	0	0	0	0	0
rm, str	0	64	64	rm, str	57	45	102	rm	8	10	18
0	0	0	0	rm	2	0	2	0	0	0	0
fac	250	350	550*	0	0	0	0	0	0	0	0
br	0	4	4	0	0	0	0	0	0	0	0
sq	0	27	27	0	0	0	0	0	0	0	0
sm	3	4	7	0	0	0	0	0	0	0	0

For oryctocenoses IV and V, the sum total of species, provided that they belong to the same community, is

$$\tau = \left(11^{\frac{1}{0.27}} + 9^{\frac{1}{0.27}} \right) 0.27 \approx 13.$$

Hence, the theoretically expected number of common species is $11+9-13=7$, which is close to the actual number (6). The allochthonous elements of these two oryctocenoses probably belong to slope vegetation. Here elements of the Nilssoniopteridi-Pityophylletum association that forms the bulk of taphocenosis III are absent, so that we may conclude that this association occupied a lower hypsometric level and were situated farther downstream, closer to the outer margin of the delta, during the time that layers IV and V were formed.

The forest with *Elatides* apparently spread over a large part of the delta plain, bordering numerous lakes and bogs, since remains of this conifer, consisting mostly of small fragments of shoots and the more transportable strobili, are found in almost all taphocenoses that were formed closer to the shoreline.

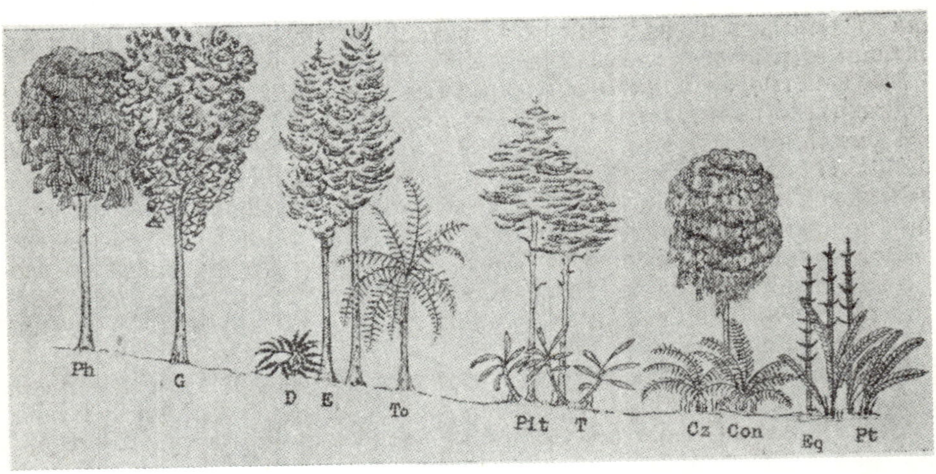

Figure 23. Example of a Mesozoic catena (Jurassic of Kamenka):

Pt – Ptilophyllum, Eq – Equisetites, Con – Coniopteris, Cz – Czekanowskia, T – Taeniopteris, Pit – Pityophyllum, To – Todites, E – Elatides, D – Dictyophyllum, G – Gingko, Ph – Phoenicopsis.

On the whole, a **deltaic catena** is reconstructed as follows (Figure 23):

1) a zone of mangroves and marshes with monospecific thickets of very large *Equisetites,* dwarf tree ferns *Coniopteris,* and the mangrove bennettite *Ptilophyllum*;

2) a zone of pioneer arboreal vegetation (Podozamito-Czekanowskietum) with the schizaeid fern *Klukia*;

3) a broad *Elatides* zone and Nilssoniopteridi-Pityophylletum bog forests with *Nilssonia* species and the bog fern royal *Cladophlebis denticulata.* Forests dominated by *Elatides* occupied the shores of the branches of the distributing water bodies of the delta plain and possibly also the lower part of the slopes descending to the plain. The underbrush features the arborescent fern *Todites, Caytonia,* and the bennettite *Ptilophyllum* (*Williamsonia*). In the clearings is the xeromorphous fern *Dictyophyllum*;

4) slope vegetation with *Gingko* and *Phoenocopsis.*

To check the accuracy of the reconstruction, let us look at other deltaic taphofloras of Jurassic age. We shall begin with the Upper Jurassic taphoflora of the Talyndzhan suite of the Bureya River, for the interpretation of which I used a similar method. A coal-bearing series with flora (intradelta facies) overlies sediments with marine fauna. In the lower part of the coal-bearing series we find remains of Limulidae, possibly indicating the presence of mangroves. The sequence of taphocenoses proceeding up the section is as follows:

1) tuffites with *Coniopteris*: well-preserved fern fronds, twigs of *Pityocladus* with brachyblasts, shoots and cones of *Elatides,* and leaves and seeds of *Pseudotorellia, Czekanowskia,* and others. The saturation coefficient is quite low. In this apparently allochthonous taphocenosis there is a mixture of material from several beds. It corresponds to the glauconitic sandstone taphocenosis of the Kamenka section;

2) siltstones and tuffites with *Czekanowskia* and *Pseudotorellia* (Pseudotorellio-Czekanowskietum): a very saturated taphocenosis with numerous foliar roofs made up of leaves, strobili, and scales of the dominant trees. The fern royal *Raphaelia* dominates in the lower story. Maceration techniques revealed very interesting mesotaphocenoses consisting of small Hepaticae, Lycopodiales, dispersed equisetum cuticles, megaspores, cuticles, leaves, snips of shoots, and seeds of *Elatides,* etc. (Krasilov, 1971). As will be seen in the following, the influence of the neighboring associations makes itself felt here. The taphocenosis as a whole corresponds to the iron ore taphocenosis (I) and partly to the brown clay taphocenosis (III). Both the autochthonous (*Czekanowskia*) and the allochthonous (*Elatides*) elements coincide here;

3) carbonaceous argillites and siltstones with *Equisetites.* This taphocenosis is repeated several times in the section. Its interpretation is obvious;

4) thin coals, siltstones, and sandstones with *Pityophyllum* and *Pityocladus.* The members are thin, with foliar roofs formed of needles of *Pityophyllum*; again repetition in the section. The members correspond to the brown clay taphocenosis (III) of Kamenka;

5) sandstones with *Phoenicopsis.* They dominate in the upper part of the Talyndzhan suite section near the mouth of the Umal'ta River. In individual layers of the thick series *Phoenicopsis* associates with *Pterophyllum, Nilssonia, Sphenobaiera,* and other genera.

We see that the paleosuccessions in the Bureya and Kamenka sections have much in common. The reconstruction of the catena *Equisetum* + *Coniopteris* → *Czekanowskia* → *Elatides* + *Pityophyllum* → *Phoenicopsis* is on the whole confirmed. However, in the Bureya catena *Ptilophyllum* is absent, in the zone of coastal forests *Pseudotorellia* is featured first and foremost (this genus is also present in the Donets basin, but in the Kamenka section described it is not encountered), *Pityophyllum* is not associated with *Nilssoniopteris,* and we can draw inferences on forests with *Elatides* only from allochthonous beds. Meanwhile, the correctness of our interpretation of the taphocenosis with *Nilssoniopteris* is corroborated by the fact that accumulations of leaves belonging to this bennettite are found in other sections of the Bureya River, plus its roots and seeds in bog facies.

The Yorkshire Jurassic taphoflora, described by the eminent paleobotanists Lindley and Hutton, Brongniart, Seward, Thomas, Harris, and others, is much richer than the Donets taphoflora, and its taphocenoses are extremely diverse. Most of the species common to Kamenka derive from the Whitby (lower series) and Gristhorpe (middle series) beds. Equivalents of the Kamenka taphocenoses here are apparently the *Equisetum* bed, the *Solenites* (*Czekanowskia*) bed with *Klukia exilis,* the *Ptilophyllum* bed, and the *Elatides* bed. In the Bathonian Tkvaroheli taphoflora studied by Delle (1967) the rather poor taphocenoses from the upper part of the porphyritic suite contain *Ptilophyllum* and *Pachypteris* – probably a mangrove association. Also found here is *Elatides,* most likely in an allochthonous bed. Several taphocenoses of the producing horizon of the coal-bearing series (deposits 2, 5, 8, and B-1) contain *Czekanowskia* in association with *Podozamites.* And, finally, in the upper part of the producing horizon *Nilssoniopteris* associates with *Pityophyllum.* But here we see species of the Donets Podozamito-Czekanowskietum associations and of the forests with *Elatides.*

Clearly, the reconstruction of a catena, which is here represented schematically, is quite a laborious process. However, some preliminary ideas may be expressed even with a sketchy knowledge of a paleosuccession. Let us take as examples deltaic catenas of various age. The Paleogene taphoflora of England has been mentioned more than once in this book. According to

Chandler's data (a series of publications the findings of which were summarized by Chandler, 1964), a belt of marshes and mangroves with *Nipa* in the outer and *Acrostichum* in the inner zone is reflected in beds of the Sparnachian strata and London Clay formation. Beyond the mangrove belt, on the delta plain and in the pre-estuarine parts of the river valleys, there grew at this time various Icacinaceae and Sapotaceae, plus, according to palynological data, *Typha, Sparganium,* and others. Judging from the Lower Bagshot taphocenoses, the lowland vegetation included *Ficus, Icacinicarya, Mastixicarpum,* and other genera. In the Boscombe series the tropical lowland forest is represented by palms (*Sabal*), *Solanispermum, Symplocos, Wardenia, Thymelaeaspermum, Eomastixia,* Lauraceae, *Rutaspermum,* etc. The vegetation of the foothills and slopes, judging from the wholly allochthonous taphocenoses of the prodelta facies and the allochthonous elements of the delta plain taphocenoses, in Thanetian time included *Pinus* spp., *Araucarites, Sequoia couttsiae, Cephalotaxus, Cupressinites, Trochodendron, Cantitilia, Dunstania, Magnolia,* and Hammamelidaceae. In the Lower Bagshot series and Bournemouth the vegetation was represented by *Araucarites, Sequoia, Protoaltingia, Carpinus,* and *Hantsia.* More temperate elements were present here. Let us note that the interpretation of *Araucarites* as an upland plant invites some doubt; in view of its high frequency and the abundance of its remains, this conifer may have been a dominant of lowland forests (araucarian forests at present grow in places along beaches: Aubréville, 1964). The vegetation of the upper belts can be visualized only by proceeding from palynological data. Present here apparently were *Picea, Cedrus,* Juglandaceae, *Carpinus,* and other genera of the Arcto-Tertiary flora. Isolated macroremains of these genera appear in the upper horizons of the Paleogene section, indicating that some shifting of the vertical zones took place.

Teichmüller (1958) used beds in the Miocene lignite series of the Lower Rhine basin to restore a deltaic catena made up of reed thickets in the foreground, a bog *Taxodium–Nyssa* forest, a bog with *Myrica,* and a sequoia forest at the foot of the slopes.

In the Paleozoic the best studied (as we have mentioned in the second and third parts of this book) is a deltaic catena consisting of five palynocenoses: *Densosporites oblatus*; *Punctatisporites obliquus*; *Thymospora pseudothiesseni*; *Lycospora*; *Lycospora–Guthoerlisporites erectus* (Habib et al., 1966). A minute analysis of the facies limitations of these taphocenoses yielded a correct interpretation of the sequence of the *Lycospora–Densospora* palynocenoses, a sequence that is characteristic for the paralic coal-bearing deposits of the Euramerican Region (Cross, 1947; Smith, 1961; Navale, 1963; Agrali, 1964; Piérart, 1968; Chaloner and Muir, 1968; etc.). Formerly this paleosuccession was linked with fluctuations of climatic conditions or with the evolution of the flora as a whole. In fact, what we have here are successive elements of a catena that alternate in the geological section due to migration of geomorphological elements of the delta.

In Carboniferous coal beds of Great Britain, Smith and Butterworth (1967) described a paleosuccession that went as follows: a *Lycospora* phase (vitrain) → a *Densospora* phase (durain) → a *Laevigatosporites* phase. In some beds a complete cycle is present, ending in a *Lycospora* phase. The omission of certain phases is also observed. The palynocenoses with *Densosporites* (pollen of *Florinites* is also found here) reflect more stable conditions, whereas *Lycospora* is more frequently encountered fissured, allochthonous coals, etc. Navale (1963) also mentions a relation between palynocenoses and certain microlithotypes of coals (which he classifies according to macerals) in Westphalian D.

Punctatisporites dominates in zones enriched in vitrite, *Densosporites* in zones with duroclarite and fusite. On the whole, the reconstruction performed by Habib, Riegel, and Spackman (see above) is corroborated.

For the Devonian we as yet do not have reliable reconstructions of a deltaic catena, although Höeg (1942) points out that *Svalbardia, Hyenia, Enigmophyton Bergeria,* and *Protolepidodendron* occur in littoral-marine facies in Spitsbergen beds, whereas taphocenoses with *Archaeopteris* (not found here) are usually confined to continental deposits.

Recently support has been found for Naumova's assumption that Devonian plants which produced small spores with simple sculpture were confined to the coastal zone (Lele and Streel, 1969; Richardson, 1967). From this viewpoint Chaloner's data (Chaloner, 1970) on the gradual enhancement of the role of large sculptured spores during the Devonian may point to the expansion of the communities in the upper zone of the catena at the expense of the vegetation in the maritime lowlands.

Examples of an **alluvial catena,** reflecting the zonal distribution of vegetation in the limits of a drainage basin, are exceptionally plentiful, and we shall mention just a few of them. In the Oligocene of Bukhtarma Rayushkina (1969) distinguishes the following groups: aquatic and coastal plants, plants of coastal and valley forests, and plants of sod and mountain forests. Macrotaphocenoses are present here due to the high relief, and they correspond to two mountain zones: *Fagus–Metasequoia* and *Picea–Pseudotsuga,* which in the same-aged flora of Ashutas are indicated only in palynocenoses.

In the Miocene of Poland the transition from lake-bog facies of an alluvial plain to slopes is accompanied by a change of the zones of aquatic plants (*Potamogeton, Lemna*); of bog forest with *Taxodium–Nyssa,* including *Liquidamabar, Typha, Juncus,* etc.; of riparian thickets with *Alnus, Cornus, Sambucus, Ulmus,* and *Platanus*; of mixed conifer forest on slopes with

Liriodendron, Gingko, Eucommia, Carya, Betula, Fagus (Lancucka-Srodoniowa, 1966; Oszast, 1967). Approximately the same catena has been established in the Neogene of Czechoslovakia (Knobloch, 1968, 1970b). In the Upper Oligocene of Hungary Andreánszky (1966) described riparian thickets with *Ulmus, Salix,* etc. and floodplain bogs with *Myrica* which were replaced by laurel forests with *Castanopsis* on the slopes. In the upper part of the section the riparian groups included *Ficus*; the forest vegetation of the lowlands had a tropical habit on the whole (*Cedrella–Symplocos*).* A deciduous forest with *Myrica* probably occupied a higher hypsometric level and was replaced by savanna in drier periods.

Vertical zonality is particularly distinctly expressed in Neogene taphocenoses of Japan. Tanai and Onoe (1961) identify here a zone of riparian forests with *Juniperus, Populus,* and *Juglans*; an intermediate zone between the riparian forest and slope vegetation (*Pterocarya, Alnus, Carpinus, Quercus, Zelkova*); a mixed conifer-broadleaf forest on the slopes with *Abies, Carya, Castanea, Fagus, Quercus, Sassafras, Hammamelis,* and others; a mountain forest including several groups; with a predominance of conifers, *Ulmus, Liriodendron,* and *Liquidamabar*; a predominance of *Magnolia*; with *Rhododendron,* etc.

We have already repeatedly mentioned paleoecological reconstructions of Cenozoic communities of North America. For instance, on the basis of the Mascall taphocenoses Chaney describes a *Taxodium–Nyssa* complex with evergreen oaks, *Carya,* etc.; areas of grasslands on terraces; hillside forest with *Acer* supp., *Amelanchier, Platanus, Populus, Quercus* spp., *Sassafras, Tilia,* and *Zelkova* (the Appalachian element is uppermost in these communities); mountain forest with fir, birch, box, *Libocedrus,* pine, spruce, etc.

In the Carboniferous of the Donets basin Shchegolev (1969 and other works) recognized a number of plant groups, some of which are components of a paralic and others of an alluvial catena. *Subsigillaria* dominated in the swamped lowlands, while the uplands were occupied by conifers in association with mesophytic pteridosperms.

An alternation of anthracophilous and anthracophobic associations (see Part IV, section 1) in continental Mesozoic and Paleozoic sections is apparently not always to be interpreted as a catena. Whereas anthracophilous communities occupied the lowest hypsometric level, the anthracophobes included on the one hand riparian groups and on the other slope vegetation. Nevertheless, in a number of cases a paleosuccession of anthracophiles–anthracophobes probably corresponds to a catena. As an example we may take

* The Mastixiaceae-Lauraceae communities were not coalifying and probably grew on high ground (Jung et al., 1971; Mitt. Bayer. Staatssaml. Paläont. hist. Geol., 11, 223–250).

Figure 24. Example of a Paleozoic catena (Pennsylvanian of Kansas, after Cridland and Morris, 1963):

S – *Sigillaria*, *Ca* – *Calamites*, *L* – *Lepidodendron*, *F* – ferns and pteridosperms, *C* – *Cordaites*, *T* – *Taeniopteris*, *D* – *Dichophyllum*, *W* – *Walchia*.

the taphocenoses described by Gothan and Gimm (1930): autochthonous taphocenoses situated directly above coal beds, with *Calamites*, *Sphenophyllum*, *Odontopteris*, and others; allochthonous taphocenoses with *Callipteris* and *Walchia*; allochthonous taphocenoses of the mixed type. The data presented by Cridland and Morris (1963) show that communities with *Walchia*–*Taeniopteris* occupied a relatively high hypsometric level (Figure 24). We shall return to this question later. Josten (1961) made a detailed study of the distribution of key genera of the European Carboniferous in a Ruhr basin section. He described several associations that are constantly repeated. Lepidophytes and pteridosperms as a rule dominate in the various associations, possibly indicating that they held different positions in the catena system (Figure 25). Havlena (1970) also established a rhythmical repetition of succession: Arthrophyta → Lepidophyta → coal seam → Pteridophyta in Namurian cyclothems.

B. Syngenesis and cenophylum

Du Rietz (1921) called the science of the phylogeny of communities syngenetics. Several syngenetic hypotheses have been advanced on the basis of

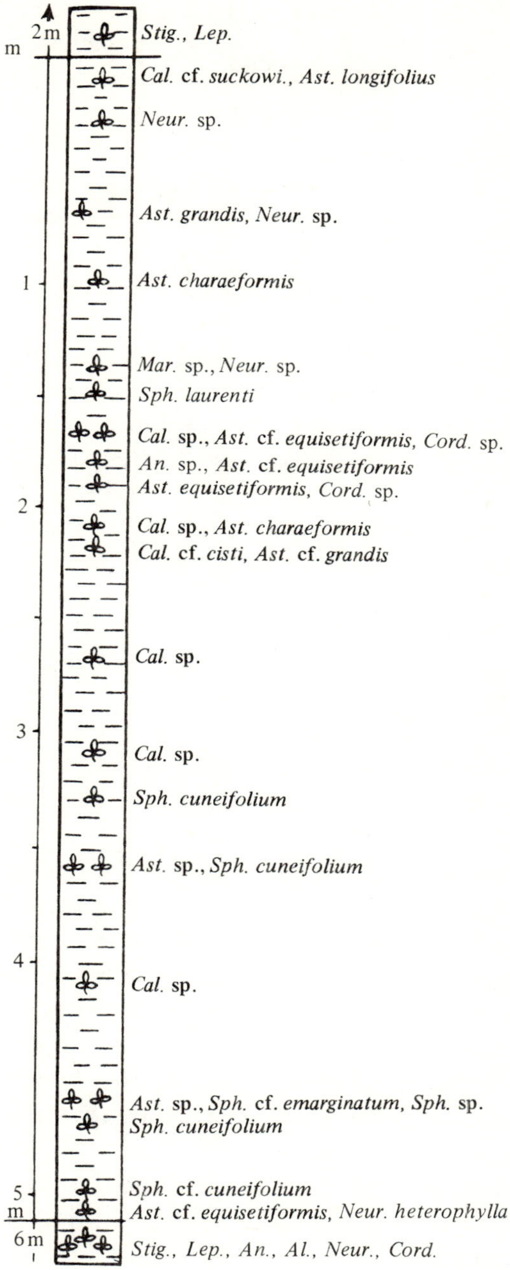

Figure 25. Distribution of *Calamites, Sphenophyllum*, and Lepidophyta in a section of the Carboniferous of the Ruhr basin, after Josten (1961); dominant species printed in bold face (*Ast.* – *Asterophyllites, Cal.* – *Calamites, Sph.* – *Sphenophyllum; Stig.* – *Stigmaria*).

paleobotanical data. Krishtofovich (1946) expounded a theory of **polychronous floras** – the "stable major floras of the past," that encompass practically the whole of the Earth's surface. Extensive geographic distribution, uniform composition over the entire territory of development, prolonged existence, and distinct differentiation (rapid replacement in time via a short series of brachychronous floras) – these, according to Krishtofovich, are the main characteristics of polychronous floras. It is worth noting that Krishtofovich saw in polychronous floras complex systems that are in a state of dynamic equilibrium ("the existence of a certain finished type of vegetation in a climax state is apparently not restricted"). The renewal of polychronous floras is related to the effect of dynamic and static factors. To the former category Krishtofovich refers marine transgressions, change of climate, orogeny, etc. These factors create new biotopes. Static (endogenous) factors include degradation of soils, swamping, the development of intrazonal communities of exposures, cliffs, solonchaks, etc. Although Krishtofovich says that the endogenous factors are "of no lesser importance," he still clearly gave more weight to the exogenous factors, since "in the life of the Earth's plant cover the cardinal moments are those during which, especially in epochs of orogeny or intensive epeirogenic movements, life conditions are substantially altered over continuous large stretches of territory." Noteworthy is the epithet "continuous", which Krishtofovich used deliberately. According to him, the stability of polychronous floras was determined precisely by homogeneity, continuity. If this system happened to break down into parts and lost its continuity, its stability was greatly diminished. He notes that geographic zonality is more distinctly expressed at the borders of polychronous floras.

Krishtofovich drew a sharp distinction between the evolution of plant formations and the evolution of the plants themselves, which proceeds very slowly in the bosom of polychronous floras. When the time comes for renewal and the creation of new biotopes, there are always enough forms ready to "burst in." Up to this crucial moment they had waited in the background in intrazonal communities. In this respect Krishtofovich's theory has much in common with the views of paleozoology theorists. For example, Westoll (1954) claims that evolutionary episodes are not directly related to tectonic movements or other exodynamic factors. "To talk of the extinction of the graptolites and the origin of land floras and tetrapods as 'results of the Caledonian mountainbuilding' is rather empty verbiage." Here he blasts the theory, going back to Buffon, Cuvier, and Brongniart, that connects the successive renewal of the organic world with catastrophes of global significance. The appearance of new groups is not to be regarded as the result of tectonic movements: they simply took advantage of the favorable situation that was thereby created. Simpson (1965) also thinks that "most of the broad features of vertebrate history might have been much the same if the Earth's crust had been static."

Potonié (1951) went further than others in refuting the role of exogenous factors of development. According to him, the evolution of vegetation is an autocyclical process. Diversification of an initial type and space filling (amplification) lead to intensified biotic pressure (Bionanke). The latter is considered a mutagenic factor. New groups arising (as in Krishtofovich's theory) inside a climax vegetation (interpolation) do not play an important part at first, but then gradually alter the environment in such a way that the whole system is reorganized (permutation), and a new vegetation structure is formed.

A concept which has grown up among American paleobotanists and which is in general similar to Krishtofovich's theory of polychronous floras is that of the **geoflora**. Geofloras are major types of vegetation that are characterized by broad territorial distribution, stability in time, and relative homogeneity ("major vegetation type... that maintained its essential identity through space and time": Axelrod, 1958). But a polychronous flora is a concept that is first and foremost chronological, embracing all the Earth's vegetation, whereas geofloras correspond to zonal vegetation (in 1947 Chaney was still calling them Floras with a capital F, while Cain and Mason rejected the idea of the homogeneity of Floras).

Axelrod's work on the evolution of the Madro-Tertiary geoflora (Axelrod, 1958) can rightly be considered one of the finest syngenetic studies. This xerophilous geoflora, characterized by sclerophyllous forms, grew up in the ecotone of the tropical and temperate zones (the Neotropico-Tertiary and the Arcto-Tertiary geofloras). Individual sclerophyllous groupings (like Krishtofovich's intrazonal element) existed in North America during the Eocene (Green River flora). The Oligocene Florissant flora contains savanna and chaparral elements (sclerophyllous species were present way back in the Cretaceous floras of this territory). Some species of the halophilous communities of the Paleogene Wilcox flora were also preadapted to arid conditions, so that the Madro-Tertiary flora was composed wholly of autochthonous elements.* Aridization in connection with the uplift of the coastal mountain ranges in the west led to the appearance of a broad adaptive zone and a stepped-up (quantum according to Simpson) evolution of the xerophilous populations. The composition of the Madro-Tertiary flora was more or less homogeneous in the Miocene. However, in the middle of the Pliocene the Rocky Mountains, the Sierra Nevada, and the Cascade and Coast Mountains had already almost attained their present-day height. Conditions abruptly became

* On the other hand, Stebbins and Major (1968, Ecol. Monogr., 35, 1–35) assume that extensive migrations occurred with the formation of the Madro-Tertiary flora.

differentiated, leading to a splitting up of the geoflora and its breakdown into productive formations. Axelrod's data (see also Mason, 1947) bear out Krishtofovich's notion on the relation between the stability of a cenogenetic system and its complexity. Productive formations, that have a less complex structure, are liable to rapid change and come to bear a closer and closer similarity to the recent sclerophyllous groupings of Mexico and the south of the USA.

In the opinion of Ralph Chaney and his adherents, the evolution of geofloras is affected mainly by migrations and autochthonous reorganizations related to climatic changes. This school pays less attention to the evolution of plants, since it is assumed that at least in the Cenozoic there were no serious evolutionary events. The change in the structure of plant communities took place almost solely on account of elimination and transpositions of already existing groups (Chaney, 1936, 1940, etc.). Axelrod (1958, 1960) thinks that the major groups of flowering plants were formed at the beginning of the Mesozoic. At the same time, the increasing continentality and contrasts of climate in the Cenozoic resulted, according to Axelrod, in the appearance of new adaptive zones (e. g. desert). On the evolutionary plane this is expressed in diversification on the lower taxonomic levels.

On the other hand, it is still thought by many that the main evolutionary episodes (the emergence and extinction of groups with higher taxonomic ranks) are related to the influence of exodynamic factors. The decisive role, in their view, is played by the appearance of new biotopes as bases for a new adaptive type. Extinction is also considered to be the consequence of the nullification of a given biotope (Newell, 1967). In paleozoology this trend is sometimes called neocatastrophism.

Among paleobotanists similar views were developed by Vakhrameev (1947), who stressed, following Golenkin, the significance of aridization of the climate in the early evolution of the angiosperms, Edwards (1955), who inveighed against Krishtofovich's theory on the emergence of new types from within a polychronous flora, and a number of other authors.

The concept of the new adaptive zone, advanced by Simpson (1944), played a marked role in the development of the evolutionary theory. Simpson links intermittent (quantum) evolution with an invasion of a new adaptive zone. More recently (Simpson, 1965) he seems to have retreated somewhat from these views. Nevertheless, the concept of the adaptive zone remains of unquestionable importance in syngenetic studies. A succession of plant groups that replace each other within one adaptive zone I call a **cenophylum**. In contrast to Sukachev's ecological-phytocenotic series, the uniting of groups into a cenophylum does not presuppose phylogenetic links. The problem of an organismal approach to a community of organisms and its relation to Gleason's individualistic theory were discussed in the preceding section.

The development of communities includes, along with coevolution, such processes as transition from one group to another, the cohabitation of species with a different evolutionary history, etc. We may speak of phylogenetic links between successive elements of a cenophylum only in the sense of autochthonous development of dominant types. The antagonist of this process is a succession of nonrelated dominant types within adaptive zones, a process that Simpson (1960) aptly named a relay.

It is to be emphasized that the adaptive zone is not identical to the biotope. For instance, if the first seed-bearing plants were wind-pollinated, for them insect pollination meant a break-through into a new adaptive zone. This zone has apparently existed continuously from the Late Paleozoic (the first bisexual flowers) to the present and has been repeatedly occupied – at first probably by the bennettites, and later by the angiosperms (relay).

Zoochory may have been of some importance in the Early Mesozoic, but only the appearance of mammals and birds at the end of the Jurassic period opened up a broad adaptive zone of zoochory. The seed plants, that had invaded this zone earlier than the others, gave rise to a new adaptive type of angiosperms (see Part III).

Such adaptive zones as mangroves, bogs, river valleys, and mountain slopes have also existed without interruption, whereas the development of cryogenic and arid deserts probably occurred only episodically.

Thus, we may speak of topical (related to landscape) and cenotic (adaptation to pollination, zoochory, etc.) and persisting and ephemeral adaptive zones, as well as primary and secondary invasions. In each case the development of the cenophylum has its own characteristic features. We shall review some of these.

1) Replacement of communities in ephemeral adaptive zones takes place mainly due to the disappearance and reappearance of the zone itself. For instance, a zone of sclerophyllous vegetation was widely distributed in the south of the USSR and in China in the Eocene (Vasil'ev, 1969; Kornilova, 1970; Tao, 1965). During the Miocene xerophilous groups were formed in the Mediterranean countries (Avakov, 1969; Negru, 1969; Palamarev, 1967) and in the west of the USA. In the Pliocene the conifer-broadleaf forests of Australia were replaced by xerophilous vegetation with *Eucalyptus* and *Acacia* (Gill, 1965). All these regions were disjunct, and the xerophilous vegetation was each time formed de novo primarily out of autochthonous elements. The connection with climate evolution is clear here.

2) In persisting adaptive zones the disappearance of a zone naturally cannot cause plant communities to be replaced. Taking such adaptive zones as

lowlands and uplands, we may demonstrate (see Part IV, section 2A for examples) that a climatic change was the chief factor in syngenesis. Transformations of plant groups (e.g. the replacement of lepidophyte formations by cordaite formations in the Namurian of the Kuznetsk basin or by conifers in the Permian of Europe) developed along the lines of a relay. Vertical migrations (catena reorganization) were decisive here. In the end they led to a lateral shift of vegetation zones (expansion of the Turgai flora at the expense of the Poltava flora at the end of the Paleogene). It may be that the substitution process involved frequent fluctuations (Andreánszky, 1966).

Only in some cases may we suspect a phylogenetic connection between dominants of successive components in a cenophylum in persisting zones. For instance, *Elatides,* the Cretaceous forms of *Parataxodium,* and the Cenozoic *Taxodium* species dominated in groups of the same zone of bog forests. They may even belong to one phylum in the family Taxodiaceae.

3) Replacements in cenotic persisting zones are also related to climate. For example, a tropical climate favors the development of zoidophily, while a temperate climate promotes anemophily (see Part III).

4) The species diversity and structural complexity of ecosystems depend on climate. When cooling occurs, tropical communities find themselves oversaturated. Under these conditions autochthonous syngenesis develops along the lines of a successive discarding of the thermophilous elements (the history of the Cenozoic floras provides a multitude of examples bearing out this theory). Allochthonous replacement proceeds by way of a frontal attack and is excellently reflected in plant beds. With the onset of warming, the communities of the temperate zone find themselves to be underpopulated. An allochthonous change takes place primarily by way of diffusion (the migration of the warmth-loving *Weichselia* and *Matonidium* to the north at the end of the Jurassic–beginning of the Cretaceous is a good example, that was described by Berry, 1945). Autochthonous development in undersaturated cenoses leads to the appearance of new groups. This consideration seems to point to a connection between the effect of exogenous factors, ecogenesis, and syngenesis. The warming that occurred at the end of the Jurassic period was apparently one of the factors promoting the development of the first angiosperms, since the Middle Jurassic groups, that existed under conditions of a colder climate (the evolution of the Jurassic climate will be discussed in greater detail in the next section), proved to be undersaturated.

5) A seasonal climate hinders the development of narrow specializations. Very complicated adaptations (e.g. to pollination), demanding stable conditions, are ineffective here. Therefore, warming and the spread of a tropical climate are conducive to the expansion of "specialists," while cooling favors the "generalists" (Valentine, 1969).

It follows from the above that the development of cenophyla has nothing to do with Potonié's autocyclical process. Contrary to Krishtofovich's ideas, plant phylogenesis is not to be separated from the development of vegetation and, consequently, from climate evolution. Matthew (1939) put forward a general theory that links megaevolution with climatic change. He based this theory on that of the American geologist Chamberlain, i.e. that there is an alternation of warm, humid thalassocratic and cold, arid geocratic epochs. The latter type of epoch sees a wide spread of races, generalization, heightened activity, in short, biological progress, whereas the abundance of food and "sluggish life" in humid epochs lead to regression. In a number of other works the rhythmicity of geological processes is also considered as the primum mobile of syngenesis and evolution. For instance, Titov (1952) thinks that forest vegetation developed during orogenic epochs, while in the intervening glyptogenic periods the vegetation was degraded, and deserts appeared (in fact, desert development is related rather to increasing contrast of relief and climatic zones that occurs during the main episodes of tectogenesis; this also contradicts analogous theories expounded by Lichkov, 1945).

If we consider the successions reptiles – mammals and gymnosperms – angiosperms as an expression of biological progress (Huxley, 1955), we see that the development of life was at a higher level in the Late Paleozoic than in the Mesozoic: in the animal world the mammalian habit dominated (reptiles, Therapsida) and in the plant world gymnosperms that were most similar to angiosperms – pteridosperms and glossopterids. In the Mesozoic the dominant forms were, among animals, the sauropsids, which were further removed from the mammals, and, among plants, the conifers, gingkos, and benettites, that were further removed from the angiosperms. I have termed this phenomenon the "Mesozoic zigzag of evolution." The most likely explanation for the zigzag lies in the fact that climatic conditions were more differentiated and the contrast of climatic zones was sharper in the Late Paleozoic than in the Mesozoic (at least from the middle of the Triassic to the first half of the Cretaceous period). What we find here is apparently the manifestation of a general pattern that was mentioned at the beginning of this book: the level of complexity of an evolving system is governed by its energetic potential and by the complexity of the environment. The structural complexity of a biocenosis is expressed both in species diversity (the variety of life forms) and in the diversity of cenotic relations. As already noted, a change of climate acts on both these indexes. In the event of cooling a community comes to be oversaturated in terms of species diversity and some species have to be eliminated (Sukachev's hologenesis) and cenotic relations simplified (for instance, the significance of zoogamy and zoochory diminishes). In the event of warming there is a trend toward enrichment in species and toward the appearance of new adaptive types (Sukachev's phylogenesis) and new forms of cenotic interaction. Evolutionary progress in the plant world, the emergence and development of the angiosperms, was apparently related to 1) enhancement of the structural complication of the lithosphere and differentiation of climates as a result of continental drift; 2) warming at the end of the Jurassic period; 3) the spread of the higher insects and mammals; 4) development of zoogamy and zoochory; 5) complication of the functional structure of biocenoses.

3. FACTORS GOVERNING PALEOSUCCESSIONS

For plants it is the climate and the soil that are the principal ecological factors determining the nature of most paleosuccessions. The interplay of climate, soil, and vegetation operates on the feedback principle: the climate exerts an effect on the vegetation and the geological substrate, the latter has its effect on the soil, the soil on the vegetation, and, finally, the vegetation influences the climate and the soil. The topography affects soil-forming processes, climate, and vegetation simultaneously. Thus, singling out one particular factor as the cause of a paleosuccession is always an arbitrary act, just as it is to subdivide paleosuccessions into climatogenic, edaphogenic, and orogenic types (in a previous paper, Krasilov, 1969b, the last were incorrectly termed topogenic).

A. Climate

A study of climate evolution using different methods shows that this process is of a periodic nature.

The hypotheses explaining the periodicity are nominally classed into two groups: geological and astronomical. In the first category the hypothesis expounded by Brooks (1949) may be considered the best elaborated. Brooks' main premise consists in that an insignificant temperature drop – of about 5° – is sufficient to bring about glaciation.

Further cooling promotes the formation of actual ice, whose chilling effect is proportional to the area of the glacier (according to Kerner's calculations, the mean January temperature at the pole in the absence of glaciation should be about 50° higher than at present). A glacier alters the pattern of atmospheric circulation: at first it is only the layers of air which are in direct contact with the glacier that cool down, but as the glacier grows in bulk, a high-pressure region is formed, and strong cold winds blow in all directions, the system of currents changes, more cloud builds up over the equator, proportionally more heat is lost on account of the reflection of solar radiation by the clouds, etc.

At first all these changes develop gradually, but once the glacier attains certain dimensions they assume catastrophic proportions.

According to Brooks, a 10% increase in area of arctic land is enough for the initial temperature drop. An increase in the height of the relief leads to the formation of mountain glaciers and heavier cloud cover.

Calculations show that approximately a 300-m increase in the average height of mountains will produce the necessary temperature drop.

On the other hand, a decrease in land height would cause Bering Strait to widen, the Gulf Stream to straighten out (Florida would sink below sea level), and the ice to melt.

At a time of violent eruptions volcanic dust reduces the amount of solar radiation reaching the Earth's surface by 15–20%, and this is also enough. Therefore, there is no lack of purely geological causes of glaciation (though Brooks does not rule out the effect of cosmic factors).

Brooks attempted to encompass the system of climatic factors in all its complexity, taking into account their interweaving, feedback mechanisms, etc. Other investigators laid weight on a particular geological factor. For instance, Heer (1860) found a link between climate evolution and changes in the system of currents. Currents influence not only temperature, but also the distribution of precipitation. In the absence of glaciation and cold currents, the climate of the western fringes of the continents in the northern hemisphere would be more humid.

Also connected with the current mechanism is Hann's hypothesis [J. Hann, Handbuch der Klimatologie, 1883], which attributes glaciation to the fact that the pole is situated in a closed arctic basin, which in the absence of circulation is transformed into a refrigerator (see also Rukhin, 1959). Ewing and Donn (1956, 1958, 1966) think that with minor depths between Labrador and Scotland a drop in the level of the World Ocean during glaciations may heavily restrict the influx of warm waters into the Arctic basin and of cold waters into the Atlantic. In their opinion, this will cause icemelt, since with the lesser water exchange the climate will be drier and cloudiness reduced. This is just one of the many autocyclical hypotheses (see Schwarzbach, 1968; Figure 26).

Another set of hypotheses imputes climatic changes to fluctuations in the composition of the atmosphere – the varying contents of water vapor, carbon dioxide (see the review of these hypotheses in Dorman, 1968), or volcanic dust. Franklin, Brooks, and many others believed volcanic eruptions to be one of the causes of glaciations. But the opposite variant is also possible: volcanic dust lowers the albedo of ice and causes thawing (Bloch, 1965).

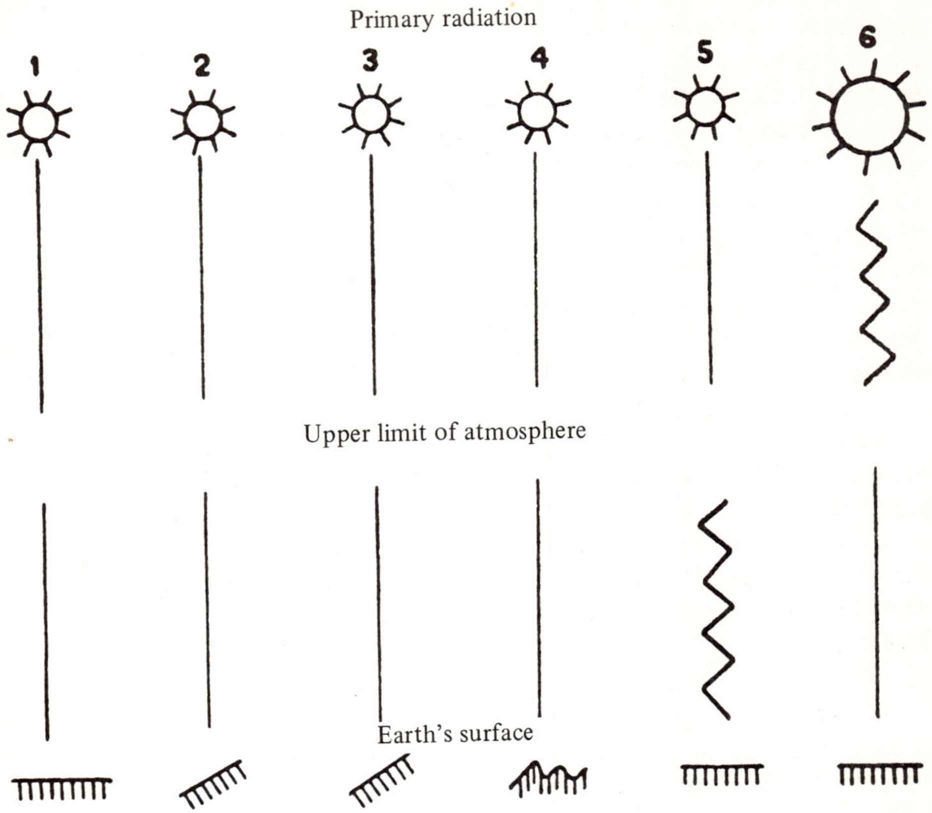

Figure 26. Hypotheses of climate evolution (after Schwarzbach, 1968). Causes of changes: 1) continental drift; 2) nonperiodic changes of obliquity of ecliptic; 3) periodic changes of obliquity, orbital eccentricity, and position of the relative center of the Galaxy; 4) changes of relief; 5) changes in the content of O_2, H_2O, CO_2, and of volcanic dust in the atmosphere; 6) changes in solar emittance.

These examples illustrate that often the same geological factor can be held responsible for diametrically opposed phenomena.

A third group of hypotheses relates climate evolution to the expansion of the Earth (Jordan, 1966) or continental drift. A change of solar emittance (Simpson, 1929–1930), 11-year cycles of sunspots, a change in the obliquity of the ecliptic (with a periodicity of about 40,400 years according to Milankovich: Malinovskii and Florenskii, 1963; Brooks, 1949; and others), variation in eccentricity with a periodicity of 21,000 years (the theories of Croll, Köppen and Wegener, Zeuner, and others: see Brooks, 1949), changes in the position of the solar system relative to the center of the Galaxy with a period of about 280 million years (Lungersgauzen, 1963; Schwarzbach, 1968, etc.) – such are the presumed cosmic causes of periodic changes of climate. We shall not go into them here. New data on marine terraces, coral reefs (Mesolella et al., 1969), the evolution of the Foraminifera, etc. show that Milankovich's theory satisfactorily explains the dynamics of Quaternary glaciations. With regard to larger-scale periodicity, many try to see the explanation in a complex interaction between geological and cosmic factors.

The main procedures and criteria adopted in an analysis of the climatogenic changes of past vegetation are described below.

a) Morphological criteria

We have already analyzed ways of interpreting morphological characters in connection with the reconstruction of life forms (Part III). In this we are limited, firstly, by the variety of adaptations to a single factor (e.g. the poikilohydrous mosses, that possess protoplasmic drought resistance, do not show morphologically expressed xeromorphous adaptations, unlike the gymnosperms: Abramov, 1969), secondly, by the ambiguousness of adaptations (xeromorphism, reduction of the leaf blade, increased number of stomata as adaptations to arid, arctic, alpine, and boggy conditions, etc.), and, thirdly, by the changes that occur in adaptations in the course of evolution. For instance, at present cauliflory is a phenomenon encountered mainly in the tropics, and therefore we might conclude that it is an indicator of tropical conditions. But in fact, in recent plants cauliflory serves as an adaptation to pollination by bats (chiropterophiles), and this is why it is found in the tropics. The adaptive significance of cauliflory in the past is not clear.

MacGinitie (1941) was skeptical of the idea that a study of leaf evolution in the dicotyledons would be able to introduce substantial corrections into Sinnot and Bailey's rules concerning the distribution among the climatic zones of simple and compound leaves, of leaves with an entire and a dissected margin, and of leaves with pinnate and palmate venation. If, for example, simple leaves preceded compound leaves in the course of evolution, then in the early stages there should be more of them in all the climatic zones. In other words, we must find out whether we are dealing with a constitutional or an epharmonic sign. It has often been suggested that the absence of growth rings in Paleozoic wood of the Euramerian Region is a constitutional feature: a seasonal growth rhythmics simply was not proper to the given evolutionary level. Restrictions of this kind apparently have minimal significance: the very first Early Cretaceous dicots had quite varied leaves, the form of which was determined by the climate (Part III, section 3). The pattern observed by MacGinitie in the evolution of the leaves of the Tertiary

Platanus, from entire to lobed, correlates perfectly with the overall climatic change. Growth rings are present in the stems of Devonian progymnosperms (Beck, 1970). In comparing this character with other climate indicators, we see that distinctness of boundaries, the width of the zones, their splitting, and so on had the same value in the past as they have at present (Daugherty, 1941; MacGinitie, 1953).

b) Tolerance

Describing the evolution of the Tertiary climates in Europe, Göppert (1855) started by noting that in the Eocene there were many palms, tropical Proteaceae, *Artocarpus,* Sterculiaceae, Sapindaceae, and others. During the Miocene, according to him, temperate conifers, Cupuliferae, Salicaceae, Aceraceae, and Juglandaceae prevailed, though *Daphnogene,* Proteaceae, and Malvaceae were still present, whereas in the Pliocene *Platanus* gained wide distribution and the tropical elements disappeared.

In order to gauge the tolerance of ancient species and their groups, we have to rely on certain methods of reconstructing climate. These may be classified into **autecological** and **synecological** methods, there being, however, no clear-cut boundary between them.

Iversen's indicator method (Iversen, 1944) is based on the idea of a **thermosphere** as the totality of thermal limits for the existence of a species. The geographic distribution of a species naturally gives an inaccurate picture of the thermosphere, since distribution is governed by humidity and other factors; still, in Europe the northern and eastern range boundaries do depend mainly on temperature conditions. What is of decisive importance here is the ratio of the mean temperatures of the warmest and the coldest month. In constructing the graph we plot the mean temperature values of the coldest month on the abscissa and those of the warmest month on the ordinate. The meteorological stations are placed inside this coordinate system (stations with destroyed vegetation are designated as substations). A curve is drawn linking the stations marking the beginning and end of the region of occurrence of the given species (Figure 27). Marked on the same graph are stations in the region of which the species occurs in the fossil state, and changes of climate are judged according to deviations from the curve. Iversen used three maritime species, *Ilex aquifolium, Hedera helix,* and *Viscum album,* whose distribution is strongly affected by winter temperatures. Station 1 on the graph characterizes the conditions in the Atlantic period (in the limits of tolerance of all three species), station 2 conditions in the Subboreal period (the winter temperatures are too low for *Ilex* and are close to the limiting values for *Hedera*), and station 3 the present-day conditions. The curve of

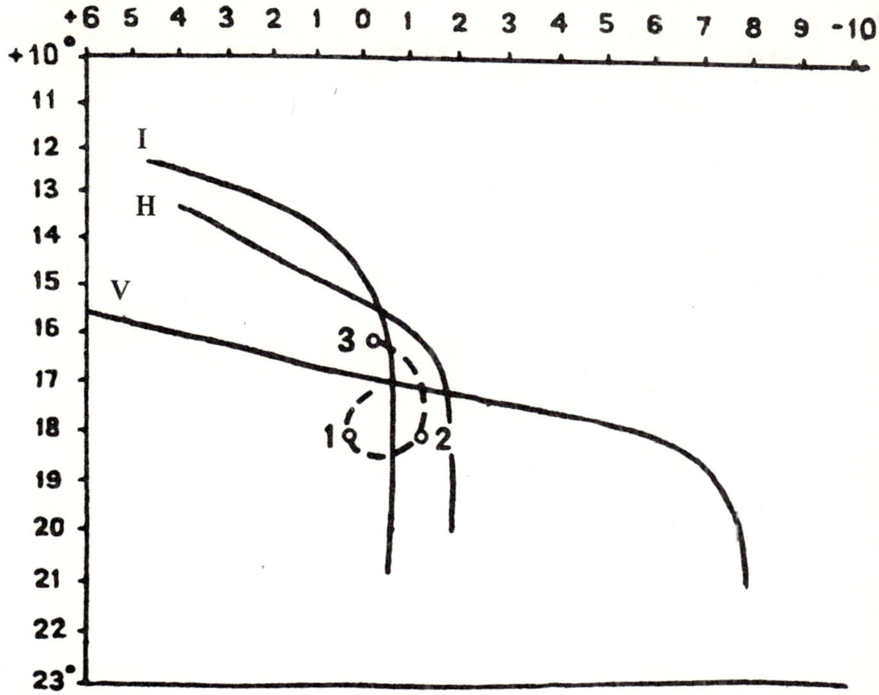

Figure 27. Determination of climatic conditions by Iversen's method (Iversen, 1944)

temperature change from the Subboreal to the present time is approximately parallel to the curve of limiting values for *Hedera*. Meanwhile, the absence of one of the indicators may be due to the fact that it has not yet reached the region in question in the course of migration.

Szafer's synecological method (Szafer, 1946) is based on mapping of the regions that are common ground for several species. If today the isoline corresponding to a particular frequency of common occurrence contours area A, but, let's say, in the Atlantic period of the Holocene it contours area B, this means that the climate of the Atlantic period in area B matched the present-day climate in area A.

Several methods of narrower application exist that take account of the common occurrence of two species and (or) their ratio. For instance, Churchill (1968) determined climate humidity on the basis of the association of pollen of *Eucalyptus diversicolor* and *E. calophylla.*

Grichuk's method (Grichuk, 1969) combines the reconstruction techniques used by Szafer and Iversen and involves the construction of two graphs, where the abscissa and ordinate represent in the first the mean January temperature and the mean July temperature and in the second the duration of the vegetative (frost-free) period and the annual precipitation total.

Axelrod and Bailey's method (Axelrod and Bailey, 1969) consists in the determination of two indexes that are crucial for vegetation – **effective temperature** (combining the mean annual temperature and the difference between the January and July temperatures) and **equability**, established by calculating the lowest temperatures and the number of hours in the year with subfreezing temperatures.

Fuji's method (Fuji, 1969) is relatively simple, being reduced to the construction of triangles the apexes of which correspond to 100% quantitative share of cryophilous (A), warm-temperate–subtropical (B), and temperate (C) genera. The distribution of the points denoting the content of the relevant groups in the taphoflora in question in comparison with the recent flora indicates the changes of the climate (Figure 28).

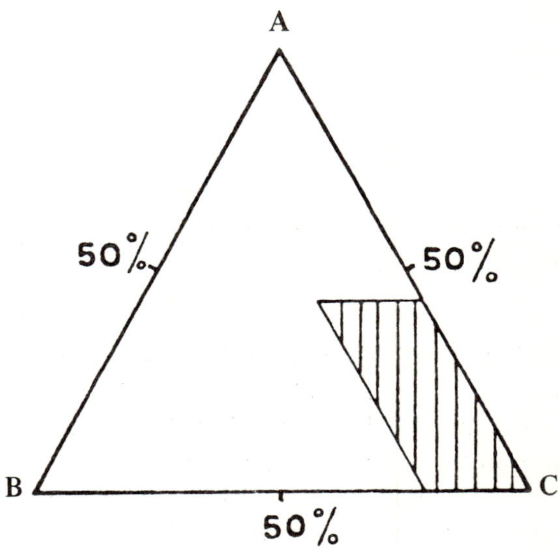

Figure 28. Climatological interpretation of a taphoflora, after Fuji, 1969:

A) cryophilous species; B) warm-temperate and subtropical species; C) temperate species. The hatched area marks the position of the recent vegetation of Central Japan.

Rougher methods of constructing temperature curves are based on calculations of the content of indicator (thermophilous, xerophilous, etc.) groups for each time interval (Polyakova, 1967; Gol'bert et al., 1968).

Chronocline. The word **cline**, widely known by now, was used by Julian Huxley to designate unidirectional geographic variability related to changes in conditions. Some of the "orthogenetic series" described by paleontologists apparently represent a succession of ecotypes belonging to one geographic cline, that is, a temporal cline (chronocline). A number of examples (*Populus eotremuloides–P. alexandri–P. trichocarpa*, etc.) can be extracted from the works of Axelrod (1941, 1948). The analogy between the geographic cline and the chronocline implies that there is an analogous change of conditions.

Actualistic methods of interpretation are of limited value for the pre-Cenozoic floras, because virtually none of the species present in them have survived to the present. Often all taxa of higher rank are considered as indicators of certain conditions (bennettites are thermophilous and xerophilous, ferns hygrophilous, conifers xerophilous, etc.), and this leads to oversimplified and inaccurate concepts of climate. Where the pre-Cenozoic floras are concerned, the limitations derive mainly from two sources.

1) The tolerance of the majority of the recent species is not known exactly. Odum (1959) makes the following additions to Shelford's principle of tolerance (with minor changes):

a) a species may have wide limits of tolerance with respect to one factor and narrow ones with respect to another;

b) if one of the factors is not in the optimum state, this limits the tolerance with respect to the other factors;

c) tolerance varies geographically, depending on the season, etc.;

d) for a number of reasons (e.g. the spread of parasites) organisms often prefer conditions that deviate from the optimal.

Potonié (1953) points out that the presence of "tropical" ferns in a taphoflora is not to be considered a sign of high temperatures, since ferns more often occur at a considerable altitude above sea level in the tropics. Moreover, the present-day conditions of existence of groups that were more widely distributed in the past give no indication at all of their tolerance.

2) From Gleason's individualistic theory (Part IV, section 1A) it follows that interspecific unions have a varying (often poor) degree of resistance. We have already given examples of the formation of new associations by way of the union of species with a different evolutionary history (Negru, 1969; Goff and Zedler, 1968; and various other publications) or by way of the breakup of associations characteristic for the Tertiary taphofloras. Associations of species that were relatively climate-resistant drifted apart in the Postglacial due to different tempos of migration. For instance, Iversen (1944) mentions that, judging from the spread of *Hedera*, the postglacial climate in Denmark was favorable for *Ilex aquifolium*. And yet this species was absent, as it had not yet managed to reach Denmark on its way from the west.

The limitations of the autecological criteria were already obvious to Reid and Chandler (1926), who claim that there is no basis for focusing attention on index species when reconstructing climate. In the taphoflora they studied (as in most of the Paleogene floras) elements of tropical and temperate vegetation are combined.

Brett (1960) also draws attention to the combination of tropical and non-tropical species of oak in the British Eocene (the similar observations made by Chaney were mentioned above). In part, such "forbidden" combinations may be considered posthumous, related to shifts of the remains from various altitudinal zones.

However, an analysis of quantitative ratios in many cases obliges us to reject this explanation. If we consider that a change of one factor entrains a change of tolerance with respect to other factors (e. g. a rise in humidity broadens the range of temperature tolerance), the "forbidden" combinations become fully explainable. Axelrod and Bailey (1969) assume that the "equability" of the Tertiary climate was responsible for the wanderings of "temperate" and "tropical" genera. The western margins of the continents (England, western Canada) had a wet-summer climate in the Paleogene. This is attributed to the different position and height of the mountains or to the absence of arctic ice and the influence of the Tethys Sea (Piel, 1971).

One of the reasons for the broad spread of some species in the past and for their having entered communities that today would be alien to them might have been the formation of **complex polyploidal assemblages**. According to Stebbins, polyploidal assemblages evolve from simple to complex and then again to simple, this probably being accompanied by a change in the limits of tolerance (Axelrod, 1941).

c) Statistics of life forms

In analyzing the dispersion of the life forms he identified, Raunkiaer (1934) joined up points where a particular form was uniformly abundant and obtained isolines which he called **biochores** (in doing so, he slightly altered the original meaning of the term as introduced by Köppen). It turned out that the biochores were more or less parallel to the isotherms. Raunkiaer spoke of phanerophyte, hemicryptophyte, and chamaetophyte climates. Explaining the climatological significance of, let's say, ferns, Raunkiaer calculated the ratio of angiosperms to ferns for the entire world flora and for a specific flora. The ratio of these two values constitutes the **pteridophyte quotient** of the flora in question. This technique may be applied to floras of any age. It is possible, for example, to identify a stigmarian climate in the Paleozoic or a cycadeoid climate in the Mesozoic. Such definitions are more specific than a commonplace statement such as "the climate was warm and humid."

d) Florogenetic method

Asa Gray was the first to show that the flora of the Atlantic states of the USA is more similar to the East Asiatic flora than the flora of the western states. He advanced a hypothesis according to which the eastern part of North America and East Asia were in the Tertiary period settled by species that had migrated from the north, bypassing the western regions of North America, where conditions were no longer suitable for them. Engler (1879) suggested that the reason for there being no East Asiatic species in the west of the USA is that they died out at the end of the Neogene. This point of view later gained more general recognition. Axelrod (1960) distinguishes European, East Asiatic, East American, and West American floristic elements in the Tertiary floras. The elimination of the East Asiatic and East American elements in Europe and the west of North America is due to the reduction in the amount of summer precipitation and to the switch from a wet-summer to a wet-winter climate. Thus, the change in the ratio of floristic elements reflects the history of the climate.

Another variant of the florogenetic method consists in calculating the proportion of local (living at present on the territory in question), exotic, and extinct (or of doubtful systematic position) genera. In North America the significance of the first group is seen to wane from the Late Cretaceous to the Eocene and then progressively wax beginning with the Oligocene, whereas the second group reaches a peak in the Early Oligocene and from then on steadily declines (Wolfe and Barghoorn, 1960). In Europe Reid, Szafer, and Nemejc (Nemejc, 1970) also used this method.

e) Phytocenological criteria

In view of the above-mentioned limitations, the methods based on the tolerance principle can only be auxiliary props when it comes to reconstructing pre-Cenozoic climates. The value of characteristics that show the overall structure of communities accordingly gains strength here. The most important of these are apparently the following:

1) The index of species diversity, indicating the size of populations:

$$\alpha = \frac{\text{number of species}}{\text{number of individuals}} \times 100.$$

The usefulness of this index for paleoclimatology lies in the fact that populations of tropical plant species are as a rule smaller than populations of temperate species. Accordingly, in the tropical floras the species are represented by a smaller number of individuals per unit area than in the temperate floras (Pires et al., 1953; Fedorov, 1966; etc.). This same circumstance also affects the likelihood of burial of a species, there being fewer chances for the tropical species. Furthermore, the differences noted in species diversity are leveled out owing to: deciduousness (greater likelihood of burial in the temperate zone), entomophily (lessens the chances of burial in the tropics), and the rate at which the material decomposes (higher in the tropics).

Despite the effect of these three leveling factors, the index of diversity is not uniform for taphofloras of different climatic zones and to some extent reflects changes of climate. For instance, it has been established according to a number of indicators that cooling occurred at the end of the Early Cretaceous epoch in the Suifun basin. The index of diversity is 4.5 for the Aptian flora, 2.4 for the Albian (Krasilov, 1969b).

Determining diversity according to the formula $\alpha = \frac{S}{N}$ is worthwhile only in a case where the number of samples is relatively small; this is usually the case, though, in paleontological collections.

If conditions for collection are good, however, one can gather so many samples that a further increase in N will no longer yield an increase in the number of species (S). In such situations it is better to determine the index of diversity on the basis of the logarithmic relationship between the number of species and the number of specimens (see Part IV, section 2): $S = \alpha \ln\ (1 + \frac{N}{\alpha})$.

2) The coefficient of mosaic structure (Dyrenkov, 1970):

$$K = \frac{S + D}{N},$$

where D is the number of dominants, S the number of subdominants, and N the total number of species.

Mono- and oligodominance are characteristic for forests of the temperate zone. Monodominant communities are found in a tropical rain forest, but relatively rarely (Walter, 1971). The trouble is that the definition of dominants and subdominants in a taphocenosis is too subjective. According to Chaney (1959), species which make up more than 6.6% in a macrotaphocenosis can be classed as abundant.

3) Jaccard's generic coefficient is the number of genera expressed as a percentage of the number of species. The significance of this coefficient is not quite clear. It may characterize the diversity of conditions (Greig-Smith, 1957) or the history of a flora, proceeding from the assumption that autochthonous development enriches a flora in genera (Malyshev, 1969).

4) The coefficient of multiformity of taphocenoses, m, that is, the number of different taphocenoses f as a percentage of the total number of taphocenoses, t:

$$m = \frac{f}{t} \times 100.$$

To me this coefficient seems very important for a paleoclimatological evaluation of major taphofloras, since within one climatic zone it has similar values in different stratigraphic horizons. If beds that are confined to the same facies are compared, we see that for the Late Cretaceous Tsagayan taphoflora of the Zeya-Bureya basin f is close to 1 (the taphocenoses in the clay lenses of the Bureya Belogor'e, Kordon Mountain, Arkhara, etc., that are more than 100 km distant from each other, are practically identical: Krasilov, 1970c). The same situation is repeated in the Miocene flora of the Zeya-Bureya basin: the beds along the Guran River west of Blagoveshchensk and along the Zeya River contain practically the same set of species. On the other hand, in the Early Jurassic flora of the Maritime Territory, studied by Shorokhova and myself, taphocenoses from the same stratigraphic horizon that are confined to the same facies and situated not more than 0.5 km apart often show a different composition.

The same picture is observed in the Early Cretaceous taphoflora of the Maritime Territory described by the author, where in 15 beds of the Lipovets suite seven different taphocenoses were discovered (m = 46.6%). These examples allow us to hope that the coefficient m will eventually find wide application in paleoecology. The difficult part about it is that any estimation of degrees of difference between taphocenoses is highly subjective. Here we could resort to the coefficient of similarity, but for all practical purposes it seems to be enough to make a rough evaluation, in which taphocenoses that differ even just in terms of one dominant or subdominant are considered to be different.

5) The ratio of different growth habits. Edwards (1936) thought that the high proportion of remains of Devonian plants (97%) in the Eocene taphoflora of the London Clay formation pointed to a tropical climate. Other authors also analyzed the spectrum of growth habits from this standpoint (Chaney, 1936; etc.). But the distortions introduced by the different likelihood of trees, shrubs, and grasses being buried are so heavy that this criterion proves to be of minimal value.

An intuitive assessment of the physiognomic characteristics of communities usually yields contradictory results. For instance, most authors consider the Carboniferous forest of the Euramerican Region to be the equivalent of the present-day tropical rain forest, whereas Potonié (1953), Jongmans (1958), and some others see here a closer similarity to the moist conifer forest of the temperate zone. Potonié emphasized the features that attest to high humidity, but, in his opinion, there is nothing to prove that the temperature conditions were tropical. On the other hand, the forest of the Euramerican Region was apparently the most thermophilous community of the Carboniferous, so that if it was not tropical by modern standards, it should still be considered tropical in relation to the conditions obtaining on the whole in the Carboniferous.

Potonié interpreted the size variability of the pinnules of *Neuropteris praedentata* along the axis as a sign of seasonal growth rhythms. In fact, seasonal fluctuations of climatic conditions are scarcely expressed in a tropical forest, although Walter (1971) notes that short dry periods commonly occur in all equatorial regions. If taphonomic observations confirm seasonal leaf fall, then this may serve as proof of an extratropical climate. However, the signs of growth periodicity described by Potonié and by Mägdefrau (1956) give no indications in this respect, since autonomous periodicity of growth, flowering, and fruiting are proper to some extent to all trees in a tropical forest (Walter, 1971; Fogie, 1947). For example, the Dipterocarpaceae display a periodicity of fruiting that in some regions equals 9–11 and in others 3–4 years (Poore, 1968). Species that regularly replace their foliage with a periodicity of 3–4, 9, 12, or 32 months occur in various regions of the tropics.

f) Migrations and the reorganization of catenas

There are those who flatly reject the possibility of whole floristic complexes undergoing migrations, and this inevitably leads them to refute extensive continental glaciations or their periodicity. It is not within the scope of this book to discuss these views. However, it can be stated categorically that cooling of the climate is apt to result in a shifting of altitudinal zones, and since altitudinal zonality (at least in the high latitudes) is analogous to latitudinal zonality, a vertical movement is in the final analysis transformed into a lateral one. When warming occurs, phytocenoses find themselves undersaturated, and the door is open for thermophilous species to filter in (see Part IV, section 2B). With all this, however, the idea of a community migrating en bloc is, of course, an oversimplification of the actual situation. Due to differences in tolerance and in migration tempos, groups of species are always being separated and restructured in the course of their wanderings, though this should not prevent us from perceiving an overall trend of migration and its relationship to changes of climate. Moreover, the parallel development of disjunct populations simulates migration (Menitskii, 1969).

There have been many paleobotanists (Gardner, Chaney, Krishtofovich, and others) who believed that in the high latitudes the plant groups that spread far south at the end of the Oligocene had become established as early as in the Eocene. However, Engler drew attention to a number of weak points in Gardner's hypothesis, and recent investigations have shown that it does not hold water. The Paleogene floras of the Arctic are in no way identical to the Neogene floras of the middle latitudes (Wolfe et al., 1966).

1) Migration of isoflors and biochores. Chaney (1940) defined isoflors as "lines connecting similar fossil floras, which are considered to represent essentially similar climatic conditions." In the first approximation they are parallel to isotherms, but deviate from these due to the varying continentality of the climate. Still, the degree of coincidence is high enough for us to be able to evaluate changes in temperature conditions from the migrations of isoflors (Figure 29).

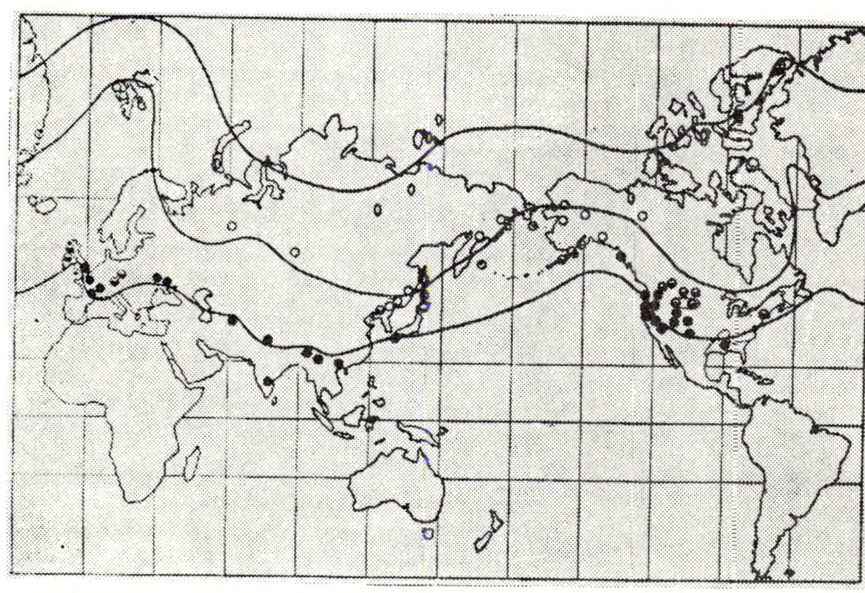

Figure 29. Eocene isoflors of the northern hemisphere (after Chaney, 1940)

A study of Raunkiaer's biochores (see above) will probably yield more reliable conclusions. But for the time being paleobotanists will have to make do with a rougher assessment of climatic changes according to the boundary shifts of the temperate and tropical zones. Dorf (1960) presented a series of maps showing that the northern boundary of the temperate zone was shifted northward by 20° and the southern boundary by 10–12° in the Eocene (at this time it coincided with the northern boundary of the USA). In the Late Oligocene and at the beginning of the Miocene the boundary of the temperate and tropical zones were displaced southward and the redwood region expanded at the expense of the subtropical formations. There was again a shift to the north in the second half of the Miocene and during the Pliocene. The migration of the northern forest boundary, established by palynological methods (the NAP/AP ratio, the rate of pollen deposition, etc.: see Part IV, section 1), has an important bearing on the interpretation of Quaternary climates.

The zone shifts may be due not only to climatic change but also to continental drift. However, the fact that the same changes are observed on different continents indicates that there is a relationship with climate (Chaney, 1940). It is very telling that the northward boundary shift in the temperate and tropical zones of the northern hemisphere in the Eocene is matched by an analogous shift to the south in the southern hemisphere (Menéndez, 1969).

2) Catena reorganization. Latitudinal migrations can be visualized as an asymmetrical expansion of an ecotone in a particular direction. This process probably does not differ in essence from the movements of components in a successional series (catena). Furthermore, we may assume that in most cases lateral migrations are realized by way of a change in the vertical structure of the catena: its elements move up or down (upland–lowland).

Many authors have expressed the opinion that the lateral migrations of vegetation zones in the Quaternary took place (albeit only partly) on account of the vertical migrations. It is hard to imagine how it could be otherwise, considering that the horizontal and altitudinal zonalities in the high latitudes show an almost total correspondence (Ahti et al., 1968):

Zone	Belt
Arctic	Alpine
Boreal	Subalpine
Boreal-meridional	Montane
Submediterranean	Hill

Since both the horizontal and the altitudinal zonalities are related here to the same thermal conditions, Ahti et al. (1968) consider it better to use the term "zone" for both, rather than distinguish "belts" for the altitudinal components. He names the alpine zone oroarctic, the subalpine oroboreal, and so on. Du Rietz, Grebenshchikov, and other phytogeographers showed that with advancement to the south, the zones situated at sea level in Northern Europe became displaced higher and higher, eventually to disappear.

We may thus be sure that climatic change in this two-dimensional system (actually, three-dimensional, as there is another climatic gradient from the coast to the inland parts of the continent) is simultaneously expressed in the vertical and lateral migration of zones.

Grosset (1967) describes the migration of vertical vegetation zones in the Crimea and Caucasus during the Würm glaciation, and Fuji (1969) makes use of the same procedure to explain the expansion of the conifers in the Pliocene of Japan. At the boundary between the Pliocene and Pleistocene the flora with *Metasequoia* species is replaced by plants groups with *Pinus koraiensis* and *Picea maximoviczii.*

Yaroshenko and Grabar' (1969) mention that in the Carpathians and the Caucasus the timberline is moving down despite a warming of the climate. But at the same time the lower boundary of the forest is moving up, so that what appears to be happening is a general degradation of the forest under the influence of Quaternary factors.

Chaney (1963, etc.) suggested that in the Paleogene, when the lowlands of the high latitudes in the northern hemisphere were occupied by a tropical flora, communities with a temperate habit already existed in the uplands. With the onset of cooling at the end of the Paleogene – beginning of the Neogene, the lowland groups shifted to the south and the upland groups took up position according to the following scheme:

Strata	Lowlands	Uplands
Eocene	Clarno	Bridge Creek
Upper Oligocene–Miocene	Bridge Creek	Mascall
Upper part of Miocene	Mascall	

Finds of remains of temperate genera (e.g. spruce or hornbeam) in allochthonous beds of an Eocene taphoflora (Chandler, 1964) in general confirm Chaney's supposition, although his assertion of a "Miocene flora on the Eocene uplands of Oregon" is no more than a metaphore that oversimplifies the situation. The Eocene lowland flora is similar to the tropical flora, but in the low latitudes, as is known, there is no such clear-cut homology between the horizontal and the altitudinal zonality as there is in the north. Therefore, the Paleogene upland flora could not have been the exact equivalent of the Neogene lowland flora of the same area. Nevertheless, it is evident that a vertical migration of zones did take place.

On the territory of the Bureya basin there were three main formations at the end of the Jurassic and in the first half of the Cretaceous: forests with *Czekanowskia–Pseudotorellia*; forests with *Phoenicopsis* (plus *Pityophyllum* and *Baiera*), and forests with *Gingkoites* (Krasilov, 1971a). Mainly on the basis of a quantitative analysis, I established that at the end of the Jurassic the first formation dominated in the lowlands, the second occupied a somewhat higher hypsometric level, and the third was distributed in the uplands. At the beginning of the Cretaceous (Urgal suite) the second formation expanded at the expense of the first, and at the end of the Neocomian the catena underwent a drastic reorganization with the incursion of the *Gingkoites* forest into the lowlands. Farther north, in the Lena basin, some authors note an increase in the content of *Gingkoites* with entire leaves at more or less the same stratigraphic level. A vertical migration probably occurred in this area too.

It is thought that Permian associations of the *Walchia–Taeniopteris* type in the USA (Cridland and Morris, 1963) or the *Walchia–Callipteris* type in Europe (Shchegolev, 1965) existed back in the Carboniferous, growing on uplands. The succession of floras at the Carboniferous-Permian boundary in this case may be related to vertical migration. However, Remy (1952) denies that *Callipteris* existed in the Carboniferous. Also uncertain is the role played by vertical migrations in the Namurian "floristic leap" that took place not only in Europe but, evidently, also in Siberia (Meien, 1968).

The limitation of climatic reconstructions based on catena reorganization lies in the fact that only an upland–lowland migration markedly alters the composition of phytocenoses (the upland communities move closer to their place of burial), whereas migrations in the opposite direction are hard to recognize: they are expressed merely in a reduced quantitative share of allochthonous remains of upland plants, which are in any case few.

g) Physical indicators; rhythmicity

Such a complex phenomenon as climate evolution naturally requires a many-sided approach. Some investigators insist that lithological indicators of climate have an advantage over paleobotanical pointers, which seem to give contradictory indications. However, as was discussed in Part II, section 11, such widely used indicators as red beds were formed under quite varied conditions. The recent conditions of formation of a laterite profile or red-sediment alluvium (Walker, 1967) cannot be considered to be the exact equivalent of the climate in which the ancient red beds were formed. Only autochthonous remains of plants that have been preserved in strata of red sediments can enable us to determine the "red-bed climate." For instance, the Mount Eden taphoflora, preserved in a series of red beds, contains groups of riparian forests with elm, walnut, plane tree, poplar, etc.; groups of lake shores or marshes (horsetails, reedmace); groups of savanna or open woodland (*Arbutus, Pinus pieperi, Quercus hannibali, etc.*); groups from the edges of deserts (*Ephedra, Lepidospartum,* and others). Axelrod (1938) thinks that the conditions here corresponded to those obtaining at present in the southwestern regions of North America. Conifer stems in the Triassic red-bed series of Arizona indicate a forest type of vegetation, while the pattern of the growth rings points to short, dry periods during the summer (Daugherty, 1941).

Brunner (1970) mentions semihumid conditions for the Miocene–Pliocene laterites of southern India, with an annual total of atmospheric precipitation in the region of 1000–2500 mm and three (maybe sometimes five) dry months. In wetter periods this territory was covered with a monsoon forest. It has recently become clear (Kudryavtsev, 1966) that even rock salt deposits do not show such a distinct relation to paleoclimate as is usually believed. There are now salt lakes not only in the arid zone but also in Greenland and on Spitsbergen (Brooks, 1949).

Plant remains in coal-bearing series, glauconitic sandstones, and so on give a much more detailed picture of paleoclimate than the series themselves.

With all this, plant communities react variously to fluctuations of climatic conditions. Their reaction is more readily perceivable in ecotone zones than in the interior of formations (McAndrews, 1966; Watts, 1969). The stability of a community depends on its degree of complexity and maturity, or, in other words, on its corresponding to a theoretical climax. In conifer forests of Siberia climatically governed cycles with a span of 240–280 years are expressed, whereas the bog vegetation in the same localities reflects periodicities of 80, 30–40, and 8–15 years (Glebov, 1963).

In a study of relatively short cycles good results are yielded by a correlation of paleobotanical data with the periodicity of lake transgressions, the dynamics of mountain glaciers, the fluctuations in the population dynamics of blue-green algae and diatoms, etc. (Shnitnikov, 1970).

An important source of information on temperature conditions is isotope thermometry. At one time it seemed that this method would supplant all others, but later it emerged that the primary isotope composition of a shell differs in different systematic groups. For example, shells of ammonoids are richer in O^{18} than those of *Inoceramus,* and they contain less C^{13} (Tourtelot and Rye, 1969).

Recrystallization of calcite and epigenetic changes in general markedly affect the reliability of measurements (Stahl and Jordan, 1969). Further, the isotope composition of a shell depends on the chemism of seawater. It has been found that Mesozoic brachiopods and cephalopods were able to live in water with abnormal salinity. The strontium/magnesium ratio in the shell may serve as an index of salinity in the case of brachiopods, while for belemnites it is the strontium/calcium ratio; however, the reliability of this criterion has not been proved. Apart from this, data on belemnites depend on the depth of habitat, on the body temperature of these active predators, and on the season in which they died (Berlin et al., 1970). Thus, the results of paleotemperature measurements must be checked by other methods.

It is interesting to compare paleothermometric and paleobotanical findings. From isotope analyses and the ratio of calcium and magnesium we observe a drop in mean annual temperatures at the beginning of the Liassic. This was followed by a warming right up to the Toarcian, cooling in the Middle Jurassic, and another rise in temperatures in the Upper Jurassic (Berlin et al., 1970; Fabricius et al., 1970; and other publications). High values of Late Jurassic paleotemperatures have been obtained for different areas of the Russian Platform, Western Europe, Greenland, and New Zealand. Cooling is seen to have occurred in the Neocomian (in West Siberia from 17–20°C in the Valanginian to 10–13° in the Hauterivian: Berlin et al., 1970). In the Late Cretaceous, after a cooling in the Cenomanian, the mean annual temperatures show a rise, reaching a maximum in the Coniacian-Santonian (Bowen, 1961a, b; Domestoy, 1967). Meanwhile, data for the Russian Platform and Siberia do not corroborate this Coniacian-Santonian maximum (Naidin, 1966; Berlin and Khabakov, 1968). A temperature drop is observed everywhere at the end of the Cretaceous period. Kobayashi (1942), basing himself on the content of cycadophytes, the degree of dissection of the leaves of Gingkoales, and other paleobotanical criteria, constructed a paleotemperature curve for East Asia which shows a warming during the Triassic and Jurassic with a peak at the end of the Jurassic and then a marked cooling that continued right up to the end of the Cretaceous.

Kobayashi mentioned secondary temperature drops for the Early and beginning of the Late Jurassic. In other areas a change in taphofloras at the Triassic-Jurassic boundary and then another one in the Middle Jurassic points to cooling. The optimum at the end of the Jurassic apparently also embraces the Berriasian. According to my data, in the Bureya basin signs of cooling (reduced amount of cycadophytes, upland–lowland migrations: Krasilov, 1971a; and others) are observed in the Valanginian (Soloni suite) and then in the Barremian-Aptian (Chagda and Chemchuko suites). In the Maritime Territory temperature conditions improved during the Aptian, to deteriorate considerably in the Early Albian (Krasilov, 1967, 1969b). My data on the Late Cretaceous taphofloras of Sakhalin do not confirm a Coniacian-Santonian optimum. Judging from the amount of entire leaves, the warmest climate was in the Campanian (the Cape Zhonkier flora), after which there was an abrupt cooling in the Maestrichtian and especially in the Danian age. On the basis of palynological data, cooling at the end of the Cretaceous is indicated for North America (Norton and Hall, 1967; Srivastava, 1970).

There are very few paleothermometric data for the Cenozoic. High values of paleotemperatures are in evidence for the Eocene and Middle Miocene, this being borne out by the successions of marine faunas (Margolis and Kennett, 1970). In accordance with the changes in taphofloras (mainly on the basis of the content of smooth-edged leaves and the ratio of local and exotic elements) quite similar paleotemperature curves have been constructed by a number of authors (Becker, 1961; Dorf, 1969; Wolfe and Barghoorn, 1960; Wolfe and Hopkins, 1967; Wolfe, 1969, 1970). They indicate a temperature maximum in the Late Eocene – Early Oligocene, followed by a marked cooling and then a second peak in the Middle Miocene. A different view is held by Axelrod and Bailey (1969), who postulate a steady drop in the effective temperature during the Cenozoic. They rightly remark that a change of a taphoflora does not adequately reflect the change of mean annual temperatures (actually, I think that this applies to effective temperatures as well), and they consider the inflections in the temperature curves drawn by other authors to be pseudofluctuations (deriving from a correlation of taphofloras situated in different latitudes or on different hypsometric levels). However, a Middle Miocene optimum is established not only in America but also in Europe (Sitár, 1967), New Zealand (Hornibrook, 1967), and Japan (Huzioka and Tanai, 1967). On the other hand, Schwarzbach (1968) believes that the fluctuations in the Cenozoic or Europe are related to a change of humidity rather than of temperature, which shows a steady drop. This question of fluctuations is very important, since a unidirectional change of climate may be attributed to continental drift alone, whereas fluctuations require that other factors be taken into account.

Just as important is the rhythmicity of the fluctuations. Such investigators as Mai and Szafer liken the Neogene climatic rhythms to the Quaternary ones (Szafer, 1961). In his analysis of spore-pollen diagrams, van der Hammen (1961) showed the importance of cycles with a duration of between 2 and 6 million years. The latter, in his opinion, define the boundaries of stages, while megarhythms of 60 and 200–250 million years correspond to geological periods and eras. His conclusions confirm the significance of climatic cycles for evolution (see p. 192).

Temperature conditions are an essential ecological factor, but we must not forget Odum's advice (Odum, 1959, p. 106): "by all means consider temperature, but do not stop there."

It is impossible to understand the differentiation of the past vegetation without considering the vital factor of the **continentality of climate**. Palynologists have established that the boundary between the Cretaceous floristic zones *Aquilapollenites* and *Normapolles* runs in a meridional, not a latitudinal direction (Vakhrameev et al., 1970). The distribution of land and sea in the Late Cretaceous epoch shows that this boundary divides regions of a marine and continental climate, the differences between which were particularly pronounced at the time. Apart from this, the differences between the Angara and Cathaysian geofloras of the Late Paleozoic were also probably related to continentality (in addition to the thermal gradient). According to palynological data, in the Cenozoic of Europe an ocean–continent gradient is not evident right up to the Pliocene but is distinctly expressed in the Pleistocene (Boulter, 1971). The boundary sliding observed in some palynozones in Europe possibly point to a progressive oceanization of the climate (Hibbert et al., 1971).

B. Edaphic conditions

Paleosuccessions in paralic series are to a considerable extent related to changes in edaphic conditions. The dynamics of associations with *Lycospora–Densosporites* (the Smith effect, after Chaloner and Muir, 1968), described in Part IV, section 2A, illustrates this.

Paleosuccessions of swamp vegetation in connection with variations in edaphic conditions have been described in many works, primarily on the basis of palynological data. Rybniček and Rybnicková (1968) reconstructed bog associations on the basis of pollen and macroremains. They established eight phases of development of swamp vegetation, from the Late Dryas to the present time: 1) a lacustrine phase with a predominance of aquatic plants; 2) a *Carex* invasion, drying, development of moss cover, and, at the end of the phase, the appearance of sphagnums and *Salix*; 3) a reduced water influx, development of sphagnums, intrusion of arboreal vegetation; 4) cessation of peat accumulation in connection with fires, renewal of vegetation; 5) at the beginning of the Atlantic period, a mesotrophic phase with associations of the *Eriophorion gracilis* alliance; 6) development toward oligotrophic and dystrophic types (mainly in connection with rapid peat bog growth); 7) an oligotrophic phase with *Eriophorum* and sphagnums and the sporadic appearance of woody plants (birch, spruce, pine). This type of vegetation is preserved up to the end of the Subatlantic period. The present-day upland bog is not formed due to insufficient humidity of the climate; 8) the recent phase: disappearance of oligotrophic peat-formers, invasion of shrubs and pine in connection with drying.

Most authors note a successive development of communities from eutrophic to oligotrophic, although under certain conditions development starts out from the oligotrophic phase. Heinselman (1970) considers that the overall trend of a succession accompanying growth of a peat bog lies not so much in a trophic change as in an enhanced diversity of the plant communities.

Thomson (1949) made a detailed study of the evolution of Tertiary communities in lignite series, where succession is related simultaneously to climatic change and to edaphic conditions. The superimposition of edaphic factors (soil impoverishment, podzolization, etc.) on climatogenic succession has been described in many publications on the Quaternary period (Karpov, 1969; Livingstone, 1968; Mania and Mai, 1969; etc.). Grosset (1967) links the expansion of *Ulmus elliptica* in the Holocene with warming and the subsequent shrinkage of the range with an impoverishment of the soils.

A more or less distinct parallelism is observed in the development of vegetation in the interglacial epochs. Von Post distinguished here three phases of development: archeo-, meso-, and telocratic.

Later this scheme was broken down further by Firbas, Iversen, Andersen, and others (see Turner and West, 1968). Andersen established an oligocratic phase, characterized by degradation of the *Quercum mixtum* broadleaf forest and by the spread of hornbeam, fir, and sometimes spruce and box in connection with soil impoverishment.

We should also note that the effect of volcanic eruptions on the development of vegetation is also partly related to soil changes (Lovelius, 1970). Thus, from annual growth rings we perceive a slowdown in the growth rate of Kamchatka larch during eruptions. This is ascribed to an overall worsening of the meteorological conditions and to the settling of ash on trunk and crown.

After eruptions, however, an increase is observed in the species diversity; this is apparently connected with an improvement in the edaphic conditions.

C. Relief

Although changes in relief are unquestionably a vital factor in paleosuccessions, it is not always possible to recognize their effect. The nature of topography and its variations are assessed according to the following indications:

1) *Diversity of ecological groups.* The paleobotanical literature abounds in observations to the effect that a combination of ecological groups in a bed (especially of plants with a tropical or temperate habit) is an indication of high, rugged relief. Intuitive evaluations of this type are not of much use when it comes to reconstructing topography. Conditions within a floodplain are quite varied, and the ecological spectrum includes a multitude of groups, from hygrophilous to xerophilous and halophilous. We know too little about the tolerance limits of the majority of recent plants (not to mention the extinct ones) to be able to assert with certainty that all the "forbidden" combinations were formed posthumously rather than during life. Even if we

manage to tie up specific taphocenoses with certain elements of a catena, this still does not give us any information on the nature of the relief. Such information can be obtained only from a detailed analysis of the ecological spectrum of the taphocenosis. For example, in the Chalk Hills Pliocene taphoflora Axelrod (1962a) described: a) riparian groups with birch, poplar, willow, etc.; b) redwood on the slopes with *Sequoiadendron, Chamaecyparis,* and sixteen others out of the twenty-four common companions of *Sequoiadendron* in the northern part of its range; c) xerophilous groups of volcanic ridges with *Rhamnus* and others. At the same time, the absence of remains from open oak woodland and chaparral shows that the material was brought in from wetter northern and western slopes.

Chandler studied a rich collection of seeds and fruits from the Pleistocene taphofloras: Barnwell and Lee Valley. At Barnwell she established the following groups: a) an arctic-alpine group – 42% of all species; b) species of broader distribution, primarily aquatic and bog forms; c) the southern element – species that at present do not occur north of 63°N; d) a calciphilous group of species growing on chalk exposures; e) a halophilous group of species growing in river estuaries or on maritime plains (*Rumex maritimus, Zannichellia pedunculata,* etc.). In the Lee Valley flora the arctic-alpine group constitutes 22%. This disparity might have been imputed to the differences in age and in the climatic conditions. But the calciphilous group is also scarcely represented in Lee Valley, suggesting that the difference is related to the topography. The Barnwell bed was apparently situated nearer to slopes with chalk exposures.*

2) *Gradient analysis.* Schaffer and Wilke (1965) made an attempt to ordinate palynospectra from 21 postglacial profiles in Michigan and Ohio (for the procedure see Part IV, section 1A). It is assumed that the distribution of the samples in the coordinate system reflects the soil change and the temperature gradient associated with increased height of relief.

Maher (1963) holds that the usual type of correlation of palynological spectra yields little information for interpreting topographic conditions. He attempted to clarify how variations in altitude above sea level affect the ratio of the amounts of different pairs of genera, e.g. the *Pinus/Picea* ratio. It was found that the logarithms of some ratios correlate with the altitude.

Oszast (1967) noted that the proportion of conifers increases in Miocene beds with increasing proximity to the Carpathians.

I made similar observations with the Early Cretaceous flora of the Maritime Territory (Krasilov, 1967). The amount of Pinaceae increases in the northern part of the Murav'ev-Amurskii peninsula, where considerable elevations were traced according to coarse deposits.

* Chandler's formulations have recently been reviewed (Bell and Dickson, 1971, New Phytol., 3, 627–638).

3) *Altitude indicators.* Taking as his starting point the *Populus pliotremuloides* and *P. alexandri* associations, with a prevalence of the former, Chaney (1938, 1959) determined the altitude reached by the Deschutes flora of East Oregon to be 3000–3500 feet. At present, an analogue of the second species, *P. trichocarpa,* grows mainly at an altitude of around 2000 feet. Higher up it associates with *P. tremuloides,* to which it is inferior in terms of abundance. Still higher, *P. tremuloides* occurs together with mountain conifers, which do not figure in the taphoflora in question. The height of the Miocene plateau Shang wang, in his opinion (Chaney, 1940), exceeded 1000 m. This is borne out by the absence of sequoia. The Miocene Mohawk formation must have been established at an altitude of less than 2000 feet, since it does not contain any mountain conifers (Axelrod, 1962b). Puri (1943, etc.) also used the indicator method to determine the altitude of the Himalayas in different epochs.

Axelrod (1962b, 1968, and other works) greatly perfected the methodology of determining altitude by utilizing effective temperature, a composite index proposed by Bailey (1960). Proceeding from the boundaries that Köppen had fixed between the tropical (mean January temperature +10°C) and polar (mean July temperature −10°) climate, Bailey designated regions of eternal winter and eternal summer in a coordinate system where the abscissa gives the difference between the July and January temperatures, i.e. the mean annual range (AR), and the ordinate the mean annual temperature (T). In the interval between these values the winter/summer ratio is defined as the effective temperature

$$\mathrm{ET} = \frac{8\mathrm{T} + 14\mathrm{AR}}{\mathrm{AR} + 8}.$$

This index gives us the temperatures at the beginning and end of the vegetative period and information on its duration. Axelrod calculates the effective temperature from the composition of the taphocenoses and compares it with the effective temperature at sea level, determined from finds of marine fauna of the same age. From the temperature difference we may calculate the altitude above sea level.

4) *Comparison of same-aged beds.* Nathorst (1888) suggested that the differences between the Pliocene Mogi taphoflora and other Japanese taphofloras of the same age are due to their having been formed at different altitudes. Yabe (1929b) calculated that the Mogi bed was situated at an altitude of 700 m. At present it is almost at sea level. Subsidence on such a scale, in his opinion, could have separated the Japanese islands from the continent.

If beds that are now situated at different altitudes contain the same plants, we may infer that the topography was less dissected at the time of their formation (e.g. the Miocene Denton Creek and Table Mountain formations with almost identical taphocenoses are at present situated at altitudes of 5200 and 2200 feet; this attests to a more gently sloping relief in the Miocene: Axelrod, 1962).

Braun (1947) demonstrated a connection between the evolution of mixed mesophilous forests in the east of the USA and the subsequent erosion cycles and the formation of peneplains.

5) *Orographic effect.* This is the term used to describe the influence of mountains on the distribution of sediments (Walter, 1971). Mountains have a marked influence on the planetary system of atmospheric circulation. The differences between the climates of Siberia and North America and between those of India and North Africa are due in good measure to the different distribution and height of the mountain ranges (Brooks, 1949).

Especially sharp climatic differences on the leeward and windward sides of mountains are observed in areas where the prevailing winds blow constantly in one direction, e.g. tradewinds.

The effect mountains exert on the distribution of ancient vegetation was first shown by Gothan (1907). He attributed the difference between the taphofloras of the Carboniferous deposits of England, Belgium, Ruhr, and the Upper Rhine region on the one hand and the "inland" regions of Bohemia, Lower Silesia, and Zwickau on the other to the effect of Hercynian mountain structures on the climate.

In North America the changes of climate and vegetation in connection with mountain uplift has been thoroughly studied on the basis of numerous beds of Tertiary floras (Chaney, 1936, 1940, 1952; MacGinitie, 1941; Dorf, 1938; and others). Differences between the vegetation of the Rocky Mountains and that of the areas lying to the west of them are noted as early as in the Eocene. The Chalk Bluffs beds investigated by MacGinitie give evidence that the Sierra Nevada existed in the Eocene, but they were not as high and scarcely affected the climate. A reduction in atmospheric precipitation related to the uplift of the Sierra Nevada is distinctly traced only in the Late Miocene (judging from the Table Mountain flora), though the southern part of the range may have risen in the middle of the Miocene.

The first signs of the uplift of the Cascade Mountains are observed at the end of the Miocene, but it was not until the Pliocene that the mountains attained a considerable height. At the end of the Pliocene or beginning of the Pleistocene the Sierra Nevada and Cascade ranges approached their present-day height. Their influence on the climate led to the disintegration of the Madro-Tertiary geoflora and the extinction of the "East-Asiatic-type" groups in the west of the USA.

In Russia the change in climate that went along with the rise of the Carpathians is expressed in the expansion of the steppe zone, which began in the

Late Miocene and progressed without interruption right up to the end of the Pliocene (Vronskii, 1970).

Beginning with the Permian period, on the areas adjacent to the Sea of Japan plant groups (vegetation zones) of the maritime and more continental types kept appearing all the time, situated parallel to each other and to the margin of the continent. In the Early Permian the Russkii Island flora (Zimina, 1969), with an "Angara" habit, refers to the inland zone.

The outer zone is represented by the Japanese Maiya flora with *Cathaysiopteris, Psygmophyllum,* and other Cathaysian genera (Asama, 1967). Presumably, the differences between the Russkii Island and Maiya floras reflect the continentality of the climate. The Rodai series, in which plant remains were found, is composed of conglomerates, black siltstones, and limestones with a marine fauna. According to Japanese geologists, the source of the clastic material was the Abe Mountain Range situated along the coast of Sino-Korea and stretching north to the Maritime Territory. The Russkii Island and Maiya beds evidently correspond to the vegetation of the inner and outer slopes of this range. In the Late Permian the vegetation of the outer zone gradually spread to the west, encroaching upon the Maritime Territory. The orographic effect was apparently weakened in connection with the destruction of the Abe Mountains.

The Triassic vegetation of the inner zone is represented by the Monguhai flora with diverse but primarily small-leaved cycadophytes (Srebrodol'skaya, 1958). To the outer zone is referred the recently discovered Tudo-Vaku flora in the Maritime Territory (Krasilov and Shorokhova, 1970) and the Japanese Nariwa and Mine floras (Huzioka, 1970). Figuring importantly here are conifers and pteridosperms, which may have lived in a coastal mangrove belt.

At the beginning of the Cretaceous period the coastal ranges were situated much farther east, separating the outer zone of the main Honshu belt with the Ryoseki flora and the inner Hida belt with the Tetori flora. In the Southern Maritime Territory the orographic effect was related to the uplift of the Murav'ev-Amurskii anticlinorium (Krasilov, 1967). The eastern slope, that faced the sea, was taken up by a moist conifer forest of the redwood type (Suchan basin). On the western slope shrub formations of various bennettites and conifers prevailed, these probably being similar to chaparral (Suifun basin). At the end of the Early Cretaceous the vegetation shows signs that the climate became more arid, due to the rise of the mountain chain in the east. The effect of these mountains on the climate was critical at the beginning of the Late Cretaceous, when thick series of red beds were formed in the Maritime Territory.

The distribution of the plant groups in the Late Cretaceous on the whole falls into three parallel structural zones. The western zone, that included

Sikhote Alin and the inner belt of southwestern Japan, was characterized by powerful superficial volcanism typical for coasts of the Andean type (Mitchell and Reading, 1969). The beds in this zone (the partisan flora of the Maritime Territory, the floras of Asuwa, Omichidani, and others of Japan: Matsuo, 1962) are contained in sediments of small lakes. The fact that most of the dicots have small leaves shows that the climate was somewhat dry. In the middle zone (West Sakhalin trough) sediments of the miogeosyncline type are distributed. The beds in the deltaic facies correspond to redwood. The marine igneous and flysch strata of the eastern zone are characteristic for eugeosynclines of island arcs.

Thus, the distribution of the Permian, Triassic, and Cretaceous plant groups gives us information on the position of the coastal mountain ranges and the structure of the continental margin. The phytogeographic elements cut across the Sea of Japan if we disregard the present shorelines. These data acquire special interest in connection with the concept of lithosphere plates and attempts at an actualistic interpretation of geosynclines. In our example they allow us to date the migration of the deepwater trough toward the ocean, that led to the formation of an island arc and an inner oceanic plate. In other cases a detailed correlation of taphocenoses on both sides of the presumed boundary of lithosphere plates introduces substantial corrections into the formulations of geophysicists. For instance, Visscher (1967) showed that the presumed "Tethys twist" is not borne out by the distribution of palynological assemblages.

Spore-pollen spectra from deepwater sediments are much poorer than coastal spectra. This enables us to establish the movement of the shoreline (Williams and Sarjeant, 1967; Lele and Chandra, 1969). Although this region has only just begun to be reconnoitred, interesting findings have already been obtained that confirm the parting of the shores of the northern Atlantic (Habib, 1970), submergence of land areas in the region of the Demerara Plateau, where unconformity of the boundaries of the African and American continents is observed, and also the recent formation of deepwater depressions in the southeastern part of the Bering Sea (Hopkins et al., 1969), in the Sea of Japan, and in other regions.

D. Activity of man

The changes that have been wrought in vegetation by anthropogenic factors are a complex field of study that requires the bringing together of paleoecological, archeological, and historical documents. We have no space here to go into a minute discussion of this problem, which has a bearing only on the last phase in the history of vegetation. We shall confine ourselves to listing the major sources of information on the ways in which man has shaped vegetation:

a) remains of plants in archeological digs. Chaney and Daugherty (1933) determined charred remains of *Cercis blackii* from a *Sinanthropus* bed in China. The Peking man evidently selected shrubs and low trees to make his fires, Kats and Kats (1969) described fruits which might have been eaten by Neolithic man living in the region of the Berendeevo bog;

b) remains of anthropochorous plants. The appearance of pollen belonging to *Triticum, Secale, Linum,* Chenopodiaceae, *Amaranthus,* and *Zea* in plentiful amounts marks the beginning of agriculture. A high content of *Calluna* pollen is evidence of forest clearings. More developed agriculture is suggested by the pollen of *Polygonum, Solanum,* and *Centaureum.* The expansion of *Plantago, Rumex, Artemisia,* and *Ranunculus* is related to clearings and cattle grazing. Some of these genera flourished at the end of the Pleistocene, then suffered a decline, and later spread out again, as ruderal forms (Faegri, 1944; Jelinek, 1966; Moore and Chater, 1969; Vuorella, 1970; Birks, 1970; and many other works);

c) changes in the composition of woody plants. Varying ratios of arboreal and nonarboreal pollen in spectra may serve as a sign of human activity, since man used fire to make clearings and, probably, as a hunting technique. Low amounts of *Ulmus* pollen are commonly thought to be due to the fact that the young shoots of this tree were used as food for cattle in the Early Neolithic (Moe, 1970). There are, however, other theories, linking the decline of *Ulmus* to the effect of climatic factors (Tauber, 1965).

In many cases it is hard to give preference to one or another factor. Godwin (1968) mentions that a reduction in the amount of oak and hazel in the presence of a constant ratio of arboreal to nonarboreal pollen may be related to selective felling for construction purposes.

Linden was used in husbandry mainly in the Neolithic, whereas later it lost its economic importance due to the cultivation of hemp, and consequently the amount of its pollen increased in palynological spectra. The importance of box increased in connection with forest clearings in the Anglo-Saxon epoch: before this it was unable to penetrate into a nonfelled oak forest. The increase in the content of alder is also ascribed to forest felling, as under these conditions the groundwater level rises and so does the soil moisture (Jacob, 1957).

4. IMPORTANCE OF PALEOECOLOGY FOR PHYTOGEOGRAPHY AND STRATIGRAPHY

Paleoecology opens the way for a causal approach to phytogeographic and stratigraphic problems. In order to make the meaning of this statement clearer, let us think of plant geography as the classification of vegetation and stratigraphy as the classification of layers in a geological section and their aggregates (stratons). There are utilitarian and analytical classifications. In the first type the multitude of objects are broken down into groups to facilitate description or memory. The principle of classification is chosen in accordance with the aim of the classification (there is no sense in classifying breeds of horse according to milk yield or breeds of cow according to trotting ability!). In more complex classifications total similarity is determined on the basis of a large number of characters so as to ensure homogeneity of the taxa to be identified. The analytical type of classification is suited to investigations into the causes of the diversity of objects and of the similarity between them. It thus rests on a causal idea. For instance, phylogenetic systematics presupposes that affiliation is the most universal cause of similarity between organisms.

In an analysis of the spatial distribution of vegetation the areas of distribution of taxa or those of plant communities assume the role of elementary units. However, one species may be included in several communities. These are grouped into floristic elements (Shmitkhyuzen, 1966) out of which arise the floristic kingdoms, regions, and provinces. The hierarchy of these units is determined by the number and taxonomic status of the endemic forms. If the division is based on total floristic similarity, it is not to be considered analytical. On the analytical plane paleoendemic and neoendemic forms have a different value, since endemism has a different cause here. Thus, an analytical floristic classification assumes the assessment of endemism in the light of a particular causal idea. It is thought that endemism reflects the history of a flora, i.e. the development of florogenetic centers and the dispersion of taxa. Without going into details, let us note that the centers of the greatest diversity of allied taxa are by no means always the same as the original genetic centers, which are practically impossible to pinpoint. Among the factors exerting an effect on dispersion great importance is attached to the distribution of land and sea. On the basis of floristic similarity conclusions are drawn on relations between continents. Back in 1867 Hooker suggested that the similarity between the floras of oceanic islands and continents is due to there having once been "bridges" that served for migrations. In 1895 Wallace somewhat modified this hypothesis, surmising that dispersion proceeded by way of a "trail" of stepping stones composed of small islands. Many biogeographers later had recourse to this hypothesis of bridges or stepping stones (see Axelrod, 1960). Others, wishing to piece together again the sundered range of a certain taxon, opted for the drift mechanism, that more or less shifted and parted continents. The following considerations will help us evaluate the reliability of such reconstructions:

1) Almost all plants are able in some way or another to overcome water barriers. In Part II, section 5 we looked at examples of dispersion via an

expanse of water stretching over hundreds and thousands of kilometers (many examples of this nature are to be found in Ridley's summary: Ridley, 1930). Carlquist (1967) perceptively remarked that the biogeographer feels surer of himself if he has even an imaginary bridge. He is willing to acknowledge that certain dryland organisms can wander across seas; it's just that this doesn't apply to those that he himself is studying. For example, bryologists reduce the free-ranging capacity of Bryophyta to a minimum and are ready to move whole continents in order to reunite fragments of the range of one genus of liverworts (see Abramov, 1969). Yet mosses and liverworts were among the first plants to appear on islands whose vegetation had been wiped out by volcanic eruptions. Furthermore, the probability of even the rarest event increases when it is multiplied by millions of years (Simpson, 1952).

2) Carlquist assumes that similar conditions on both sides of a water barrier have more importance for colonization than the width of the barrier. Antevs (1947) specified the following subordination of barriers according to their influence on the dispersion of plants: climatic, edaphic, biotic, and geographic. In Axelrod's opinion (Axelrod, 1952), plants easily overcome water barriers 200–250 miles wide.

3) Climate exerts a much stronger effect on floristic similarity than water barriers. For instance, the flora of East Asia is more similar to the east American than to the west American (Li, 1952). This phenomenon is due to the elimination of certain elements of the integral Arcto-Tertiary geoflora in the west of the USA in connection with a change of climate (Axelrod, 1960) and has nothing to do with the trans-Atlantic or trans-Bering links of continents.

4) The floristic links between the northern and southern hemispheres were apparently governed by the climate, not by a water barrier (the Tethys). Edwards (1955) observes that the Carboniferous flora of North Africa (the southern shore of the Tethys) is practically identical to that of Europe. In his opinion, the obstacle that the plants came up against was not the Tethys but the arid zone (also the equatorial zone for the temperate-climate plants). Wesley (1965) described the Jurassic flora of Veneto, that was situated on the southern shore of the Tethys. It differs from the Yorkshire flora in the absence of such gymnosperms as *Nilssonia, Ctenis,* and *Anomozamites.* Since these plants belong among the anthracophiles and the Veneto vegetation displays signs of aridization, we may assume that their absence here is due to their sensitivity to the climatic conditions, not to water barriers. Yet another example confirms this idea: the Mesozoic floras of the northern and southern hemispheres differ mainly in the composition of conifers (which do not grow in the tropical lowlands) and Gingkoales (which are primarily plants of the temperate zone), but they have practically an identical composition of thermophilous Bennettitales.

5) The most intensive migrations over the "Bering bridge" (a "splinter" of it is thought to be St. Lawrence Island with Miocene beds of *Metasequoia*: Chaney, 1940; Graham, 1946) coincide with the thermal maxima (e.g. at the end of the Eocene: Simpson, 1965) and are probably related to the temperature conditions rather than to the existence of a strait (actually, water barriers are more serious for animals than for plants).

6) Plant formations do not have equal chances of appearing in the paleobotanical record: lowland formations are represented more fully, the mangrove groupings being favorites in this sense. For this reason their specific weight is very high in the vegetation of an ancient continent, and since mangroves of all coasts are very similar, the similarity of floras comes to be overstated.

7) Floristic similarity is not a proof of drift because there are always enough "stepping stones" to account for their spread without having to bring in this mechanism. According to Axelrod (1952, 1960), Australia, New Zealand, Tasmania, Antarctica, and South America with the islands between them and the shallows that periodically rose above sea level constituted a "dry" enough tract for the dispersion of the Tertiary plants. With the discovery of a glossopterid flora in Antarctica, we were obliged to reckon with the possibility of Paleozoic migrations along this stretch.

Those who consider drift to be the sole mechanism responsible for sea-separated ranges are forced to have recourse to arbitrary datings of evolutionary episodes. For instance, Melville (1967, 1969) considers that all the families of dicots common for Africa and South America emerged before the drift, that is, at least in the Triassic. This contradicts paleobotanical facts.

In its classical version, the continental drift hypothesis was based mainly on the fact that the two coasts of the South Atlantic fit together (as Francis Bacon noted) and on the Paleozoic glaciation of the southern continents. This was in essence a meteorological hypothesis, since Wegener and Köppen extended the laws of atmospheric circulation to the Earth's crust without taking into account the differences in the viscosity of the material or the structure of the sea floor, which had not been studied in their time. Recent achievements in geophysics and marine geology have made it possible to bring the drift hypothesis up to date. Drift is now considered to be the consequence of ocean-floor spreading. Continents are passively moved by the lithosphere plates, while this entire "conveyer belt" undergoes a spatial shift as a result of migration of mid-oceanic ridges (rifts) and troughs (Le Pichon, 1968; Wilson, 1968; Isacks et al., 1968; etc.). Certain laws of the distribution of ancient vegetation are confirmed by geophysicists' formulations (Sullivan, 1967; Meien, 1969; Krasilov, 1970a), yet there is no reason to believe that drift is the principal mechanism responsible for floristic similarity. As shown above, it is determined to a much greater degree by climate. Water barriers only underline climatic differences, and in the absence of the latter, they do not have any appreciable influence on the spread of vegetation (their influence may prove considerable at a certain stage, but it is smoothed out in the course of time). As for plant communities, the relationship between their geographic distribution and climate is evident ("a good vegetational map is a very practical map of climates": Odum, 1959; this conviction is shared by many authors).

We thus reach the conclusion that floristic units based on the ranges of species and units of vegetation based on the ranges of communities are causally related to climate and, at least on the higher classification levels, necessarily coincide (see Shmitkhyuzen, 1966). As Popov (1949) wrote: "the boundaries of solar zones, that seem to be completely imaginary lines, ... are lines of the most profound biogeographic changes: on these lines both floras and plant communities (formations) replace each other with unusual rapidity" (see also Matthew, 1939).

By now quite a number of schemes have been conceived for "zoning" the vegetation of different geological epochs, based on floristic statistics. However, all the classifications that have stood the test of time have a climatological basis.

An analytical classification of ancient regions cannot take total floristic similarity as a point of departure. The incompleteness of the geological record reduces the value of this criterion to a minimum. Only an analysis of causal links makes it possible to weigh individual differences. This is where paleoecology steps in. For example, the species composition of the Czekanowskiales in Vakhrameev's Siberian and Indo-European regions (Vakhrameev, 1970) is practically identical, but only in the first do these plants acquit themselves as chief forest-formers. Nor do the Bennettitales include any genera that are confined to a particular region, but it is apparently only to the south of their boundary that there existed mangrove formations with *Ptilophyllum* and "Mesozoic chaparral" consisting of bennettites and conifers. Stigmariae are absent in the Carboniferous flora of Siberia. The significance of this fact for phytogeography is determined by the connection of the stigmariae with the Paleozoic mangrove formation (Meien, 1970). Recently paleozoogeographists have been increasingly stressing the importance of the climatic factor (Valentine, 1963; Yuferev, 1969).

Paleoecology plays the same role in stratigraphy as it does in biogeography, in that it provides the possibility for a causal analysis. However, many authors believe that the paleoecological method is useful for resolving only certain specific and generally minor problems and in any case cannot be compared in value with the phylogenetic method. The popular notions on the biostratigraphic line of investigation can be summarized as follows. Darwin discovered that species change under the influence of natural selection. Dollo showed that this change is irreversible. Consequently, phylogenesis can be converted into a time scale. Animals and plants evolve at different tempos, and as a result we have two scales, one for animals and the other for plants ("paleophyte," "mesozophyte," "cenophyte"). In order to reduce them all to a common denominator, we have to sacrifice the plants and give priority to the animals (and among these, to the "archistratigraphic groups"). The temporal boundaries corresponding to phylogenetic stages intersect the natural boundaries of geological bodies. The latter are determined in the main by tectonic movements, which do not coincide in time in different areas and hardly have any effect on phylogenesis. There is no point in listing the works that push this set of views – they are legion.

Let us now return to the alternative point of view, the basis for which has been set out in different sections of this book.

In performing a correlation of geological sections, we place on one level strata that contain the most similar taphocenoses. In other words, similarity of taphocenoses serves as a means of synchronizing strata. This actually makes it possible to consider stratigraphy as a classification, that is, a discipline, in which the assessment of similarity is the main working method. Let us recall that in 1816 Smith spoke not of synchronizing but of identifying strata (his work is in fact entitled "Strata identified by organized fossils"). His approach was thus utilitarian rather than analytical; he did not seek the causes of similarity, or rather he saw them in immanent properties of nature, that apportions to each class its own special stratum. Following Werner, Smith postulated that each layer was laid down everywhere, and all we have to do is to recognize it.

Subsequently the goals and tasks of stratigraphy changed, the method, however, remaining practically the same. We measure the similarity of taphocenoses. We synchronize strata according to degree of similarity. It must be, we say, that synchronousness is the cause of similarity (and in exactly the same way we determine affinity according to degree of similarity, since affinity is a cause of similarity). As Huxley showed, here we are standing on unsteady ground: naturally, synchronousness cannot per se be a cause of similarity. The real causal basis of a stratigraphic correlation is the nonidentity of conditions obtaining in successive geological periods (Nairn, 1965). Let's say that Jurassic taphocenoses are on the whole similar and differ from Cretaceous ones because conditions were different in the Jurassic and Cretaceous periods and the taphocenoses mirror these differences. It may be argued that conditions are always changing, whereas substantial changes of fauna and flora are confined strictly to certain boundary lines. A study of the dynamics of recent communities shows that in most cases they possess a high capacity for regeneration. Johnson (1964) presents examples where communities of marine benthos have been destroyed as a result of heavy frosts, changes in salinity, the epidemic that hit *Zostera* at the beginning of the thirties, etc. As a rule, numbers are quickly restored. He drew the conclusion that only a change of overall significance can lead to extinction.

The major stratigraphic boundaries are determined by switchovers of dominant types (dinosaurs and mammals, gymnosperms and angiosperms: see Huxley, 1955) that mark a fundamental reorganization of ecosystems. They are accompanied by dying out on a catastrophic scale. The switchover of the dominant types takes place more or less simultaneously in the animal and plant kingdoms (mammals appeared in the Jurassic a little earlier than angiosperms; in the Cretaceous the latter were fairly diverse, but played a subordinate role relative to the conifers; only after the sweeping extinction that occurred in the Cretaceous did both the mammals and the angiosperms come out into the foreground as the main dominants: Krasilov, 1972) and is confined to a transitional zone which I have named stratoecotone (Krasilov, 1970a).

The periodicity of macroevolutionary phenomena coincides with that of tectogenesis (the development cycle of lithosphere plates is equal to about 250 million years: Dewey and Horsfield, 1970). Activization of certain rifts (Dietz and Holden, 1970) inevitably galvanizes the entire system of lithosphere plates into motion. This accounts for the coincidence of tectonic episodes in different regions. Concomitantly with the evolution of lithosphere and climate occurs a change of biotopes, the appearance of new adaptive zones, breakdown of the dynamic equilibrium of ecosystems, and a modification of their structure. It is this factor which decisively affects sedimentation, and this is why biostratigraphic boundaries coincide with the natural boundaries of geological bodies.

Although extinction occurs relatively rapidly and is of a catastrophic nature, it may be related to a change of conditions that does not go beyond the limits of tolerance for a given factor. According to paleothermometry and paleobotanical data (Norton and Hall, 1965; Srivastava, 1967; Krasilov, 1970c), the cooling that set in at the end of the Cretaceous period was not too considerable and was hardly accompanied by glaciation on a scale comparable to the Quaternary glaciation. It did, however, destroy the balance of the Mesozoic ecosystems and was apparently the chief cause of the massive Cretaceous dying out.

The situation is complicated by the fact that overall changes are manifested as a tendency against a background of minor fluctuations. For instance, the absence of conifers in the Miocene Shang wang flora may be imputed to the effect of local topographic and edaphic factors (Chaney and Hu, 1940). The change in the proportion of conifers in the Miocene floras of the USA may be associated with the rise of mountain ranges in the west of the country. But if analogous changes occur simultaneously in the Miocene floras of

Europe and Japan, we have to suspect a more general cause. And once we have gathered all the data on the variation of the structural indexes of Miocene taphocenoses, on the displacements of the boundaries of climatic zones, on vertical migrations, and on paleothermometry, we shall be able to unravel the cause. Hence, the paleoecological approach to solving stratigraphic problems is far more complex than the simple noting of the presence or absence of particular leading forms, but it offers hope for transforming stratigraphy into an analytical science.

Like phytogeography, stratigraphy operates with units of two types: 1) ranges of the vertical distribution of species and higher taxa; 2) ranges of the vertical distribution of associations and higher syntaxa. The phylogenetic method makes use of the first type of units – acrozones, teilzones, eidozones, acmezones, etc.

However, such evolutionary episodes as the appearance and disappearance of a species are omitted from the paleontological record due to the small size of the initial and definitive populations. "Heyday," established according to increased abundance, is a purely ecological phenomenon.

Specific phylogeneses, described by paleontologists (Waagen's races, Neumayr's series, orthogenesis in the family Equidae, the evolutionary phases of the Neogene Stipeae: Elias, 1942) and forming the basis for stratigraphic formulations, are in fact successions of ecotypes or nonallied species in connection with changes of conditions (Kaufmann, 1933; Stebbins, 1947; Rensch, 1954; Simpson, 1960). Consequently, on the stratigraphic plane what we have here is a succession of biomeres, i.e. units corresponding to "nonevolutionary changes" in Palmer's definition (Palmer, 1965).

The paleobotanist more often operates with generic, rather than with specific units. Plant genera as a rule had a long life span. Their acrozones and even teilzones are too large for the purposes of detailed stratigraphy. They often intersect the time boundaries of geofloras. Apart from this, very few genera are truly cosmopolitan, and this limits the possibilities of correlation on a universal scale. The vertical range of an association (cenozone) is usually considerably smaller than the generic zone.

A change in the general structure of a geoflora affects all the phytocenoses and, as was shown in the preceding section, only individual fragments of associations make the transition from one geoflora to another, so that associations more accurately reflect nonidentity of conditions than do individual taxa. Moreover, the systems of vicarious communities embrace the entire vegetation of the Earth and thus make it theoretically possible to carry out correlations on a universal scale.

The stratigraphy of Quaternary deposits shows the potentials of the paleoecological method. Climatic cycles are distinctly expressed in the glaciation dynamics, in the migration of the northern forest boundary, in the eustatic fluctuations of sea level, and in the alternation of dry and wet (pluvial) epochs in the low latitudes. It is easy to show that these changes affect not only individual groups of animals and plants but also the entire bios (Shackleton, 1969). A change in the hydrographic regime (e.g. the submergence of the Florida peninsula during transgressions: Valentine, 1963), in the isotope composition of seawater (Emiliani, 1970), and in carbon content (Arrhenius, 1952) leads to upheavals in marine biocenoses. In the tropics dry periods go along with the expansion of savanna and the splitting up of the forest massif into separate outliers, which are reunited in the next wet period. The fluctuations in the size of populations that accompany this process have a marked influence on speciation (Haffer, 1969). It is sometimes observed that the cycles do not coincide in different regions (Ericson and Wollin, 1970), that transgressions occur during cooling, etc. These may be related in part to taphonomic conditions. For instance, a reduction in the vegetation of a maritime plain during transgressions and the concomitant enhancement of the role of montane plants in palynological spectra may simulate cooling.

A less distinct but still fairly evident rhythmicity is established in the development of pre-Quaternary climates. Some authors (Hammen, 1961; Stockmans and Willière, 1963) consider that the boundaries of all the stages in the Tertiary system are determined by rhythmic fluctuations of mean annual temperatures (see Klimaänderungen..., 1969).

Turner and West (1968) suggest that the phases of an interglacial period (cryocratic, protocratic, mesocratic, telocratic) be considered as **cenozones.**

Theoretically, therefore, climate evolution, traced by means of paleoecological investigations, should form a most reliable basis for correlating strata from different facies on a universal scale. However, the practical side of the matter has still been little studied. Properly speaking, we have two practical devices: the identification of cenozones (assemblage zones) and the construction of cliseres.

Von Post identified a number of palynozones and correlated them with climatic periods on the Blytt-Sernander scale. It was assumed that changes of climate exert a broad and synchronous influence on vegetation and as a result, palynozones have a chronostratigraphic significance. On the whole, von Post's method can be very fruitfully applied to Quaternary geology (and presumably to more ancient sediments as well: Krasilov, 1970a, 1971a); yet when it comes to identifying more detailed palynozones and applying radiocarbon datings, we see that analogous changes of pollen spectra are not always synchronous. This has led to a revision of the concept of the palynozone, which is now equated with the cenozone (assemblage zone), that is, a stratigraphic unit determined by a characteristic combination of organisms and (or) their numerical ratios. Beginning in 1961, Cushing and his school defined palynozones as biostratigraphic units of local or regional distribution, characterized by a certain pollen spectrum and not bearing any paleoecological or climatological connotation (Cushing, 1967; West, 1970; Birks, 1970; and many others); let us note that earlier too, some palynologists regarded pollen spectra as index fossils of a special kind, a point of view that was vehemently opposed by Faegri and Iversen (1950, p. 112). The asynchronousness of cenozones is attributed to the varied reactions of different phytocenoses to climatic changes, to the different migration tempos (e.g. the successive emergence of *Betula, Corylus,* and *Pinus* may indicate the rate of their migration in the presence of stable climatic conditions: Smith, 1965; Birks, 1970), and to the time it takes for arboreal vegetation to be restored (birch groves appear after 100–200 years on bare moraine). On the other hand, many palynological boundaries are synchronous (according to radiocarbon datings) over large areas. In some cases asynchronousness may derive from inaccurate radiocarbon datings (their accuracy in general does not exceed ±1000 years: Ogden, 1967).

Examples may be cited of very rapid reorganization of vegetation: the speedy replacement of spruce by pine or oak, indicating migration for a distance of about 600 miles (Ogden, 1967; both genera are considered slow wanderers), the rapid replacement of boreal forest by mixed conifer-broadleaf groups that occurred simultaneously over a large area (Watts, 1967), and so on. In cases where the appearance of certain genera proves nonsimultaneous, a reason can almost always be found that hampered spreading. For instance, the time-staggered rise in the content of alder and hazel pollen in the west and east of Europe is apparently due to the fact that oceanization of the climate spread in this direction (Hibbert et al., 1971). There are no weighty proofs that different migration tempos were in themselves responsible for the heterochronousness of vegetation cenozones. If heterochronousness is due mainly to climatic factors, it does not rule out the use of cenozones for synchronization (provided that a detailed paleoecological analysis is performed). Many paleobotanists studying Tertiary floras support the theory of "migration waves" coming from the north. It is thought, for example, that the Eocene communities of the Arctic were identical to the Miocene communities of the middle latitudes. If this is the case, then the age of cenozones would change from north to south from the Eocene to the Miocene. But in fact, as mentioned in the preceding section, there is no question of the communities being identical in the given instance. We have no real reason to doubt the synchronousness (at least in the time units used in pre-Quaternary geology) of cenozones in the limits of their distribution. These limits may be very broad, but even so a global correlation is possible only with the aid of cliseres.

A **clisere** is a succession of plant groups that reflect changes of the climate.* Axelrod (1948) seems to have been the first to apply cliseres practically for correlation purposes (he correlated Neogene floras that have hardly any species in common). Analysis of a clisere enables us to construct a curve showing the variation of certain characteristics of the climate (or of vegetation characteristics related to climate). Such graphs are then used as a basis for correlation. For interpreting a clisere it is desirable to use the whole set of techniques described in Part IV, section 3A. Of course, in practice this is highly complicated, so that usually just one particular method is selected, which leads to various interpretations of cliseres (the differences of opinion among American paleobotanists illustrate this state of affairs: Axelrod and Bailey, 1969; Wolfe and Barghoorn, 1960; Wolfe and Hopkins, 1967). Cliseres are also useful for correlations of continental and marine sediments. In this case the paleoecological interpretation of a clisere may be verified by means of isotopic paleothermometry.

* Clements understood by clisere a succession of climaxes in relation to climatic changes; however, in view of the vagueness apparent in the interpretation of climaxes, I prefer the more general formulation (V. K.).

BIBLIOGRAPHY

AARIO, L. Pflanzentopographische und paläogeographische Mooruntersuchungen in N-Satakunta. – *Commentat. for.,* Vol. 17, 1932.

ABRAMOV, I. I. *Aspects of Endemism in the Musci.* – Leningrad, "Nauka," 1969. 55 pp. (Russian)

AFANAS'EV, B. L. *Paleotectonics and Presence of Coal.* – Moscow, "Nedra," 1968. 155 pp. (Russian)

AGER, D. V. Principles of Paleoecology. An Introduction to the Study of How and Where Animals and Plants Lived in the Past. – *Intern. Series on the Earth Sci.,* New York – San Francisco – Toronto – London, 1963. 352 pp.

AGNEW, A. D. G. and J. E. C. FLAX. Plant Dispersal by Hares (*Lepus capensis L.*) in Kenya. – *Ecology,* Vol. 51, No. 4, 1970.

AGRALI, B. Valeur stratigraphique des genres *Densisporites* et *Lycosisporites* et leur utilisation pour l'établissement de subdivisions palynologiques dans le houiller d'Amasza. – *Annls. Soc. géol. N.* **84**, No. 1 (1964), 9–17.

AHTI, T., L. HÄMET-AHTI, and J. JALAS. Vegetational Zones and their Sections in Northwestern Europe. – *Ann. Bot. Fenn.,* **5**, No. 3 (1968), 169–204.

ALBRITTON, Cl. C. Uniformity, the Ambiguous Principle. – In: Cl. C. Albritton (editor). *Uniformity and Simplicity; Spec. Pap. Geol. Soc. Am.,* Vol. 89 (1967), 1–2.

ALEKSANDROVA, V. D. *Classification of Vegetation.* – Leningrad, "Nauka," 1969. (Russian)

ALFVÉN, H. and G. ARRHENIUS. Two Alternatives for the History of the Moon. – *Science,* Vol. 165 (1969), 11–17.

ALLEN, J. R. L. The Classification of Cross-Stratified Units, with Notes on their Origin. – *Sedimentology,* Vol. 2 (1962), 93–114.

ALLEN, J. R. L. Petrology Origin and Deposition of the Highest Lower Old Red Sandstone of Shropshire, England. – *J. Sedim. Petrol.,* Vol. 32 (1962), 657–697.

ALLEN, J. R. L. Henry Clifton Sorby and the Sedimentary Structures of Sands and Sandstones in Relation to Flow Conditions. – *Geologie Mijnb.,* No. 7 (1963), 223–228.

ALLEN, J. R. L. Primary Current Lineation in the Lower Old Red Sandstone (Devonian), Anglo-Welsh Basin. – *Sedimentology,* **3**, No. 2 (1964), 89–108.

ALLEN, J. R. L. A Review of the Origin and Characteristics of Recent Alluvial Sediments. – *Sedimentology,* Vol. 5 (1965), 91–180.

ALLEN, J. R. L. Studies in Fluviatile Sedimentation: A Comparison of Fining-Upwards Cyclothems, with Special Reference to Coarse Member Composition and Interpretation. – *J. Sedim. Petrol.* **40**, No. 1 (1970), 298–323.

ANAN'EV, A. R. Psilophyta, General. – In: *Osnovy paleontologii.* Moscow, Akad. Nauk SSSR, 1963. (Russian)

ANAN'EV, A.R. and S.A. STEPANOV. A Find of Spore-Bearing Organs in *Psilophyton princeps* Dawson emend. Halle in the Lower Devonian of the South Minusinsk Basin. – *Trudy Tomsk. Univ.,* Vol. 202 (1968), 30–46. (Russian)

ANAN'EV, G. S. and I. A. KAREVSKAYA. Distribution of Pollen and Spores in Recent Continental Deposits of the East Slope of the Southern and Central Urals. – *Bot. Zh.* **55**, No. 6 (1970), 865–869. (Russian)

ANDERSEN, Sv. Th. Tree Pollen Rain in Mixed Deciduous Forest in South Jutland (Denmark). – *Rev. Palaeobot. and Palynol.,* Vol. 3 (1967), 267–275.

ANDREÁNSZKY, G. On the Upper Oligocene Flora of Hungary. Analysis of the Site at the Wind Brickyard, Eger. – *Stud. biol. hung.,* Vol. 5. Budapest, 1966. 151 pp.

ANTEVS, E. Biogeographical Principles. – *8th Ann. Biol. Colloq., Oregon State Coll.,* pp. 7–9, 1947.

ARBER, A. *Natural Philosophy of Plant Form.* – Cambridge, 1950.

ARCHANGELSKY, S. Studies on Triassic Fossil Plants from Argentina. IV. The Leaf Genus *Dicroidium* and its Possible Relation to *Rhexoxylon* Stems. – *Palaeontology* **11**, No. 4 (1968), 500–512.

ARRHENIUS, S. On the Influence of Carbonic Acid in the Air upon the Temperature of the Ground. – *Phil. Mag.,* Vol. 41 (1896), 237–276.

ARRHENIUS, G. Sediment Cores from the East Pacific. – *Rep. Swed. Deep Sea Exped. (1947–1948)* **5**, fasc. 1 (1952). 89 pp.

ASAMA, K. Two Types of Evolution in *Sphenophyllum.* – *Bull. Natn. Sci. Mus.,* Tokyo **9**, No. 4 (1966), 577–608.

ASAMA, K. Permian Plants from Maiya, Japan, I. *Cathaysiopteris* and *Psygmophyllum.* – *Bull. Natn. Sci. Mus.,* Tokyo **10**, No. 2 (1967), 149–153.

ASAMA, K. Evolution of *Nilssonia orientalis.* – *Mem. Natn. Sci. Mus.,* No. 1 (1968), 33–36.

ASH, S. R. *Pagiophyllum simpsoni,* a New Conifer from the Chinle Formation (Upper Triassic) of Arizona. – *J. Paleont.,* Vol. 44, No. 5 (1970), 945–952.

ATTIWILL, P. M. A Method for Estimating Crown Weight in *Eucalyptus* and Some Implications of Relationships between Crown Weight and Stem Diameter. – *Ecology* **47**, No. 5 (1966), 796–804.

AUBREVILLE, A. Les reliques de la flore des conifères tropicaux en Australie et en Nouvelle-Calédonie. – *Adansonia,* Vol. 4 (1964), 482–492.

AVAKOV, G. S. Geochronological Importance of the Karagan-Konskaya Flora of Georgia. – *Sakartvelos SSR, Akad. Moambe, Soobshch. Akad. Nauk Gruz. SSR* **53**, No. 1 (1969), 245–248. (Russian)

AXELROD, D. I. A Pliocene Flora from the Mount Eden Beds, Southern California. – *Publs. Carnegie Instn.* 476, 1938. 125 pp.

AXELROD, D. I. The Concept of Ecospecies in Tertiary Paleobotany. – *Proc. Natn. Acad. Sci.* **27**, No. 12 (1941), 545–551.

AXELROD, D. I. Climate and Evolution in Western North America during Middle Pliocene Time. – *Evolution* **2**, No. 2 (1948), 127–144.

AXELROD, D.I. Variables Affecting the Probabilities of Dispersal in Geologic Time. – *Bull. Am. Mus. Nat. Hist.* **99**, Art. 3 (1952), 177–188.

AXELROD, D.I. Evolution of the Madro-Tertiary Geoflora. – *Bot. Rev.* **24**, No. 7 (1958), 433–503.

AXELROD, D.I. The Evolution of Flowering Plants. – In: S. Tax (ed.). *Evolution after Darwin,* Vol. 1, pp. 227–305. Univ. Chicago Press, 1960.

AXELROD, D.I. Post-Pliocene Uplift of the Sierra Nevada, California. – *Bull. Geol. Soc. Am.,* Vol. 72 (1962), 183–198.

AXELROD, D.I. A Method for Determining the Altitudes of Tertiary Floras. – *Palaeobotanist,* Vol. 14, Nos. 1–3, 1965 (1966).

AXELROD, D.I. Tertiary Floras and Topographic History of the Snake River Basin, Idaho. – *Bull. Geol. Soc. Am.* **79**, No. 6 (1968), 713–734.

AXELROD, D.I. and H.P. BAILEY. Paleotemperature Analysis of Tertiary Floras. – *Palaeogeogr. Palaeoclimatol. Palaeoecol.,* Vol. 6 (1969), 163–195.

BACHMAYER, Fr. Ein bemerkenswerter Fund: *Myrica*-Früchte im Flyschharz. – *Annln. naturh. Mus. Wien,* Vol. 72 (1968), 639–643.

BAIDIN, S.S. Recent Delta-Forming Processes in the Mouth of the Terek River. – In: "*Gidrologiya i gidrokhimiya morei i ust'ev rek,*" pp. 49–59. Moscow, Gidrometeoizdat, 1970. (Russian)

BAILEY, H.P. A Method of Determining the Warmth and Temperateness of Climate. – *Geogr. Annls.* **42**, No. 1 (1960), 1–16.

BAILEY, I.W. and E.W. SINNOTT. The Climatic Distribution of Certain Types of Angiosperm Leaves. – *Am. J. Bot.,* Vol. 3 (1916), 24–39.

BAIRD, A.M. A Contribution to the Life History of *Macrozamia reidlei.* – *J. Proc. R. Soc. West. Aust.,* Vol. 25 (1938), 153–175.

BAKKER, R.T. Ecology of *Brontosaurus.* – *Nature* **229**, No. 5281 (1971), 172–174.

BALUKHOVSKII, N.F. *Geological Cycles.* – Kiev, "Naukova Dumka," 1966. 168 pp. (Russian)

BANDEL, K. Trace Fossils from Two Upper Pennsylvanian Sandstones in Kansas. – *Paleont. Contr. Univ. Kans.,* Vol. 18, 1967.

BANKS, H.P. and M.R. DAVIS. – *Crenaticaulis,* a New Genus of Devonian Plants Allied to *Zosterophyllum,* and its Bearing on the Classification of Early Land Plants. – *Am. J. Bot.* **56**, No. 4 (1969), 436–449.

BARLETT, A.S., E.S. BARGHOORN, and R. BERGER. Fossil Maize from Panama. – *Science,* Vol. 165 (1969), 389–391.

BARREL, J. – *Am. J. Sci.,* ser. 4, Vol. 36 (1913), 429–472.

BATTEN, D.J. Probable Dispersed Spores of Cretaceous *Equisetites.* – *Palaeontology,* Vol. 11 (1968), 633–642.

BAXTER, R.W. The Growth Habit of *Lepidodendron serratum* Felix. – *Palaeobotanist* **14**, Nos. 1–3 (1965, 1966), 1–4.

BECK, C.B. Reconstructions of *Archaeopteris* and Further Considerations of its Phylogenetic Position. – *Am. J. Bot.,* Vol. 49 (1962), 373–382.

BECK, C.B. The Appearance of Gymnospermous Structure. – *Biol. Rev.,* Vol. 45 (1970), 379–400.

BECKER, H. F. Paleobotanical Record of Solar Change. – *Ann. N. Y. Acad. Sci.,* Vol. 95 (1961), 684–687.

BECKING, L. G. M. Notes on the Determined and the Undetermined in Biology. – *Acta biotheor.,* Vol. 8 (1946), 18–41.

BECKING, R. W. Vegetational Response to Change in Environment and Change in Species Tolerance with Time. – *Vegetatio* **16**, Nos. 1–4 (1968), 135–158.

BEERBOWER, J. R. Cyclothems and Cyclic Depositional Mechanisms in Alluvial Plain Sedimentation. – *Bull. Kans. Univ. Geol. Surv.,* No. 169 (1964), 31–42.

BEERBOWER, J. R. Interpretation of Cyclic Permocarboniferous Deposition in Alluvial Plain Sediments in West Virginia. – *Bull. Geol. Soc. Am.,* Vol. 80 (1969), 1843–1848.

BELENKO, N. G. Conditions of Formation of Upper Permian Coal-Bearing Deposits in the Kuznetsk Basin. – *Litol. i Polez. Iskop.,* No. 5 (1970), 98–106. (Russian)

BELOUSSOV, V. V. Against the Hypothesis of Ocean Floor Spreading – *Tectonophysics,* Vol. 9 (1970), 489–511.

BEREZINA, N. A. and S. N. TYUREMNOV. On the Degree of Preservation of Pollen in Holocene Deposits. – *Vest. Moskov. Univ. Biol. i Pochvoved.,* No. 4 (1967), 74–86. (Russian)

BERG, F. A Review of Some Theories of the Thermal History of Earth. – *Izv. Akad. Nauk SSSR, geol. ser.,* No. 1 (1966), 39–56. (Russian)

BERGER, W. Pflanzenreste im Flysch. – *Mitt. geol. Ges. Wien,* Vol. 58 (1965, 1966), 233–245.

BERKNER, L. V. and L. S. MARSHALL. Oxygen and Evolution. – *New Scient.,* No. 11 (1965), 415–419.

BERLIN, G. S. and A. V. KHABAKOV. Results of Paleotemperature Determinations according to Belemnites by the Calcium-Magnesium Method. – *Byull. Moskov. Obshch. Ispyt. Prirody* **43**, No. 1 (1968), 17–30. (Russian)

BERLIN, T. S., E. L. KIPRIKOVA, D. P. NAIDIN, I. D. POLYAKOVA, V. N. SAKS, R. V. TEIS, and A. V. KHABAKOV. Some Problems of Paleotemperature Analysis (according to Belemnite Rostra). – *Geologiya i Geofizika,* No. 3 (1970), 36–43. (Russian)

BERLIN, G. S., D. P. NAIDIN, V. N. SAKS, R. V. TEIS, and A. V. KHABAKOV. Climates in the Jurassic and Cretaceous Periods in the Northern USSR according to Paleotemperature Determination. – *Geologiya i Geofizika,* No. 10 (1966), 17–31. (Russian)

BERLIN, G. S., S. I. PASTERNAK, and A. V. KHABAKOV. Temperature Determination according to the Calcium/Magnesium Ratio in Calcite of Upper Cretaceous Fossils of the Volyn-Podolian Platform and the Lvov Syncline. – *Geokhimiya,* No. 9 (1968), 1128–1131. (Russian)

BERRY, E. W. The Environment of the Early Vertebrates. – *Am. Nat.* **59**, No. 663 (1925), 354–362.

BERRY, E. W. Revision of the Lower Eocene Wilcox Flora of the Southeastern States. – *Prof. Pap. U.S. Geol. Surv.,* 156, 1930.

BERRY, E. The Origin of Land Plants and Other Papers. – *Johns Hopkins Univ. Stud. Geol.,* No. 14. Baltimore, 1945.

BEWS, J. W. Studies in the Ecological Evolution of Angiosperms. – *New Phytol.,* Vol. 26 (1927), 1–21, etc.

BHATNAGAR, S. P. and G. MEENAKSHI. Structure and Ontogeny of Stomata in *Mimusops elengi.* – *J. Indian Bot. Soc.* **48**, Nos. 3–4 (1969), 362–367.

BIRKS, H. H. Studies in the Vegetational History of Scotland. – *J. Ecol.* **58**, No. 3 (1970), 827–846.

BISHOP, D. J. and E. R. FORCE. The Reliability of Graded Bedding as an Indicator of the Order of Superposition. – *J. Geol.* **77**, No. 3 (1969), 303–318.

BLACK, M. Drifted Plant-Beds of the Upper Estuarine Series of Yorkshire. – *Q. Jl. Geol. Soc. Lond.,* Vol. 85, Pt. 4, 1929.

BLOCH, M. R. A Hypothesis for the Change of Ocean Levels Depending on the Albedo of the Polar Ice Caps. – *Palaeogeogr. Palaeoclimatol. Palaeoecol.* **1**, No. 2 (1965), 127–142.

BOCH, M. S. Attempt to Apply Statistical Analysis for Identifying Layers in Peat Deposits that Are Homogeneous in Terms of Botanical Composition. – *Bot. Zh.* **55**, No. 8 (1970), 1142–1149. (Russian)

BONAMO, P. M. and H. P. BANKS. *Tetraxylopteris schmidtii*: its Fertile Parts and Relationships within the Aneurophytales. – *Am. J. Bot.* **54**, No. 6 (1967), 755–768.

BOND, T. A. Permian Palynological Assemblage from the Wellington Formation, Kay County, Oklahoma. – *Pollen et spores* **10**, No. 2 (1968), 385–393.

BONNEFILLE, R. Analyse pollinique d'un sédiment récent: vases actuelles de la rivière Aouache (Ethiopie). – *Pollen et spores* **11**, No. 1 (1969), 7–16.

BORTENSCHLAGER, S. Funde afrikanischer Pollen in den Alpen. – *Naturwissenschaften* **52**, No. 24 (1965), 663–664.

BOTVINKINA, L. N. *Stratification of Sedimentary Rocks.* – Moscow, Akad. Nauk SSSR, 1962. (Russian)

BOTVINKINA, L. N. On the Textures of Sedimentary Rocks and Series. – *Litol. i Polez. Iskop.,* No. 3 (1970), 37–52. (Russian)

BOUCOT, A. J. Life and Death Assemblages among Fossils. – *Am. J. Sci.,* Vol. 251 (1953), 25–40.

BOULOUARD, C. and H. DELAUZE. Analyse palynoplanctologique de sédiments prélevés par le bathyscaphe "Archimède" dans la fosse du Japon. – *Mar. Géol.* **4**, No. 6 (1966), 461–466.

BOULTER, M. C. A Palynological Study of Two of the Neogene Plant Beds in Derbyshire. – *Bull. Br. Mus. Nat. Hist. A. Geology* **19**, No. 7 (1971), 361–410.

BOUMA, A. H. A Graphic Representation of the Facies Model of Salt Marsh Deposits. – *Sedimentology,* Vol. 2 (1963), 122–129.

BOWEN, R. Oxygen Isotope Paleotemperature Measurements on Cretaceous Belemnoidea from Europe, India and Japan. – *J. Paleont.* **35**, No. 5 (1961a), 1077–1084.

BOWEN, R. Paleotemperature Analysis of Mesozoic Belemnoidea from Germany and Poland. – *J. Geol.,* Vol. 69 (1961b), 75–83.

BOWEN, R. Oxygen Isotope Paleotemperature Measurements on Mesozoic Belemnoidea and their Importance in Paleoclimatic Studies (Abstract). – *Intern. Meet. Organic Processes Geochem.* Milan, Oxford, Pergamon Press, 1962.

BOWER, F.O. *Size and Form in Plants with Special Reference to the Primary Conducting Tracts.* – London, 1930.

BRADLEY, W.H. Origin and Microfossils of the Oil Shale of the Green River Formation of Colorado and Utah. – *Prof. Pap. U.S. Geol. Surv.,* 168, 1931.

BRADLEY, W.H. Eocene Algae and Plant Hairs from the Green River Formation. – *Am. J. Bot.,* Vol. 57 (1970), 782–785.

BRAUN, E.L. Development of the Deciduous Forests of Eastern North America. – *Ecol. Monogr.,* Vol. 17 (1947), 213–219.

BRETT, D.W. Fossil Oak Wood from the British Eocene. – *Palaeontology,* Vol. 13 (1960), 86–92.

BRINKMANN, P. Zur Geschichte der Moore, Marschen und Wälder Nordwestdeutschlands. – *Engl. Bot. Jahrb.,* Vol. 66, 1934.

BRÖNNIMANN, P. Palaeontologische Überlieferung und Faunenschnitte. – *Archs. Sci.,* Vol. 19 (1966), 1–21.

BROOKS, C.E.P. *Climate through the Ages.* – London, 1949, 395 pp.

BROOKS, J. and G. SHAW. The Role of Sporopollenin in Palynology. – *III Intern. Palynol. Conference,* Novosibirsk, pp. 1–23, 1971.

BRUNNER, H. Pleistocene Klimaschwankungen im Bereich des östlichen Mysore-Plateaus (Südindien). – *Geologie* **10**, No. 1 (1970), 72–82.

BRUSH, G.S. The Absence of Pollen and Spores in Some Triassic Sediments. – *J. Paleontol.* **40**, No. 5 (1966), 1241–1243.

BRZYSKI, B. Struktury trankowe skamienialych szczątków flory karbonskiej (numur A) z Górnośląskiego zaglębia węglowego. – *Acta palaeobot., Cracow* **1**, No. 1 (1969), 98 pp.

BURLINGAME, L. The Morphology of *Araucaria brasiliensis,* 3. – *Bot. Gaz.* **59**, No. 1 (1915), 1–34.

BURMAN, G. Inkohlung und mechanische Deformation. Abgehandelt am Erhaltungszustand organischer Mikrofossilien. – *Z. angew. Geol.* **15**, No. 7 (1969), 355–363.

BYKOV, B.A. *Introduction to Phytocenology.* – Alma-Ata, "Nauka," 1970. (Russian)

CARLQUIST, Sh. Island Life. – *Publ. Am. Mus. Nat. Hist.,* 1965. 452 pp.

CARLUCCIO, L.M., F.M. HUEBER, and H.P. BANKS. *Archaeopteris macilenta,* Anatomy and Morphology of its Frond. – *Am. J. Bot.,* Vol. 53 (1966), 719–730.

CHALONER, W.G. Paleoecology of Fossil Spores. – In: *Evolution and Environment,* pp. 125–138. New Haven–London, 1968.

CHALONER, W.G. The Rise of the First Land Plants. – *Biol. Rev.,* Vol. 45 (1970), 353–378.

CHALONER, W.G. and M. MUIR. Spores and Floras. – In: *Coal and Coal-Bearing Strata,* pp. 127–146. Edinburgh, 1968.

CHANDLER, M.E.J. The Lower Tertiary Flora of Southern England. I. Palaeocene Floras, London Clay Flora (Suppl.). – *Bull. Br. Mus. Nat. Hist.,* 1961. 354 pp.

CHANDLER, M.E.J. The Lower Tertiary Floras of Southern England. II. Flora of the Pipe-Clay Series of Dorset (Lower Bagshot). – *Bull. Br. Mus. Nat. Hist.,* 1963. 169 pp.

CHANDLER, M. E. J. The Lower Tertiary Floras of Southern England. IV. A Summary and Survey of Findings in the Light of Recent Botanical Observations. – *Bull. Br. Mus. Nat. Hist.,* 1964. 151 pp.

CHANDLER, M. E. J. *The Arctic Flora of the Gam Valley at Barnwell, Cambridge.* 1964a.

CHANEY, R. W. Quantitative Studies of the Bridge Creek Flora. – *Am. J. Sci.,* Vol. 8 (1924), 127–144.

CHANEY, R. W. A Sequoia Forest of Tertiary Age on St. Lawrence Island. – *Science,* Vol. 72 (1930), 653–654.

CHANEY, R. W. Plant Distribution as a Guide to Age Determination. – *J. Wash. Acad. Sci.* **26**, No. 8 (1936), 313–324.

CHANEY, R. W. The Deschutes Flora of Eastern Oregon. – *Publs. Carnegie Instn.,* 476, 1938. 105 pp.

CHANEY, R. W. Tertiary Forests and Continental History. – *Bull. Geol. Soc. Am.,* Vol. 51 (1940), 469–488.

CHANEY, R. W. Tertiary Centres and Migration Routes. – *Ecol. Monogr.* **17**, No. 2 (1947), 139–148.

CHANEY, R. W. Conifer Dominants in the Middle Tertiary of the John Day Basin, Oregon. – *Palaeobotanist,* Vol. 1 (1952), 105–115.

CHANEY, R. W. Miocene Floras of the Columbia Plateau. – *Publs. Carnegie Instn.,* 617, 1959. 228 pp.

CHANEY, R. W. and L. H. DAUGHERTY. The Occurrence of *Cercis* Associated with the Remains of *Sinanthropus.* – *Bull. Geol. Soc. China* **12**, No. 3 (1933), 323–326.

CHANEY, R. W. and M. K. ELIAS. Late Tertiary Floras from High Plains. – *Publs. Carnegie Instn.* **476**, No. 1 (1936), 1–72.

CHANEY, R. W. and HU HSEN-HSU. A Miocene Flora from Shantung Province, China. – *Publs. Carnegie Instn.,* 507, 1940. 140 pp.

CHAPMAN, V. I. *Salt Marshes and Salt Deserts of the World.* – New York, Interscience Publisher, 1960. 400 pp.

CHIGURYAEVA, A. A. Palynology and Apomixis. – In: "*Apomiksis i selektsiya,*" pp. 80–86. Moscow, "Nauka," 1970. (Russian)

CHURCHILL, D. M. The Distribution and Prehistory of *Eucalyptus diversicolor* F. Muell, *E. marginata* Donn ex Sm. and *E. calophylla* R. Br. in Relation to Rainfall. – *Aust. J. Bot.,* Vol. 16, No. 1, 1968.

CHURCHILL, D. M. The Fossil Occurrence of *Lygodium* in Australia. – *Neues Jb. Geol. Paläont. Mh.,* No. 5 (1969), 257–265.

CLAPHAM, W. B. Nature and Paleogeography of Middle Permian Floras of Oklahoma as Inferred from their Pollen Record. – *J. Geol.* **78**, No. 2 (1970a), 153–171.

CLAPHAM, W. B. Permian Miospores from the Flowerpot Formation of Western Oklahoma. – *Micropaleontology* **16**, No. 1 (1970b), 15–36.

CLARKE, G. L. *Elements of Ecology.* – New York – London, 1954. 534 pp.

CLARKE, L. D. and N. J. HANNON. The Mangrove Swamp and Salt Marsh Communities of the Sydney District. 1. Vegetation, Soils and Climate. – *J. Ecol.* **55**, No. 3 (1967), 753–772.

CLARKE, L. D. and N. J. HANNON. The Mangrove Swamp and Salt Marsh Communities of the Sydney District. 2. – *J. Ecol.* **57**, No. 1 (1969), 213–234.

CLARKE, R. H. Quaternary Sediments off South East Devon. – *Q. Jl. Geol. Soc. Lond.*, Vol. 125, Pt. 3, No. 499 (1970), 277–319.

CLEMENS, W. A. Origin and Early Evolution of Marsupials. – *Evolution* **22**, No. 1 (1968), 1–18.

CLEMENTS, F. E. Plant Succession. – *Publs. Carnegie Instn.*, 242, 1916.

CLEMENTS, F. E. and V. E. SHELFORD. *Bio-ecology.* – New York – London, 1939.

CLOWES, F. A. L. *Apical Meristems.* – Oxford, 1961. 217 pp.

COHEN, A. D. An Allochthonous Peat Deposit from Southern Florida. – *Bull. Geol. Soc. Am.*, pp. 2477–2482, 1970.

COMANOR, P. L. Forest Vegetation and the Pollen Spectrum: an Examination of the Usefulness of the R Value. – *Bull. N. Y. Acad. Sci.*, Vol. 13 (1968), 7–19.

CONDIT, C. The San Pablo Flora of West Central California. – *Publs. Carnegie Instn.*, 476 (1938), 217–268.

CORNER, E. J. H. Debunking the New Morphology. – *New Phytol.* **65**, No. 3 (1966), 398–404.

CORNER, E. J. H. *Ficus* in the Solomon Islands and its Bearing on the Post-Jurassic History of Melanesia. – *Phil. Trans. R. Soc.* **253 B**, No. 783 (1967), 23–159.

CORREIA, M. Relations possibles entre l'état de conservation des éléments figurés de la matière organique émanant d'hydrocarbures. – *Revue Inst. fr. Pétrole* **22**, No. 9 (1967), 1285–1306.

COTTA, B. von. *Grundrisse der Geognosie und Geologie.* – Dresden – Leipzig, 1846.

CRIDLAND, A. A. The Fungi in Cordaitean Rootlets. – *Mycologia* **54**, No. 3 (1962), 230–234.

CRIDLAND, A. A. *Amyelon* in American Coal-Balls. – *Palaeontology* 7, Pt. 2 (1964), 189–209.

CRIDLAND, A. A. and J. E. MORRIS. *Spermopteris, a New Genus of Pteridosperms from the Upper Pennsylvanian Series,* 1960.

CRIDLAND, A. A. and J. E. MORRIS. *Taeniopteris, Walchia* and *Dichophyllum* in the Pennsylvanian System of Kansas. – *Kans. Univ. Sci. Bull.* **44**, No. 4 (1963), 71–85.

CROSS, A. T. Spore Floras of the Pennsylvanian of West Virginia and Kentucky. – *J. Geol.* **55**, No. 3 (1947), 285–308.

CROSS, A. T., G. THOMPSON, and J. B. ZAITZEFF. Source and Distribution of Palynomorphs in Bottom Sediments, Southern Part of the Gulf of California. – *Mar. Geol.* **4**, No. 6 (1966), 467–524.

CURTIS, J. T. *The Vegetation of Wisconsin.* – Madison Univ. Wisconsin Press, 1959. 657 pp.

CURTIS, J. T. and R. P. McINTOSH. An Upland Forest Continuum in the Prairie Forest Border Region of Wisconsin. – *Ecology,* Vol. 32 (1951), 476–496.

CUSHING, E. J. Problems in the Quaternary Phytogeography of the Great Lakes Region. – In: H. E. Wright et al. (eds.). *The Quaternary of the United States.* Princeton, 1965.

CUSHING, E. J. Late-Wisconsin Pollen Stratigraphy and the Glacial Sequence in Minnesota. – In: Cushing, E. J. and H. E. Wright Jr. (eds.). *Quart. Paleoecol., Proc. VII Congr. Intern. Assoc. Quart. Res.,* Vol. 7 (1967), 59–88.

DABER, R. A *Weichselia – Stiehleria* – Matoniaceae Community within the Quedlinburg Estuary of Lower Cretaceous Age. – *J. Linn. Soc., Botany,* Vol. 61 (1968), 75–85.

DALE, M. B. Systems Analysis and Ecology. – *Ecology* **51**, No. 1 (1970), 2–16.

DALE, M. B. and D. WALKER. Information Analysis of Pollen Diagrams. – *Pollen et spores* **12**, No. 1 (1970), 21–39.

DARBINYAN, G. A. Manifestation of Xeromorphism in Leaves of Different Layers of the Same Plants. – *Izv. Akad. Nauk Arm. SSR* **3**, No. 10 (1950), 893–904. (Russian)

DARRAH, E. L. A Remarkable Branching *Sphenophyllum* from the Carboniferous of Illinois. – *Palaeontographica,* Vol. 121B (1968), 87–101.

DARWIN, C. R. *Origin of Species by Means of Natural Selection.* – Oxford Univ. Press, 1859.

DAUGHERTY, L. H. – In: Daugherty, L. H. and H. R. Stanger. The Upper Triassic Flora of Arizona. – *Publs. Carnegie Instn.,* 526, 1941.

DAVIES, D. The Ecology of Westphalian and Lower Part of the Staffordian Series of Clydach Vale and Gilfach Goch (East Glamorgan). – *Q. Jl. Geol. Soc. Lond.,* Vol. 77 (1921), 30–74.

DAVIS, M. B. On the Theory of Pollen Analysis. – *Am. J. Sci.,* Vol. 261 (1963), 897–912.

DAVIS, M. B. Pollen Deposition in Lakes as Measured by Sediment Traps. – *Bull. Geol. Soc. Am.,* Vol. 78 (1967), 849–858.

DAVIS, M. B. Late-Glacial Climate in Northern United States: a Comparison of New England and the Great Lakes Region. – In: Cushing, E. J. and H. E. Wright Jr. (eds.). *Quart. Paleoecology Proc. VII Congr. Assoc. Quart. Res.,* Vol. 7 (1967), 11–43.

DAVIS, M. B. Climatic Changes in Southern Connecticut Recorded by Pollen Deposition. – *Ecology* **50**, No. 3 (1969), 409–421.

DAVIS, R. B. Pollen Studies of Near-Surface Sediments in Maine Lakes. – In: Cushing, E. J. and H. E. Wright Jr. (eds.). *Quart. Paleoecology, Proc. VII Congr. Intern. Assoc. Quart. Res.,* Vol. 7 (1967), 143–174.

DEHM, R. Klima und Wirbeltier-Entfaltung in Südafrika während Perm und Trias. – *Senckenberg. leth.,* Vol. 46 (1965), 35–44.

DELEVORYAS, Th. Ontogenetic Studies of Fossil Plants. – *Phytomorphology,* Vol. 14 (1964), 299–314.

DELEVORYAS, Th. Investigations of North American Cycadeoids: Microsporangiate Structures and Phylogenetic Implications. – *Palaeobotanist,* Vol. 14, Nos. 1–3, 1966.

DELEVORYAS, Th. Investigations of North American Cycadeoids: Structure, Ontogeny and Phylogenetic Considerations of Cones of Cycadeoidea. – *Palaeontographica,* Vol. 12113 (1968), 122–133.

DELEVORYAS, Th. Some Aspects of Cycadeoid Evolution. – *J. Linn. Soc., Botany* **61**, No. 384 (1968), 137–146.

DELLE, G.V. Middle Jurassic Flora of the Tkvarcheli Coal-Bearing Basin. – *Trudy Bot. Inst. Akad. Nauk SSSR*, ser. 8, No. 6, 1967. (Russian)

DEWEY, J.F. and B. HORSFIELD. Plate Tectonics, Orogeny and Continental Growth. – *Nature*, Vol. 225(1970), 521–526.

DICE, L.R. What Is Ecology? – *Scient. Mon., Lond.* **80**, No. 6 (1955), 346–351.

DIETZ, K.S. and J.C. HOLDEN. Reconstruction of *Pangea*: Breakup and Dispersion of Continents, Permian to Present. – *J. Geophys. Res.*, Vol. 75 (1970), 4939–4956.

DIETZ, R.S. Ocean-Basin Evolution by Sea-Floor Spreading. – In: S.K. Runcorn (ed.). *Continental Drift.* New York–London, Academic Press, 1962.

DIJKSTRA, S.J. Carboniferous Megaspores in Tertiary and Quaternary Deposits of South-East England. – *Ann. Mag.*, No. 34 (1950), 865–875.

DILCHER, D.L. Chlorophyll in der Braunkohle des Geiseltales. – *Natur. Mus.*, Vol. 97, No. 4, 1967.

DOKHMAN, G.I. On a Set of Diagnostic Units of Vegetation. – *Bot. Zh.* **45**, No. 5 (1960), 637–648. (Russian)

DOLIANITI, E. A flora do Gondwana inferior em Santa Caterina. IV. O gênero *Vertebraria.* – *Notes Serv. Geol.* Brasil, Vol. 81, 1954.

DOLITSKII, A.V. and I.A. KIIKO. On the Causes of Deformation of the Earth's Crust. – In: *Problemy planetarnoi geologii.* Moscow, 1963. (Russian)

DOMESTOY, G. Der Einfluss der Paläotemperaturen auf die Ökologie der Rudisten während der Kreidezeit. – *Mitt. geol. Ges. Wien,* Vol. 60, 1967 (1968).

DONSELOAR, J. van and F.P. JONKER. A Palynological Investigation of the Lower Peat in the Province of Friesland, the Netherlands. – *Meded. bot. Lab. Herb. Rijks-Univ. Utrecht,* No. 112 (1952), 257–267.

DORF, E. Late Tertiary Floras from Southwestern Idaho. – *Publs. Carnegie Instn.*, 476 (1938), 73–124.

DORF, E. Paleobotanical Evidence of Mesozoic and Cenozoic Climatic Changes. – *Proc. North Amer. Paleontol. Convent.*, Pt. D, pp. 323–346, 1969.

DORMAN, F.H. Australian Tertiary Paleotemperatures. – *J. Geol.* **74**, No. 1 (1966), 49–61.

DORMAN, F.H. Some Australian Oxygen Isotope Temperatures and Theory for a 30-Million-Year World-Temperature Cycle. – *J. Geol.* **76**, No. 3 (1968), 297–313.

DOROFEEV, P.I. On the Neogene Flora of Zhitkovichi Village in the South of Belorussia. – *Dokl. Akad. Nauk SSSR* **2**, No. 8 (1967), 719–723. (Russian)

DOUBINGER, J. and G.P. KRUSEMAN. Sur la flore du Permien de la région de Lodève ("zone de transition" et Saxonien inférieur). – *Bull. Soc. géol. Fr.* **7**, No. 4 (1965, 1966), 511–548.

DRUDE, O. *Ökologie der Pflanzen.* – Braunschweig, 1913. 308 pp.

DULHUNTY, J.A. and R. DULHUNTY. Notes on Microspore Types in Tasmanian Permian Coals. – *Proc. Linn. N.S.W.*, Vol. 74 (1949), 132–139.

DU RIETZ, G. E. Zur methodologischen Grundlage der modernen Pflanzensoziologie. – *Akad. Afh. Uppsala,* pp. 1–272, 1921.

DUTTA, S. K. and S. C. D. SAH. Palyno-Stratigraphy of the Tertiary Sedimentary Formations of Assam: 5. Stratigraphy and Palynology of South Shillong Plateau. – *Palaeontographica* **131**, Nos. 1–4 (1970), 1–72.

DYAKOWSKA, J. Researches on the Rapidity of the Falling Down [sic!] of Pollen of Some Trees. – *Extrait Bull. l'Acad. Polonaise de Sci. et des Lett., Cl. Sci. Math. Natur.,* sér. B, pp. 156–168, 1937.

DYAKOWSKA, J. and J. ZURZYCKI. Gravimetric Studies on Pollen. – *Bull. Acad. pol. Sci. Cl. II. Sér. Sci. biol.* **7**, No. 1 (1959), 11.

DYMINA, G. D. On the Structure of the Plant Cover. – *Uchen. Zap. Dal'nevostoch. Gos. Univ., ser. pochv.-bot.,* Vol. 25 (1969), 153–166. (Russian)

DYRENKOV, S. A. Quantitative Evaluation of the Structural Complexity of Forest Phytocenoses. – *Bot. Zh.* **55**, No. 6 (1970), 777–786.

EAMES, A. J. *Morphology of the Angiosperms.* – N. Y., McGraw-Hill, 1963.

EDWARDS, W. N. Flora of the London Clay. – *Proc. Geol. Ass. Lond.,* Vol. 47, 1936.

EDWARDS, W. N. The Geographical Distribution of Past Floras. An Address Delivered at the Annual Meeting of the British Association for the Advancement of Science. – *Advmt. Sci.* **12**, No. 46 (1955), 165–176.

EFREMOV, I. A. Taphonomy and the Geological Record. – *Trudy Paleontol. Inst. Akad. Nauk SSSR,* Vol. 34, No. 1, 1950. (Russian)

EGGERT, D. A. The Ontogeny of Carboniferous Arborescent Lycopsida. – *Palaeontographica,* Vol. 108 (1961), 43–92.

EGGERT, D. A. The Ontogeny of Carboniferous Arborescent Sphenopsida. – *Palaeontographica,* Vol. 110 B (1962), 99–127.

EICKE, R. Elektronmikroskopische Untersuchungen an verkieselten Coniferen. – *Palaeontographica* **97B**, Nos. 1–2 (1959), 36–46.

EISENACK, A. Erhaltung von Zellen und Zellkernen aus dem Mesozoikum und Paläozoikum. – *Natur und Mus.* **95**, No. 11 (1965), 473–477.

ELIAS, M. K. Tertiary Prairie Grasses and Other Herbs from the High Plains. – *Spec. Pap. Geol. Soc. Am.,* Vol. 41, 1942.

ELLENBERG, H. Wege der Geobotanik zum Verständnis der Pflanzendecke. – *Naturwissenschaften* **55**, No. 10 (1968), 462–470.

EL'YANOV, M. D. and V. K. TETERYUK. Finds of Microspores in Primary Kaolins of the Northwestern Part of the Ukrainian Shield. – *Dokl. Akad. Nauk SSSR* **193**, No. 6 (1970), 1374–1375. (Russian)

EMILIANI, C. Pleistocene Paleotemperatures. – *Science* **168**, No. 3933 (1970), 822–825.

EMILIANI, C. The Latest Interglacial. – *Science* **171**, No. 3971 (1971), 571–573.

ENDO, S. On the Relation between the Climatic Zones and the Leaf Margins of Dicotyledon Trees. – *J. Geogr. Soc. Tokyo,* Vol. 43 (1931), 231–234.

ENDO, S. The Climatic Condition of the Eocene *Woodwardia* Zone, Ishikari Coalfield, Hokkaido, with Description of *Ficus eowightiana* Endo n. sp. – *Proc. Japan Acad.* **40**, No. 6 (1964), 416–421.

ENDO, S. The Flora from the Eocene *Woodwardia* Formation, Ishikari Coalfield, Hokkaido, Japan. – *Bull. Natn. Sci. Mus., Tokyo,* Vol. 11, No. 4, 1968.

ENGLER, A. *Versuch einer Entwicklungsgeschichte der Pflanzenwelt, insbesondere der Florengebiete seit der Tertiärperiode.* – Leipzig, 1879–1882.

ERDTMAN, G. E. Pollen Grains Recovered from the Atmosphere over the Atlantic. – *Meddelanden från Göteborgs botaniska trädgard,* Vol. 12 (1937), 185–196.

ERDTMAN, G. *An Introduction to Pollen Analysis.* – Stockholm, 1954. 239 pp.

ERHART, H. *La genèse des sols en tant que phénomène géologique.* – Paris, 1956.

ERICSON, D. B. and G. WOLLIN. Pleistocene Climates in the Atlantic and Pacific Oceans: a Comparison Based on Deep-Sea Sediments. – *Science* **167**, No. 3924 (1970), 1483–1484.

EVENARI, M. The Physiological Anatomy of the Transpiratory Organs and the Conducting System of Certain Plants Typical of the Wilderness of Judea. – *J. Linn. Soc., Botany,* Vol. 51 (1938), 389–407.

EWING, M. and W. L. DONN. A Theory of Ice Ages. – *Science,* Vol. 123 (1956), 1061–1066; Vol. 127 (1958), 1159–1162; Vol. 152 (1966), 751–755.

FABRICIUS, F., H. FRIEDERICHSEN, and V. JACKOBSHAGEN. Paläotemperaturen und Paläoklima in Obertrias und Lias der Alpen. – *Geol. Rdsch.,* Vol. 9, No. 2, 1970.

FAEGRI, K. On the Introduction of Agriculture in Western Norway. – *Geol. För. Stockh. Förh.* **66**, No. 3, pp. 449–462.

FAEGRI, K. and J. IVERSEN. *Textbook of Pollen Analysis.* – Munkái, Copenhagen, 1964 (1950). 168 pp.

FAEGRI, K. and L. VAN DER PIJL. *The Principles of Pollination Ecology.* – Pergamon Press, 1966. 248 pp.

FAGERLIND, F. The Real Significance of Pollen Diagrams. – *Bot. Notiser,* No. 2 (1952), 185–224.

FAGESTROM, J. Fossil Communities in Paleoecology, their Recognition and Significance. – *Bull. Geol. Soc. Am.* **75**, No. 12 (1964), 1197–1216.

FAHN, A. Some Anatomical Adaptations of Desert Plants. – *Phytomorphology,* Vol. 14, No. 1, 1964.

FEDOROV, A. A. The Structure of the Tropical Rain Forest and Speciation in the Humid Tropics. – *J. Ecol.* **54**, No. 1 (1966), 1–11.

FEDOROVA, R. V. Transportation of Pollen and Spores by Flowing Water. – *Trudy Inst. Geogr. Akad. Nauk SSSR,* pp. 46–72, 1952. (Russian)

FERGUSON, L. Sedimentary Evidence for the Allochthonous Origin of *Stigmaria,* Carboniferous, Nova Scotia: Discussion. – *Bull. Geol. Soc. Am.,* Vol. 81 (1970), 2531–2534.

FIRBAS, F. *Spät- und nacheiszeitliche Waldgeschichte Mitteleuropas nördlich der Alpen, I.* – Jena, 1949. 480 pp.

FISK, H. N. and E. McFARLAN Jr. Late Quaternary Deltaic Deposits of the Mississippi River. – In: *Crust of the Earth,* Symposium edited by A. Poldervaart. *Geol. Soc. Am. Spec. Paper 62,* 1955.

FISK, H. N., E. McFARLAN, C. R. KOLB, and L. J. WILBERT. Sedimentary Framework of the Modern Mississippi Delta. – *J. Sedim. Petrol.,* Vol. 24 (1954), 76–99.

FISUNENKO, O.P. Patterns of Development of the Carboniferous Flora and Stratigraphic Zonation of the Middle Carboniferous of the Donets Basin. – In: *Geologiya uglenosnykh formatsii i stratigrafiya karbona SSSR,* pp. 199–207. Moscow, "Nauka," 1965. (Russian)

FISUNENKO, O.P. Patterns in the Formation of the Middle Carboniferous Flora of the Donets Basin. – In: "*Voprosy paleogeograficheskogo raionirovaniya v svete dannykh paleontologii,*" pp. 107–118. Moscow, "Nedra," 1967. (Russian)

FISUNENKO, O.P. On the Causes of Plant Extinction at the Boundary of the Middle and Late Carboniferous. – In: "*Materialy po geologii Donetskogo basseina,*" pp. 75–80. Moscow, "Nedra," 1968. (Russian)

FLAHAUT, Ch. and C. SCHRÖTER. Rapport sur la nomenclature phytogéographique. – *Actes Congr. Intern. Bot. Bruxelles,* Vol. 1 (1910), 131–162.

FLORENSOV, N.A. Some Features of the Basins of Major Lakes in Southern Siberia and Mongolia. – In: *Mezozoiskie i kainozoiskie ozera Sibiri,* pp. 59–73. Moscow, "Nauka," 1968. (Russian)

FLORIN, R. Studien über die Cycadales des Mesozoikums. – *K. svenska Vetensk. Akad. Handl.,* Vol. 12, 1933.

FLORIN, R. Die Koniferen des Oberkarbons und des unteren Perms, 1. – *Palaeontographica,* Vol. 85B, 1938.

FLORIN, R. Evolution in Cordaites and Conifers. – *Acta Horti Bergiani* **15**, No. 11 (1956), 285–388.

FLORIN, R. On Jurassic Taxads and Conifers from Northwestern Europe and Eastern Greenland. – *Acta Horti Bergiani* **17**, No. 10 (1958), 1–402.

FOGGIE, A. Some Ecological Observations on a Tropical Forest Type on the Gold Coast. – *J. Ecol.,* Vol. 34 (1947), 88–106.

FOSBERG, F.R. Succession and Condition of Ecosystems. – *J. Indian Bot. Soc.* **46**, No. 4 (1967), 151–154.

FOX, W. Quantitative Paleoecologic Analysis of Fossil Communities in the Richmond Group. – *J. Geol.* **76**, No. 6 (1968), 613–640.

FRIEDRICH, W.L. Tertiäre Pflanzen im Basalt von Island. – *Meddr. dansk geol. Foren.* **18**, Nos. 3–4 (1968), 265–276.

FRÜH, J. Kritische Beiträge zur Kenntnis des Torfes. – *Jahrb. geol. Reichsanstalt Wien,* Vol. 35 (1885), 677–725.

FUJI, N. Fossil Spores and Pollen Grains from the Neogene Deposits in Noto Peninsula, Central Japan, 3. – *Trans. Proc. Palaeont. Soc. Japan,* No. 76, 1969.

GALL, J.C. and L. GRAUVOGEL. A propos d'une végétation fossile conservée in situ dans le Buntsandstein supérieur des Vosges. – *C. r. somm. Séanc. Soc. géol. Fr.,* No. 7 (1967), 301–303.

GALL, J.C. and L. GRAUVOGEL. Présentation d'un profile paléoécologique du Grés à Voltzia (Buntsandstein supérieur) des Vosges. – *Bull. Serv. Carte géol. Als. Lorr.* **21**, No. 4 (1968), 305–313.

GALTIER, J. Observations sur les structures foliaires et caulinaires de *Botryopteris antiqua* Kidston. – *C. r. hebd. Séanc. Acad. Sci., Paris,* 1969.

GEKKER, R.F. *Introduction to Paleoecology.* Moscow, Gosgeoltekhizdat, 1957. (Russian)

GENSEL, P. A., A. KASPER, and H. N. ANDREWS. *Kaulangiophyton,* a New Genus of Plants from the Devonian of Maine. – *Bull. Torrey Bot. Club* **96**, No. 3 (1961), 265–276.

GIJZEL, P. van. Autofluorescence of Fossil Pollen and Spores with Special Reference to Age Determination and Coalification. – *Leid. geol. Meded.,* Vol. 40 (1967), 263–317.

GIJZEL, P. van, Th. van der HAMMEN, and T. A. WIJMSTRA. Long-Distance Correlation of Pleistocene Deposits of South America, Spain and the Netherlands. – *Palaeogeogr., Palaeoecol. Palaeoclimatol.* 3, No. 1 (1967), 65–72.

GILL, E. D. The Palaeogeography of Australia in Relation to the Migrations of Marsupials and Men. – *Trans. N. Y. Acad. Sci.* 28, No. 1 (1965), 5–14.

GILL, J. E. Origin of Continents. – *Geotektonika,* No. 6, pp. 48–55, 1965.

GITERMAN, R. E. Correlation of the Stages of Development of the Quaternary Vegetation of Northern Asia, Europe, and North America. – *Trudy Geol. Inst. Akad. Nauk SSSR,* No. 177 (1968), 237–252. (Russian)

GITERMAN, R. E., L. V. GOLUBEVA, E. D. ZAKLINSKAYA, E. V. KORENEVA, O. V. MATVEEVA, and A. A. SKOBA. *Main Stages in the Development of the Vegetation of Northern Asia in the Quaternary.* – Moscow, "Nauka," 1968. 270 pp. (Russian)

GLEASON, H. A. The Individualistic Concept of Plant Association. – *Am. Midl. Nat.,* Vol. 21 (1926), 92–110.

GLEBOV, F. Z. *Bogs and Paludal Forests of the Left-Bank Yenisei Forest Zone.* – Moscow, "Nauka," 1969. (Russian)

GLUZBAR, E.A. Distribution Patterns of Reworked Carboniferous Spores in the Upper Triassic and Jurassic of the Donets Basin. – In: *"Stepanovskie chteniya,"* pp. 24–25. Artemovsk, 1968. (Russian)

GODWIN, H. Studies on the Post-Glacial History of British Vegetation, XV. Organic Deposits of Old Backenham Meere, Norfolk. – *New Phytol.* **67**, No. 1 (1968), 95–108.

GOFF, F. G. and P. H. ZEDLER. Structural Gradient Analysis of Upland Forest in the Western Great Lakes Area. – *Ecol. Monogr.,* Vol. 38 (1968), 65–86.

GOL'BERT, A. V., L. G. MARKOVA, I. D. POLYAKOVA, V. N. SAKS, and Yu. V. TESLENKO. *Paleolandscapes of the West Siberian Territory in the Jurassic, Cretaceous and Paleogene.* – Moscow, "Nauka," 1968. 150 pp. (Russian)

GOLDYREV, G. S. and A. A. VYKHRISTYUK. Bottom Sediments of the Middle and Northern Baikal. – In: *Mezozoiskie i kainozoiskie ozera Sibiri,* pp. 98–129. Moscow, "Nauka," 1968. (Russian)

GOLENKIN, M. I. Victors in the Struggle for Existence. – Moscow, *Assots. Nauchno-Issled. Inst. pri Fiz.-Mat. Fak. Moskov. Gos. Univ.,* 1927. (Russian)

GÖPPERT, H. R. De floribus in statu fossili commentatio. – *Nova Acta Acad. Leop. Car. Nat. Cur.,* Vol. 18 (1836), 547–572.

GÖPPERT, H. R. *Die tertiäre Flora von Schossnitz in Schlesien.* – Görlitz, 1855. 52 pp.

GOTHAN, W. Pflanzengeographisches aus der paläozoischen Flora. – *Naturw. Wschr.,* No. 38 (1907), 594–599.

GOTHAN, W. and O. GIMM. Neuere Beobachtungen und Betrachtungen über die Flora des Rothliegenden von Thüringen. – *Arb. Inst. Paläobot.,* Vol. 2 (1930), 29–74.

GOTHAN, W. and H. WEYLAND. – *Lehrbuch der Paläobotanik.* – Berlin, 1954. 535 pp.

GRAHAM, A. Origin and Evolution of the Biota of Southeastern North America. Evidence from the Fossil Plant Record. – *Evolution* **18**, No. 4 (1964), 571–589.

GRAVE, M. K., V. S. GUNOVA, E. I. DEVYATOVA, M. A. LAVROVA, R. M. LEBEDEVA, L. Ya. SAMSONOVA, and E. A. CHEREMISINOVA. Mikulino Interglacial in the South-East of the Kola Peninsula. – In: "*Osnovnye problemy geomorfologii i stratigrafii antropogena Kol'skogo poluostrova,*" pp. 25–56. Leningrad, "Nauka," 1969. (Russian)

GREEN, B. H. Factors Influencing the Spatial and Temporal Distribution of *Sphagnum imbricatum* Hornsch. ex Russ. in the British Isles. – *J. Ecol.,* Vol. 56, No. 1, 1968.

GREGUSS, P. Ramification of *Sigillaria* and *Lepidodendron* and the Telome Theory. – *Phytomorphology* **11**, No. 3 (1961), 243–248.

GREIG-SMITH, P. *Quantitative Plant Ecology.* – London, Butterworths Sci. Publ., 1957.

GRICHUK, M. P. Attempt at a Reconstruction of Some Climatic Elements of the Northern Hemisphere in the Atlantic Period of the Holocene. – In: "*Golotsen.*" Moscow, "Nauka," 1969. (Russian)

GRICHUK, M. P. Patterns of Formation of Recent Spore-Pollen Spectra as the Basis for Interpreting Fossil Spore-Pollen Spectra. – In: *Istoriya razvitiya rastitel'nosti vnelednikovoi zony Zapadno-Sibirskoi nizmennosti v pozdnepliotsenovoe i chetvertichnoe vremya; Trudy Inst. Geol. i Geofiz.,* pp. 12–19. Moscow, "Nauka," 1970. (Russian)

GRICHUK, M. P., V. S. VOLKOVA, G. F. BUKREEVA, M. R. VOITAKH, T. P. LEVINA, V. P. POLESHCHUK, and A. I. STRIZHOVA. Evolution of Vegetation. – In: *Istoriya razvitiya rastitel'nosti vnelednikovoi zony Zapadno-Sibirskoi nizmennosti v pozdnepliotsenovoe i chetvertichnoe vremya; Trudy Inst. Geol. i Geofiz. Sibirsk. Otdel. Akad. Nauk SSSR,* pp. 12–19. Moscow, "Nauka," 1968. (Russian)

GRIGOR'EV, Yu. S. On Some General Patterns of Xerophilization in the Higher Plants. – In: *Problemy osvoeniya pustyn',* No. 5 (1968), 3–13. (Russian)

GRISEBACH, A. Über den Vegetationscharakter von Hardanger in Bergens Stift. – *Wiegmanns Archiv für Naturgesch.,* pp. 1–28, 1844.

GRISEBACH, A. *Die Vegetation der Erde nach ihrer klimatischen Anordnung.* – Leipzig, 1872.

GROOT, J. J. Some Observations on Pollen Grains in Suspension in the Estuary of the Delaware River. – *Mar. Geol.,* Vol. 4 (1966), 409–416.

GROOT, J. J. and C. R. GROOT. Marine Palynology Possibilities, Limitations, Problems. – *Mar. Geol.,* Vol. 4 (1966), 387–395.

GROOT, J. J. and C. R. GROOT. Pollen Spectra from Deep-Sea Sediments as Indicators of Climatic Changes in Southern South America. – *Mar. Geol.* **4**, No. 6 (1966), 525–537.

GROSSET, G.E. Migration Paths and Schedule of Crimean-Caucasian Forest Species on the Russian Plain and the Subsequent Changes of their Ranges in Connection with Landscape Evolution. – *Byull. Moskov. Obshch. Ispyt. Prirody, Otdel Biol.* **72**, No. 5 (1967), 47–76. (Russian)

GROSSGEIM, A.A. The Xeromorphogenesis Theory and Some Aspects of Floral Evolution. – In: *Problemy botaniki,* No. 1, 1950. (Russian)

GRUSHVITSKII, I.V. and N.T. SKVORTSOVA. On a New Type of Compound Leaf. The Fasciculate Compound Leaf of Species of the Genus *Schefflera* Forst. et Forst., F. (Araliaceae). – *Bot. Zh.* **55**, No. 4 (1970), 525–536. (Russian)

GULISASHVILI, V.Z. *Origin of the Woody Vegetation of Subtropical and Temperate Climates and Development of its Hereditary Features.* – Tbilisi, "Metsniereba," 1967. 215 pp. (Russian)

GÜMBEL. Bildung von Kohlen an Ort und Stelle und für die durch Transport hervorgegangenen. Beiträge zur Kenntnis der Texturverhältnisse der Mineralkohlen. – *Sber. bayer. Akad. Wiss., Math.-physik. Klasse,* 1883.

GUNTAU, M. Bemerkungen zum Aktualismus in der Geologie. – *Ber. geol. Ges.,* Vol. 8, No. 4, 1963.

GUPTA, K.M. *Williamsonia* Fructifications from the Jurassic of Rajmahal Hills – their Preservation and Plan of Construction. – *J. Palaeont. Soc. India,* Vol. 3, 1958.

HABIB, D. Spores, Pollen and Microplankton from the Horizon Beta Outcrop. – *Science* **162**, No. 3861 (1968), 1480–1481.

HABIB, D., W. RIEGEL, and W. SPACKMAN. Relationship of Spore and Pollen Assemblages in the Lower Kittaning Coal to Overlying Faunal Facies. – *J. Paleont.,* Vol. 40, No. 3, 1966.

HAFFER, J. Speciation in Amazonian Forest Birds. – *Science,* Vol. 165 (1969), 31–138.

HAJÓS, M. Magyarországi neogén Phytolithariál. – *Évi Jelent. magy. K. földt. Intéz.,* pp. 225–240, 1968.

HALL, J.B. and J. JENIK. Observations on the African Cycad in Ghana. – *Niger. Fld.* **32**, No. 2 (1967), 75–81.

HALL, J.W. and N.J. NORTON. Palynological Evidence of Floristic Change across the Cretaceous-Tertiary Boundary in Eastern Montana. – *Palaeogeogr. Palaeoclimatol. Palaeoecol.* **3**, No. 1 (1967), 121–131.

HALLIK, R. Organogene Serie einer pleistozänen Warmzeit vom Typ Bederkesa. – *Eiszeitalter Gegenw.,* Vol. 19 (1968), 244–249.

HAMILTON, L.H., A.R. RAMSDEN, and J.F. STEPHENS. Fossiliferous Graphite from Undercliff, New South Wales. – *J. Geol. Soc. Aust.* **17**, No. 1 (1970), 31–38.

HAMMEN, Th. van der. Upper Cretaceous and Tertiary Climatic Periodicities and their Causes. – *Ann. N. Y. Acad. Sci.,* Vol. 95 (1961), 440–449.

HANSEN, H. Further Pollen Studies of Peat Bogs on the Pacific Coast of Oregon and Washington. – *Bull. Torrey Bot. Club* **71**, No. 6 (1944), 627–686.

HARPER, J.L. A Darwinian Approach to Plant Ecology. – *J. Ecol.* **55**, No. 2 (1967), 247–270.

HARRIS, T.M. The Fructification of *Czekanowskia* and its Allies. – *Phil. Trans. R. Soc.* **225**, No. 625 (1951), 483–508.

HARRIS, T.M. The Zonation of the Yorkshire Jurassic Flora. – *Palaeobotanist,* Vol. 1 (1952), 207.

HARRIS, T.M. The Fossil Plant Cuticle. – *Endeavour,* Vol. 15, 1956.

HARRIS, T.M. Liasso-Rhaetic Flora in South Wales. – *Proc. R. Soc.,* ser. B, Vol. 147, pp. 289–308.

HARRIS, T.M. Misplaced Plant Microfossils. – *J. Palaeont. Soc. India,* Vol. 3 (1958), 145–148.

HARRIS, T.M. Plant Ecology of Ancient Periods. – *Nature,* Vol. 183 (1959), 1370–1371.

HARRIS, T.M. The Yorkshire Jurassic Flora. II. Caytoniales, Cycadales and Pteridosperms. – *Bull. Br. Mus. Nat. Hist. A. Geology,* 1964. 185 pp.

HARRIS, T.M. Dispersed Cuticles. – *Palaeobotanist* **14**, Nos. 1–3 (1965, 1966), 101–105.

HARRIS, T.M. The Yorkshire Jurassic Flora. III. Bennettitales. – *Bull. Br. Mus. Nat. Hist. A. Geology,* No. 675 (1969), 201 pp.

HARTIGAN, D.T. Some Present and Prospective Foliage Diseases of Australian Grown Pines. – *Aust. For.* **32**, No. 1 (1968), 39–48.

HAVLENA, V. Das Grundproblem der Grenze Karbon-Perm Tschechoslawakei. – *C. r. 4ème congr. av. et strat. geol. Carbon. Heerlen,* No. 1 (1960), 227–285.

HAVLENA, V. Ecology and Mode of Deposition of the Ostrava Formation (Namurian a). – *Věst. Ústřed. Úst. geol.* **45**, No. 5 (1970), 291–294.

HEDGPETH, J.W. Concepts of Marine Ecology. – In' J.W. Hedgpeth (ed.). *Treatise on Marine Ecology and Paleoecology,* Vol. 1; *Mem. Geol. Soc. Am.,* Vol. 67, 1957. 1296 pp.

HEDGPETH, J.W. Evolution of Community Structure. – In: J. Imbrie, N. Newell (eds.). *Approaches to Paleoecology,* pp. 11–18. New York–London–Sydney, 1964.

HEER, O. *Die Tertiäre Flora der Schweiz,* Vol. 1, pp. 1–118. Winterthur, 1855.

HEER, O. *Die Tertiäre Flora der Schweiz,* Vol. 3, pp. 101–155. Winterthur, 1859.

HEER, O. *Untersuchungen über das Klima und die Vegetations-verhältnisse des Tertiärlandes.* – Winterthur, 1860.

HEER, O. *Die Urwelt der Schweiz,* Parts 1–2. – Zürich, 1864–1865.

HEER, O. *Über den versteinerten Wald von Atanekerdluk in Nordgrönland,* pp. 259–280, 1886.

HEEZEN, C., et al. *Floors of the Oceans.* 3 vols., 1959.

HEINSELMAN, M.L. Landscape Evolution, Peatland Types and the Environment in the Lake Agassiz Peatlands Natural Area, Minnesota. – *Ecol. Monogr.* **40**, No. 2 (1970), 235–261.

HESLOP-HARRISON, J. The Unisexual Flower – a Reply to Criticism. – *Phytomorphology,* Vol. 8 (1958), 177–184.

HESSELMAN, H. Über die Verbreitungsfähigkeit des Waldbaumpollens. – *Meddn. St. SkogsförsAnst,* 1919.

HIBBERT, F.A., V.R. SWITSUR, and R.G. WEST. Radiocarbon Dating of Flandrian Pollen Zones at Red Moss, Lancashire. – *Proc. R. Soc.,* Vol. 1773 (1971), 161–176.

HIRMER, M. and L. HOERCHAMMER. Morphologie, Systematik und geographische Verbreitung der Fossilen und Rezenten Matoniaceen. – *Palaeontographica,* Vol. 81 B, 1936.

HIRST, J. M. and O. J. STEDMAN. Long-Distance Spore Transport: Vertical Sections of Spore Clouds over the Sea. – *J. Gen. Microbiol.,* Vol. 48 (1967), 357–377.

HÖEG, O. A. The Downtonian and Devonian Flora of Spitzbergen. – *Skrifter,* No. 83. Oslo, 1942. 228 pp.

HOOKER, J. D. Insular Floras. – *Gdnrs' Chron.,* Vol. 43, 1867.

HOPKINS, D. M., et al. Cretaceous, Tertiary and Early Pleistocene Rocks from the Continental Margin in the Bering Sea. – *Bull. U.S. Geol. Surv.,* Vol. 80 (1969), 1471–1480.

HORNIBROOK, N. de B. New Zealand Tertiary Microfossil Zonation, Correlation and Climate. – *11th Pacif. Sci. Congr.,* Vol. 4 (1967), 8.

HOROWITZ, A. Recent Pollen Sedimentation in Lake Kinneret, Israel. – *Pollen et Spores,* Vol. 11 (1969), 353–384.

HUBER, N. The Environmental Control of Sedimentary Iron Minerals. – *Econ. Geol.,* Vol. 53 (1958), 123–140.

HUEBBERT, M. K. Critique of the Principle of Uniformity. – In: Cl. C. Albritton (ed.). *Uniformity and Simplicity,* pp. 3–34. New York, 1967.

HUEBER, F. M. *Hepaticites devonicus,* a New Fossil Liverwort from the Devonian of New York. – *Ann. Mo. Bot. Gdn.* **48**, No. 2 (1961), 125–128.

HUGHES, N. F. and J. C. MOODY-STUART. – *Rev. Palaeobot. Palynol.,* Vol. 1 (1967), 259–268.

HUGHES, N. F. and J. SMART. Plant-Insect Relationships in Palaeozoic and Later Time. – In: *The Fossil Record, Geol. Soc.,* pp. 107–117, 1967.

HUMBOLDT, A. von. *Ideen zu einer Physiognomik – der Gewächse.* – Tübingen, 1806.

HUTCHINSON, J. *Evolution and Phylogeny of Flowering Plants.* – London, New York Academic Press, 1964. 717 pp.

HUTCHINSON, T. C. The Occurrence of Living and Sub-Fossil Remains of *Betula nana* L. in Upper Teesdale. – *New Phytol.,* Vol. 65, No. 3, 1966.

HUXLEY, J. The Evolutionary Process. – In: J. Huxley et al. (eds.). *Evolution as a Process.* London, 1954.

HUXLEY, J. *Evolution. The Modern Synthesis.* – London, 1955. 645 pp.

HUZIOKA, K. A New Species of *Sagenopteris* from Nariwa, Southwest Honshu, Japan. – *Trans. Proc. Palaeont. Soc. Japan,* No. 77 (1970), 229–234.

HUZIOKA, K. and T. TANAI. Climatic Implication of Tertiary Floras in Japan. – *11th Pacif. Sci. Congr.,* Vol. 4 (1967), 15.

IBRAHIM, A. C. *Sporenformen des Ägirhorizonts des Ruhr-Riviers.* – Diss. Berlin, 1932. Würzburg, 1933.

IDYLL, C. P. The Everglades: a Threatened Ecology. – *Sci. J.* **5 A**, No. 2 (1969), 66.

IL'INSKAYA, I. A. Fossil Monotopic and Polytopic Floras and Assemblages. – *Dokl. Akad. Nauk SSSR,* Vol. 119, No. 4, 1958. (Russian)

IMBRIE, J. and N. D. NEWELL. The Viewpoint of Paleoecology. – In: *Approaches to paleoecology,* pp. 1–7, 1964.

ISACHENKO, T. I. Structure of the Plant Cover and Mapping. – In: "*Geobotanicheskoe kartografirovanie,*" pp. 20–33. Leningrad, "Nauka." (Russian)

ISACKS, B., J. OLIVER, and L. R. SYKES. Seismology and the New Global Tectonics. – *J. Geophys. Res.,* Vol. 73 (1968), 5855–5900.

IVERSEN, J. *Viscum, Hedera* and *Ilex* as Climate Indicators. – *Geol. För. Stockh. Förh.* **66**, No. 3 (1944), 463–483.

JACOB, H. Waldgeschichte Untersuchungen im Tharandter Gebiet. – *Reprium. nov. Spec. Regni veg.,* Vol. 137 (1957), 184–279.

JANSSEN, C. R. Recent Pollen Spectra from the Deciduous and Coniferous-Deciduous Forests of Northeastern Minnesota: a Study in Pollen Dispersal. – *Ecology* **47**, No. 5 (1966), 804–825.

JANSSEN, C. R. Myrtle Lake: a Late- and Postglacial Pollen Diagram from Northern Minnesota. – *Can. J. Bot.* **46**, No. 11 (1968), 1397–1408.

JELINEK, A. L. Correlation of Archeological and Palynological Data. – *Science* **122**, No. 3728 (1966), 1507–1509.

JOHNSON, R. Y. The Community Approach to Paleoecology. – In: I. Imbrie and N. Newell (eds.). *Approaches to Paleoecology,* pp. 107–134. New York–London–Sydney, 1964.

JONGMANS, W. J. Carboniferous Floras round the Paleozoic Mediterranean. – *Proc. Int. Bot. Congr.,* p. 591, 1958.

JONKER, F. P. A Plea for the Standardization of Pollen Diagrams. – *Taxon* **1**, No. 1 (1951), 89–91.

JOPLING, A. V. Hydraulic Studies on the Origin of Bedding. – *Sedimentology,* Vol. 2 (1963), 115–121.

JORDAN, P. Die Expansion der Erde. Folgerungen aus der Diracschen Gravitationshypothese. – *Wissenschaft,* Vol. 12 (1966), 1–182.

JOSTEN, K.-H. Pflanzensoziologische Beobachtungen an Steinkohlenbohrungen im Ruhrgebiet. – *Palaeontographica* **108 B**, Nos. 1–2 (1961), 39–42.

JUX, U. Torfe des rheinischen Tertiärs im Vergleich mit heutigen Bildungen an der amerikanischen Ostküste. – *Z. dt. geol. Ges.,* Vol. 18 (1968), 69–101.

KAGEMORI, N. Chemical Composition of Fossil Wood Collected from Cenozoic Strata in the Vicinity of Osaka. – *J. Geol. Soc. Japan* **75**, No. 1 (1969), 43–50.

KAMYSHEV, N. S. Principles of Naming and Identifying Plant Associations in Nature. – *Bot. Zh.* **51**, No. 10 (1966), 1391–1400. (Russian)

KARPOV, V. G. Experimental Analysis of Competition Mechanisms Displayed by Spruce (*Picea excelsa*) and Pedunculate Oak (*Quercus robur*) in Biogeocenoses of the Southern Taiga. – In: *Eksperimental'noe izuchenie biogeotsenoza taigi,* pp. 80–104. Leningrad, "Nauka," 1969. (Russian)

KASHYAP, S. R. The Relationships of Liverworts Especially in the Light of Some Recently Discovered Himalayan Forms. – *Proc. Asiat. Soc. Beng.,* Vol. 2, No. 15, 1919.

KASK, M. *Vegetation of the Avaste Bog.* – Tartu, 1965. (Russian)

KATS, N. Ya. Bogs of the Sub-Antarctic and the Cold-Temperate Zone of the Southern Hemisphere. – *Pochvovedenie,* No. 2 (1966), 65–74. (Russian)

KATS, N. Ya. and S. V. KATS. Natural and Anthropochorous Vegetation of the Habitat of Neolithic Man in the Berendeevo Bog. – In: *Golotsen.* Moscow, "Nauka," 1969. (Russian)

KAUFMANN, L. Variationsstatistische Untersuchungen über die "Artabwandlung" und "Artumbildung" an der oberkambrischen Trilobitengattung *Olenus.* – *Abh. geol.-palaeont. Inst. Greifswald,* Vol. 10, 1933.

KAYE, C. A. Fossiliferous Bauxite in Glacial Drift, Martha's Vineyard, Massachusetts. – *Science* **177**, No. 3792 (1962), 1035–1037.

KAZARINOV, V. P. et al. *Weathering and Lithogenesis.* – Moscow, "Nedra." 1966. 456 pp. (Russian)

KEDVES, M. Palynologische Untersuchungen der Eozän-Braunkohlen-Schichten des Doroger Beckens im Gebiet von Csolnok. – *Grana palynol.* **6**, No. 2 (1965), 290–296.

KEDVES, M., L. ENDREDI, and Z. SZELEY. Problèmes palynologiques concernant le remaniement des sédiments paléo- et mésozoiques dans les bassins du Pannonien supérieur de Hongrie. – *Pollen et spores* 8, No. 2 (1966), 315–336.

KEDVES, M. and M. KARÁCSONYI. The Recycled Sporomorphs of the Boring No. Ny-1. in Kecskemét. – *Acta biol. Szeged.* **13**, Nos. 3–4 (1967), 11–13.

KELLER, B. A. Anatomy of Drought- and Heat-Resistant Leaves. – *Sovetskaya Botanika,* No. 2 (1933), 9–13. (Russian)

KESTEVEN, G. L. and R. R. INGPEN. Representation of the Structure of Biotic Systems. – *Aust. J. Sci.,* Vol. 29, 1966.

KHAIN, V. E. Metamorphic Regeneration of Regions of Repeated Orogeny and the Problem of Neogenesis of Oceanic Depressions. – *Geotektonika,* No. 3 (1969), 31–38. (Russian)

KHAIN, V. E. Conditions of Deposition and Main Stages in the Development of the Mediterranean Geosyncline Zone. – *Vest. Moskov. Gos. Univ.,* No. 2 (1970), 36–72. (Russian)

KHAITSER, L. L. On the Kaolinization of Shells in Permian Deposits of the Pechora Basin. – *Litologiya i Poleznye Iskopaemye,* No. 5 (1970), 157–160. (Russian)

KHOTINSKII, N. A., A. L. DEVIRTS, and N. G. MARKOVA. Age and History of the Formation of Bogs on the Eastern Outskirts of Vasyugan'e. – *Byull. Moskov. Obshch. Ispyt. Prirody* **75**, No. 5 (1970), 82–91. (Russian)

KING, J. E. Modern Pollen Rain and Fossil Pollen in Soils in the Sandia Mountains, New Mexico. – *Pap. Mich. Acad. Sci. Arts and Letters,* Vol. 52 (1967), 31–41.

KIRCHHEIMER, F. Mikrofossilien aus Salzablagerungen des Tertiärs. – *Palaeontographica* **90B**, Nos. 4–6 (1946), 127–160.

KISELEVA, K. V., S. M. RAZUMOVSKII, and A. P. RASNITSYN. Boundaries of Plant Communities and Vegetation Dynamics. – *Zhurn. Obshch. Biol.* **30**, No. 2 (1969), 123–131. (Russian)

KIZEL'SHTEIN, L. Ya. and N. T. SEREBRYAKOVA. Some Variable Features of the Structure and Composition of Coal Beds. – *Litologiya i Poleznye Iskopaemye,* No. 5 (1970), 107–114. (Russian)

KIZEL'SHTEIN, L. Ya. and V. G. SHCHIROV. Recent Peat Accumulation and Some Aspects of the Formation of Coal Beds. – *Izv. Akad. Nauk SSSR, geol. ser.*, No. 8 (1970), 101–107. (Russian)

KLAAUW, C. J. van der. Zur Aufteilung der Ökologie in Autökologie und Synökologie im Lichte der Ideen als Grundlage der Systematik der zoologischen Disciplinen. – *Acta biotheor.* **2**, Pt. 3 (1936), 195–240.

KLAUS, W. Vorbericht über pollenanalytische Untersuchungen von Sedimenten aus des Schlenken-Durchgangshöhle a. d. Tauge (Salzburg). – *Anz. öst. Akad. Wiss. Math.-naturwiss. Klasse* **104**, Nos. 1–14 (1967), 379–380.

KLEOPOV, Yu. D. Main Features of Development of the Flora of the Broadleaf Forests of the European SSSR. – In: "*Materialy po istorii flory i rastitel'nosti SSSR*," No. 1, 1941. (Russian)

Klimaänderungen im Tertiär aus paläobotanischer Sicht. – *Abh. zentr. geol. Inst.*, No. 10. Berlin, 1967. 230 pp.

KLOPOTOVSKAYA, N. B. Methods of Applying Spore-Pollen Analysis under Mountain Conditions. – *Sakartvelos SSSR, Metsnier. Akad. Moambe, Soobshch. Akad. Nauk GruzSSR* **50**, No. 2 (1968), 401–406. (Russian)

KLUSEMANN, H. and R. TEICHMÜLLER. Begrabene Wälder im Ruhrkohlbecken. – *Natur Volk*, Vol. 84 (1954), 373–382.

KNOBLOCH, E. Nové rostlinné nálezy z neogénu slovenské části Oravské pánve. – *Čas. Miner. Geol.* **13**, No. 4 (1968), 469–476.

KNOBLOCH, E. Ermöglicht die Paläokarpologie Aussagen zur Genese Biostratigraphie der jungen Flyschsedimente? – *Mitt. Bayer. Staatssamml. Paläont. hist. Geol.*, No. 10 (1970a), 297–308.

KNOBLOCH, E. The Tertiary Floras of Moravia (Czechoslovakia). – *Paläontol. Abh.* **3**, Nos. 3–4 (1970b), 381–390.

KNOCHE, H., P. ALBRECHT, and G. OURISSON. Organische Verbindungen in fossilen Pflanzen (*Voltzia brongniarti*, Coniferales). – *Angew. Chem.* **80**, No. 16 (1968), 666–667.

KOBAYASHI, T. On the Climatic Bearing of the Mesozoic Floras of Eastern Asia. – *Jap. J. Geol. Geogr.*, Vol. 18 (1942), 157–196.

KOKAWA, S. Morphometric Reconstruction of the Compressed Seed Remains of *Menyanthes* of Japan. – *J. Inst. Polytech. Osaka Cy. Univ.*, Vol. 11D (1960), 79–89.

KOLAKOVSKII, A. A. The Plant World of Colchis. – Moscow, *Moskov. Gos. Univ.*, 1961. (Russian)

KOLAKOVSKII, A. A. A Representative of the Oligocene Mastic Flora of Europe in the Pliocene of the Colchis Refuge. – *Sakartvelos SSR Metsnier. Akad. Moambe, Soobshch. Akad. Nauk GruzSSR* **55**, No. 3 (1969), 737–740. (Russian)

KOLISHCHUK, V. G. On the Morphological Evolution from Trees to Grasses in the Series of Prostrate Plant Forms. – *Bot. Zh.* **53**, No. 8 (1968), 1029–1042. (Russian)

KONOE, R. Preliminary Study on the Optimum Temperature for the Growth of *Metasequoia* and Related Genera. – *J. Inst. Polytech. Osaka Cy. Univ.*, Vol. 11D, 1967.

KONONOV, Yu. I. and L. P. KARAULOVA. On the Distribution of Pollen and Spores in Recent Sediments of Rivers of the Northern Sikhote-Alin Range. – *Proceedings 14th Scientific Conference of the Far Eastern State University,* pp. 51–54. Vladivostok, 1965. (Russian)

KONONOVA, M. M. Some Controversial Points Relating to Soil Humus. – *Izv. Akad. Nauk SSSR, biol. ser.,* No. 3 (1970), 369–374. (Russian)

KORENEVA, E. V. *Spores and Pollen from Bottom Sediments of the West Pacific.* – Moscow, "Nauka," 1969. (Russian)

KORNILOVA, V. S. Floral Succession in Kazakhstan during the Paleogene and Neogene. – *Paläontol. Abh.* **B3**, No. 3–4 (1970), 391–397. (Russian)

KOSYGIN, Yu. A. and V. A. SOLOV'EV. Geological Formations and Geotectonics. – *Geologiya i Geofizika,* No. 1, 1969. (Russian)

KOTSERUBA, L. A. and N. O. RYBAKOVA. Palynological Description of Different Facies of Alluvial Deposits in the Lower Ob Floodplain. – *Vest. Moskov. Gos. Univ., Geologiya,* No. 3 (1965), 75–79. (Russian)

KOZO-POLYANSKII, B. M. *Introduction to the Phylogenetic Systematics of the Higher Plants.* – Voronezh, 1922. (Russian)

KRASILOV, V. A. *Early Cretaceous Flora of the Southern Maritime Territory and its Importance for Stratigraphy.* – Moscow, "Nauka," 1967. 248 pp. (Russian)

KRASILOV, V. A. Study of Fossil Plants from the Group Czekanowskiales. – In: "*Rasteniya mezozoya.*" Moscow, "Nauka," 1968. (Russian)

KRASILOV, V. Z. Reconstruction of Extinct Plants. – *Paleontol. Zh.,* No. 1 (1969a), 4–12. (Russian)

KRASILOV, V. A. Types and Causes of Paleofloristic Successions. – *Paleontol. Zh.,* No. 3 (1969b), 7–23. (Russian)

KRASILOV, V. A. Paleoecosystems. – *Izv. Akad. Nauk SSSR, geol. ser.,* No. 4 (1970a), 144–150. (Russian)

KRASILOV, V. A. On the Origin and Homology of the Reproductive Organs of Flowering Plants. – *Zhurn. Obshch. Biol.* **31**, No. 6 (1970b), 679–689. (Russian)

KRASILOV, V. A. New Data on the Tsagayan Deposits and their Taphoflora. – *Izv. Akad. Nauk SSSR,* No. 12 (1970c), 97–101. (Russian)

KRASILOV, V. A. Paleoclimates and Correlation of Lower Cretaceous Deposits of the Far East and Northern Yakutia. – *Geologiya i Geofizika,* No. 8, 1971a. (Russian)

KRASILOV, V. A. New Data on the Structure and Evolution of the Reproductive Organs of Ancient Angiosperms. – *Dokl. Akad. Nauk SSSR,* Vol. 196, No. 6, 1971b. (Russian)

KRASILOV, V. A. On the Coincidence of the Lower Cenozoic-Cenophytic Boundaries. – *Izv. Akad. Nauk SSSR, geol. ser.,* No. 3, 1972. (Russian)

KRASILOV, V. A. and S. A. SHOROKHOVA. New Triassic Plants from the Iman River Basin (Maritime Territory) and Some Morphogenetic Aspects of Mesozoic Pteridophylls. – In: "*Triasovye bespozvonochnye i rasteniya vostoka SSSR,*" pp. 98–121. Vladivostok, 1970. (Russian)

KRASSILOV, V. A. Approach to the Classification of Mesozoic "Gingkoalean" Plants from Siberia. – *Palaeobotanist,* No. 1 (1969), 12–19.

KRÄUSEL, R. Die Gingkophyten der Trias von Lunz in Nieder-Österreich und von Neue Welt bei Basel. – *Palaeontographica,* Vol. 87 B (1943), 59–93.

KRÄUSEL, R. and H. WEYLAND. Die Flora des böhmischen Mitteldevons. – *Palaeontographica,* Vol. 78 B, 1933.

KRÄUSEL, W. Wesen und Wandel der Fossilen. – *Der Aufschluss* **19**, Nos. 7–8 (1968), 177–184.

KREMP, G. O. W. and N. O. FREDERIKSEN. Recognition of Coal Lithotypes by Palynological Methods. – *Int. Congr. Coal Petrol.,* Heerlen, 1956.

KRISHTOFOVICH, A. N. Evolution of the Plant Cover in the Geological Past and the Major Factors Governing It. – *Mater. po Istorii Flory i Rastitel'nosti SSSR,* No. 2, pp. 21–86. Leningrad, Akad. Nauk SSSR, 1946. (Russian)

KRISHTOFOVICH, A. N. Development of Botanico-Geographical Regions of the Northern Hemisphere from the Beginning of the Tertiary Period. – In: "*Voprosy geologii Azii,*" Vol. 2 (1955), 824–844. (Russian)

KRISHTOFOVICH, A. N. *Paleobotany.* – Leningrad, Gostoptekhizdat, 1957. 650 pp. (Russian)

KROPOTKIN, P. N. Paleomagnetism, Paleoclimates, and the Problem of Major Horizontal Movements of the Earth's Crust. – *Sovetskaya Geologiya,* No. 5, 1961. (Russian)

KROPOTKIN, P. N. The Problem of Continental Drift (Mobilism). – *Fizika Zemli,* No. 3 (1969), 3–18. (Russian)

KROPOTKIN, P. N. and Yu. A. TRAPEZNIKOV. Variations in the Angular Velocity of the Earth's Rotation, Fluctuations of the Pole, and Rate of Drift of the Geomagnetic Field, and their Possible Relationship with Geotectonic Processes. – *Izv. Akad. Nauk SSSR, geol. ser.,* No. 11, 1963. (Russian)

KRYUKOV, M. N. Distribution of Deposition Depths in Various Types of Peat Bog. – *Torfyanaya Promyshlennost',* No. 7, 1956. (Russian)

KUDRYAVTSEV, N. A. On the Patterns of Accumulation of Mineral Salts. – *Sovetskaya Geologiya,* No. 7 (1966), 17–35. (Russian)

KUKAL, Z. Diagnostic Features of Paleozoic Deltaic Sediments of Central Bohemia. – *Sedimentology,* Vol. 3 (1964), 109–113.

KUMMEROW, E. Die aktualistische Methode in der Geologie. – *Z. dt. geol. Ges.,* Vol. 84 (1932), 563–565.

KUTYREV, E. I. *Conditions of Formation and the Interpretation of Cross Bedding.* – Moscow, "Nedra," 1968. (Russian)

LANCUCKA-SRODONIOWA, M. Tortonian Flora from the "Gdów Bay" in the South of Poland. – *Acta palaeobot., Cracow* 7, No. 1 (1966), 135 pp.

LAVEINE, J.-P. A propos de la classification des Neuropteridées. – *C. r. hebd. Séanc. Acad. Sci., Paris,* Vol. 262 D (1966), 1680–1683.

LAWRENCE, D. R. Taphonomy and Information Losses in Fossil Communities. – *Bull. Geol. Soc. Am.* **79**, No. 10 (1968), 1315–1330.

LAWRENCE, D. R. The Nature and Structure of Paleoecology. – *J. Paleont.,* Vol. 45, No. 4, 1971.

LECLERQ, S. Les Psilophytales représentent-elles le creuset des plantes vasculaires? – *L'ann. Biol.,* sér. 3, Vol. 28, Nos. 5–6, 1952.

LELE, K. M. and A. CHANDRA. Palynological Reconnaissance of the Marine Beds at Umaria and Manendrgarn (India). – *Science and Culture* **35**, No. 2 (1969), 65–67.

LELE, K. M. and M. STREEL. Middle Devonian (Givetian) Plant Microfossils from Goé (Belgium). – *Annls. Soc. géol. Belg.* **22**, No. 1 (1969), 80–121.

LEMOIGNE, Y. Les tissus vasculaires et leur histogenèse chez les lépidophytales arborescentes du Paléozoique. – *Annls. Sci. natur., Bot. et biol. végét., sér. 12,* **7**, No. 3 (1966), 445–474.

LEMOIGNE, Y. Paléoflore à Cupressales dans le Trias-Rhétien du Cotentin. – *C. r. hebd. Séanc. Acad. Sci., Paris* **264 D**, No. 5 (1967), 715–718.

LEMOIGNE, Y. Contribution à la connaissance du gamétophyte de *Rhynia gwynevaughanii* Kidston et Lang; problème des protubérances et processus de ramification. – *Bull. mens. Soc. linn. Lyon,* Vol. 38, No. 4, 1969.

LEOPOLD, E. B. *Reconstruction of Quaternary Environments Using Palynology.* – Taos, N. Mexico, Symposium on Paleoecology, 1964.

LE PICHON, X. Sea-Floor Spreading and Continental Drift. – *J. Geophys. Res.,* Vol. 73 (1968), 3661–3697.

LEVIN, B. Yu. Thermal History of the Earth and Moon and the Physical Basis for the Differentiation of their Mineral Resources and the Volcanism on their Surface. – In: "*Vul'kanizm i glubinnoe stroenie Zemli.*" Vol. 3. Moscow, "Nauka," 1966. (Russian)

LEWARNE, G. and I. M. PALLOT. Mesozoic Plants from Fissures in the Carboniferous Limestone of South Wales. – *Ann. Mag. Nat. Hist.,* ser. 12, Vol. 10 (1957), 72–79.

LI, HUI-LIN. Floristic Relationships between Eastern Asia and Eastern North America. – *Trans. Am. Phil. Soc.* **42**, Pt. 2 (1952), 371–429.

LICHKOV, B. L. – *Zhurn. Obshch. Biol.* **6**, No. 3 (1945), 159–182. (Russian)

LICITIS-LINDBERGS, R. Branch Abcission and Disintegration of the Female Cones of *Agathis australis* Salisb. – *Phytomorphology,* Vol. 6, No. 2, 1956.

LIVINGSTONE, D. A. Some Interstadial and Postglacial Pollen Diagrams from Eastern Canada. – *Ecol. Monogr.* **38**, No. 2 (1968), 87–125.

LOCK, M. *Equisetites lyelli* (Mantel) at a New Horizon in the Wadhurst Clay near Pemburg, Kent. – *Proc. Geol. Ass. Lond.,* Vol. 64 (1953), 31–32.

LOVELIUS, N. V. Effect of Volcanic Eruptions on the Vegetation of Kamchatka. – *Bot. Zh.* **55**, No. 11 (1970), 1630–1633. (Russian)

LUNEV, B. S. Differentiation of Sediments in Recent Alluvium. – *Ucher. Zap. Permsk. Univ.,* No. 174. Perm, 333 pp. (Russian)

LUNGERSGAUZEN, G. F. On the Periodicity of Geological Phenomena and Climatic Changes in Past Epochs. – In: *Problemy planetarnoi geologii,* Moscow, 1963. (Russian)

LUT, B. F. and B. P. AGAFONOV. Morphology of the Basins of Some Recent Lakes of Transbaikalia. – In: *Mezozoiskie i kainozoiskie ozera Sibiri,* pp. 98–124. Moscow, "Nauka," 1968. (Russian)

LYELL, Ch. Review of Transactions of the Geological Society of London, Vol. 1. – *Quart. Rev.,* Vol. 34 (1826), 507–540.

LYON, A.G. Probable Fertile Region of *Asteroxylon mackieki* K. and L. – *Nature,* Vol. 203 (1964), 1082–1083.

LYUSTIKH, S.N. The Hypothesis of Differentiation of the Earth's Mantle and Geotectonic Generalizations. – *Sovetskaya Geologiya,* No. 6, 1961. (Russian)

LYUSTIKH, S.N. Neomobilism and Convection in the Earth's Mantle. – *Byull. Moskov. Obshch. Ispyt. Prirody,* Vol. 40 (1965), 5–27. (Russian)

MacGINITIE, H.D. A Middle Eocene Flora from the Central Sierra Nevada. – *Publs. Carnegie Instn.,* 534 (1941), 178 pp.

MacGINITIE, H.D. Fossil Plants of the Florissant Beds, Colorado. – *Publs. Carnegie Instn.,* 599 (1953), 1–198.

McALESTER, A.L. Animal Extinctions, Oxygen Consumption and Atmospheric History. – *J. Paleont.* **44**, No. 3 (1970), 405–409.

McANDREWS, J.H. Postglacial History of Prairie, Savanna and Forest in Northwestern Minnesota. – *Mem. Torrey Bot. Club,* Vol. 22 (1966), 1–72.

McELHINNY, M.W. and G.R. LUCK. Paleomagnetism and Gondwanaland. – *Science* **168**, No. 3933 (1970), 830–832.

McINTOSH, R.P. The Continuum Concept of Vegetation. – *Bot. Rev.,* Vol. 33, No. 2, 1967.

McKEE, E.D. Inorganic Sedimentary Structures. – In: *Approaches to paleoecology.* New York–London–Sydney, 1964.

McLEROY, C.A. and R.Y. ANDERSON. Lamination of the Oligocene Florissant Lake Deposits, Colorado. – *Bull. Geol. Soc. Am.,* Vol. 77 (1966), 605–618.

McNAUGHTON, S.J. Relationships among Functional Properties of Californian Grassland. – *Nature (Engl.),* Vol. 216, No. 5111, 1967.

MAEKAWA, F. Historical Sketch on the Theories Concerning the Stem-Leaf Complex. – *J. Japan Bot.* **40**, No. 9 (1965), 257–263.

MÄGDEFRAU, K. *Paläobiologie der Pflanzen.* – Jena, 1956. 443 pp.

MAGNITSKII, V.A. The Origin and Paths of Development of Continents and Oceans. – *Voprosy kosmogonii,* Vol. 6, 1958. (Russian)

MAHER, L.J. Pollen Analysis of Surface Materials from Southern San Juan Mountains, Colorado. – *Bull. Geol. Soc. Am.,* Vol. 74 (1963), 1485–1504.

MAHESHWARI, H.K. Studies in the *Glossopteris* Flora of India, 31. Some Remarks on the Genus *Glossopteris* Sternb. – *Palaeobotanist* **14**, Nos. 1–3 (1966), 36–45.

MAHESHWARI, P. and V. VASIL. *Gnetum.* – New Delhi, 1961. 142 pp.

MAL'GINA, E.A. and E.G. MAEV. Spore-Pollen Spectra of Bottom Sediments in the Caspian Sea. – *Izv. Akad. Nauk SSSR, geogr. ser.,* No. 2 (1966), 61–70. (Russian)

MALINOVSKII, Yu.M. and P.V. FLORENSKII. Cyclic Astronomic Phenomena in Relation to Climatic Fluctuations and the Rhythmicity of Sedimentary Strata. – In: *Problemy planetarnoi geologii.* Moscow, 1963. (Russian)

MALYSHEV, L.I. Relationship between Floristic Wealth and Environmental and Historical Factors. – *Bot. Zh.* **54**, No. 8 (1969), 1137–1147. (Russian)

MAMAEV, B.M. *Evolution of Gall-Formers – Midges.* – Leningrad, "Nauka," 1968. 236 pp. (Russian)

MANIA, D. and D. H. MAI. Warmzeitliche Mollusken und Pflanzenreste aus dem Mittelpleistozän des Geiseltals (südlich von Halle). – *Geologie* **18**, No. 6 (1969), 674–680.

MANTEN, A. A. Palynology and Environmental Geology. – *Paleogeogr. Paleoclimatol. Palaeoecol.,* Vol. 3 (1967), 7–15.

MARGOLIS, S. V. and J. P. KENNETT. Antarctic Glaciation during the Tertiary Recorded in Sub-Antarctic Deep-Sea Cores. – *Science* **170**, No. 3962 (1970), 1085–1087.

MARKOVSKII, B. P. *Methods of Biofacies Analysis.* – Moscow, "Nedra," 1966. (Russian)

MARTIN, A. R. H. Pollen Analysis of Groenvlei Lake Sediments, Knysna (South Africa). – *Rev. Palaeobot. Palynol.* **7**, No. 1 (1968), 107–148.

MASON, H. L. Evolution of Certain Floristic Associations in Western North America. – *Ecol. Monogr.* **17**, No. 2 (1947), 201–210.

MATHIEU, G. Reflexions sur le milieu biologique terrestre permo-carbonifère. – *Troisième congr. carbon. Heerlen,* 1951, pp. 443–452, 1952.

MATSUO, H. A Study on the Asuwa Flora (Late Cretaceous Age) in the Hokuriku District, Central Japan. – *Sci. Rep. Kanazawa Univ.* 8, No. 1 (1962), 177–250.

MATTHEW, W. D. Climate and Evolution. 2nd edition. – *Spec. Publs. N. Y. Acad. Sci.,* 1939, 223 pp.

MEEUSE, A. D. J. Again: the Growth Habit of the Early Angiosperms. – *Acta bot. neerl.* **16**, No. 2 (1967), 33–41.

MEHRA, P. H. and O. HANDOO. Morphology of *Anthoceros erectus* and *A. himalayensis* and the Phylogeny of the Anthocerotales. – *Bot. Gaz.* 114, No. 4 (1953), 371–382.

MEIEN, S. V. On Some Methods of Reconstructing the Ecology of Ancient Plants. – In: "*Voprosy paleogeograficheskogo raionirovaniya v svete dannykh paleontologii*"; *Proceedings 9th Session of All-Union Paleontological Society.* Moscow, "Nedra," 1967. (Russian)

MEIEN, S. V. The Age of the Ostrog Suite of the Kuznetsk Basin and Analogues of the Namurian in Continental Deposits of Northern Asia. – *Dokl. Akad. Nauk SSSR* **180**, No. 4 (1968), 944–947. (Russian)

MEIEN, S. V. The Continental Drift Hypothesis from the Viewpoint of Paleofloristics in the Carboniferous and Permian. – *Geotektonika,* No. 5 (1969), 3–16. (Russian)

MEIEN, S. V. Carboniferous Floras. – In: V. A. Vakhrameev et al. *Paleozoiskie i mezozoiskie flory Evrazii i fitogeografiya etogo vremeni,* pp. 43–110. Moscow, "Nauka," 1970a. (Russian)

MEIEN, S. V. Some Theoretical Problems of Recent Paleobotany. – *Paleontol. Zh.,* No. 4 (1970b), 3–15. (Russian)

MELCHIOR, R. C. and J. W. HALL. A Calamitean Shoot Apex from the Pennsylvanian of Iowa. – *Am. J. Bot.,* Vol. 48 (1961), 811–815.

MELVILLE, R. A New Theory of the Angiosperm Flower. – *Nature,* Vol. 188 (1960), 14–18.

MELVILLE, R. The Distribution of Land around the Tethys Sea and its Bearing on Modern Plant Distribution. – In: *Aspects Tethyan Biogeogr.*, pp. 291–312. London, 1967.

MELVILLE, R. Leaf Venation Patterns and the Origin of the Angiosperms. – *Nature*, Vol. 224 (1969), 121–125.

MENARD, H.W. Sediment Movement in Relation to Current Velocity. – *J. Sediment. Petrol.*, Vol. 20 (1950), 148–160.

MENÉNDEZ, C.A. Die fossilen Floren Südamerikas. – *Biogeogr. Ecol. S. Amer.*, Vol. 2, pp. 519–561. The Hague, 1969.

MENITSKII, Yu. L. Ancestors and Evolution of Caucasian and West Asian Oaks. – *Bot. Zh.* **54**, No. 11 (1969), 1675–1686. (Russian)

MENNER, V.V. *Biostratigraphic Principles of Correlating Marine, Lagoonal and Continental Formations.* – Moscow, Akad. Nauk SSSR, 1962. (Russian)

MERKER, H. Analyse der Rhynien – Basis und Nachweis des Gametophyten. – *Bot. Notizen*, Vol. 112 (1959), 441–452.

MERKER, H. Entwurf zur Lebenskreis-Rekonstruction der Psilophytales nebst phylogenetischem Ausblick. – *Bot. Notizen*, Vol. 114 (1961), 88–102.

MERKLIN, R.L. Life Forms and their Importance for Paleoecological Analysis. – In: "*Problemy paleontologii*," pp. 18–26. Moscow, "Nauka," 1968. (Russian)

MERKLIN, R.L. A Method for Paleosynecological Investigation of Geosynclinal Strata. – In: "*Biostratigrafiya, fauna i flora kainozoya severo-zapadnoi chasti Tikhookeanskogo podvizhnogo poyasa*," pp. 140–149. Moscow, "Nauka," 1969. (Russian).

MESERVEY, R. Topological Inconsistency of Continental Drift on the Present-Sized Earth. – *Science*, Vol. 166 (1969), 609–611.

MESOLELLA, K.J., R.K. MATTHEWS, W.S. BROECKER, and D.L. THURBER. The Astronomic Theory of Climatic Change: Barbados Data. – *J. Geol.* **77**, No. 3 (1969), 275–288.

MIKI, S. On the Change of Flora in Eastern Asia Since the Tertiary Period. 1. The Clay or Lignite Beds Flora in Japan with Special Reference to the *Pinus trifolia* Beds in Central Hondo. – *Jap. J. Bot.* **11**, No. 3 (1941), 238–301.

MIRKIN, B.M. Criteria for Dominants and Determinants in the Classification of Phytocenoses. – *Bot. Zh.* **53**, No. 6 (1968), 767–778. (Russian)

MIRKIN, B.M. Successions of River Floodplain Vegetation. – *Bot. Zh.* **55**, No. 10 (1970), 1405–1418. (Russian)

MISRA, R. Form, Function and Factors in Ecology. – *J. Indian Bot. Soc.* **46**, Nos. 2–3 (1967), 143–153.

MITCHELL, A.H. and H.G. READING. Continental Margins, Geosynclines and Ocean Floor Spreading. – *J. Geol.* **7**, No. 6 (1969), 629–646.

MOE, D. A Pollen Analysis of an Occurrence of Elm in Beiarn, Nordland County, Northern Norway. – *Årbok Univ. Bergen. Matematisk-naturvitenskapelig serie*, No. 2, 1970.

MÖHN, E. Eine neue Gallmücke aus niederrheinischen Braunkohle, *Sequoiomyia krauseli* n.g., n. sp. (Diptera, Itonididae). – *Senckenberg. leth.*, Vol. 41 (1960), 513–522.

MOKEEVA, E. A. Procedures for Studying Leaf Anatomy. – *Izv. Akad. Nauk UzbSSR*, No. 1 (1948), 18–26. (Russian)

MONK, C. D. An Ecological Study of Hardwood Swamps in North-Central Florida. – *Ecology* **47**, No. 4 (1966), 649–654.

MOONEY, H. A. and E. L. DUNN. Convergent Evolution of Mediterranean Climate Evergreen Sclerophyll Shrubs. – *Evolution*, Vol. 24 (1970), 292–303.

MOORE, G. T. Interaction of Rivers and Oceans. Pleistocene Petroleum Potential. – *Bull. Am. Ass. Petrol. Geol.* **53**, No. 12 (1970), 2421–2430.

MOORE, P. D. and E. H. CHATER. – The Changing Vegetation of West-Central Wales in Light of Human History. – *J. Ecol.*, Vol. 57 (1969), 361–379.

MOREY, P. R. and E. D. MOREY. Senckenberg Lignite: a Lignified Wood with Apparently Original Cellulose and Lignin. – *Science* **164**, No. 3881 (1969), 836–838.

MORGAN, J. The Morphology and Anatomy of American Species of the Genus *Psaronius*. – *Illinois Biol. Monogr.*, Vol. 27, 1959.

MOWBRAY, Th. B. and H. J. OOSTING. Vegetation Gradients in Relation to Environment and Phenology in Southern Blud Ridge Gorge. – *Ecol. Monogr.* **38**, No. 4 (1968), 309–344.

MÜLLER, A. H. Die praekambrische Lebewelt. Erscheinung und Probleme. – *Biol. Rundsch.*, Vol. 2 (1964), 53–67.

MULLER, J. Palynology of Recent Orinoco Delta and Shelf Sediments: Reports of the Orinoco Shelf Expedition. – *Micropaleontology*, Vol. 5 (1959), 1–32.

NAIDIN, D. P. Isotopic Paleotemperatures Determined according to Belemnoids (Author's Summary of Reports Presented 9 April 1965). – *Byull. Moskov. Obshch. Ispyt. Prirody, Otdel Geol.* **40**, No. 6 (1965), 148–149. (Russian)

NAIRN, A. E. M. Uniformitarianism and Environment. – *Palaeogeogr. Palaeoclimatol. Palaeoecol.*, Vol. 1 (1965), 5–11.

NALIVKIN, D. V. *Theory of Facies*, Vols. 1–2. – Akad. Nauk SSSR, 1956. (Russian)

NALIVKIN, D. V. Flysch – Continental Deposits. – *Dokl. Akad. Nauk SSSR*, Vol. 141, No. 4, 1961. (Russian)

NATHORST, A. G. Zur fossilen Flora Japan's. – *Paläont. Abh.*, Vol. 4 (1888), 195–256.

NAUGHTON, S. J. and L. L. WOLF. Dominance and the Niche in Ecological Systems. – *Science* **167**, No. 3915 (1970), 131–139.

NAVALE, G. K. B. Palynological Studies of Merlabach Coal in Conjunction with Petrographic Structure. – *Palaeobotanist* **12**, No. 3 (1963, 1964), 232–249.

NEEDHAM, H. D., D. HABIB, and B. C. HEEZEN. Upper Carboniferous Palynomorphs as a Tracer of Red Sediment Dispersal Pattern in the Northwest Atlantic. – *J. Geol.* **77**, No. 1 (1969), 113–120.

NEGRU, A. G. Miocene Flora of Bursuk Village in Moldavia. – *Bot. Zh.* **59**, No. 11 (1968), 1727–1733. (Russian)

NEISHTADT, M. I. *History of Forests and Paleogeography of the USSR in the Holocene.* – Moscow, Akad. Nauk SSSR, 1957. (Russian)

NEKHOROSHEV, V. P. Main Hypotheses of Tectogenesis and their Contradictions. – In: *Materialy po obshchei i regional'noi tektonike; Trudy Vses. Nauchno-Issled. Geol. Inst., nov. ser.*, No. 85, 1963. (Russian)

NEMEJC, F. The Taxonomic Significance of the Term Psilopsida and the Evolutionary Trends of this Plant Phylum. – *Palaeobotanist,* Vol. 9 (1962), 17–20.

NEMEJC, F. – *Paläont. Abh.* **3**, Nos. 3–4 (1970), 553–562.

NEUHÄSL, R. Draft Proposals for Nomenclature Principles in Floristic Phytosociology. – *Folia geobot. phytotax. Bohemoslov.* **3**, No. 1 (1968), 47–55.

NEWELL, N. D. Revolutions in the History of Life. – In: Cl. C. Albritton (ed.). *Uniformity and Simplicity; Spec. Pap. Geol. Soc. Am.,* Vol. 89 (1967), 63–89.

NICHOLS, H. The Suitability of Certain Categories of Lake Sediments for Pollen Analysis. – *Pollen et spores* **9**, No. 3 (1967), 615–626.

NIKITIN, V. P. Quaternary Floras of Western Siberia (Seeds and Fruits). – In: *Istoriya razvitiya rastitel'nosti vnelednikovoi zony Zapadno-Sibirskoi nizmennosti v pozdnepliotsenovoe i chetvertichnoe vremya,* pp. 245–311. Moscow, "Nauka," 1970. (Russian)

NIKOLAEVSKII, V. S. Anatomico-Morphological Structure of the Leaves of Woody Plants in Connection with their Gas Resistance. – *Zap. Sverdl. Otdel. Vses. Bot. Obshch.,* No. 4 (1966), 115–120. (Russian)

NIKOLAEVSKII, V. G. Comparative Studies of Xeromorphous and Mesomorphous Characters in the Leaf Structure of Grasses. – *Bot. Zh.* **55**, No. 10 (1970), 1442–1450. (Russian)

NIKOLAEVSKII, V. G. and L. D. NIKOLAEVSKAYA. Anatomy of the Conducting Bundles in the Stems of Various Ecological Types of Grasses. – *Bot. Zh.* **53**, No. 9 (1968), 1226–1232. (Russian)

NITSENKO, A. A. Criteria for Identifying Plant Associations. – *Bot. Zh.,* Vol. 51, No. 8, 1966. (Russian)

NORIN, B. N. On Some Concepts of Phytocenology and Geobotany (in their Application to the Plant Cover of the Far North. – *Bot. Zh.* **53**, No. 9 (1968), 1286–1299. (Russian)

ODUM, E. P. *Fundamentals of Ecology.* – Philadelphia, 1967. 546 pp.

ODZAKI, K. Notes on the Transportation of Leaves of Some Plants by the Hayama and Sakawa Rivers. – *Scient. Rep. Yokohama Natn. Univ.* **2**, No. 15 (1969), 95–108. (Japanese)

OGDEN, J. Y. III. Radiocarbon and Pollen Evidence for a Sudden Change in Climate in the Great Lakes Region Approximately 10,000 Years Ago. – In: E. L. Cushing and H. E. Wright, Jr. (eds.). *Quart. Paleoecol., Proc. VII Congr. Intern. Assoc. Quart. Res.,* Vol. 7 (1967), 117–127.

OLIVER, E. W. Foreign Pollen in the Ovules of *Ginkgo* and of Fossil Plants. – *New Phytol.,* Vol. 14 (1915), 149–151.

OLTZ, D. F. Numerical Analyses of Palynological Data from Cretaceous and Early Tertiary Sediments in East Central Montana. – *Palaeontographica* **128 B**, Nos. 3–6 (1969), 90–166.

ORTLAM, D. Fossile Böden als Leithorizonte für die Gliederung des Höheren Buntsandsteins im nördlichen Schwarzwald und südlichen Odenwald. – *Geol. Jber.,* Vol. 84 (1967), 485–590.

OSZAST, J. Miocencka roślinność zloza siarkowego w Piasecznie kolo Tarnobrzega. – *Acta palaeobot., Cracow,* Vol. 8, No. 1, 1967.

OVERBECK, F. and H. SCHMITZ. Zur Geschichte der Moore, Marschen und Wälder N.W. Deutschlands. – *Mitt. Prov. Zeilst. Natur.,* Vol. 3, No. 1, 1931.

PAEPE, R. and R. VANCHOORNE. The Stratigraphy and Palaeobotany of the Late Pleistocene in Belgium. – *Mém. expl. cartes géol. et min. de Belgique,* No. 8, 1967. 96 pp.

PAIJMANS, K. An Analysis of Four Tropical Rain Forest Sites in New Guinea. – *J. Ecol.* **58**, No. 1 (1970), 77–101.

PALAMAREV, E. Xerotherme Elemente in der Tertiärflora Bulgariens und Aspekte zum Problem der Formierung der mediterranen Flora aus der Balkanhalbinsel. – *Abh. zentr. geol. Inst.,* No. 10 (1967), 165–175.

PÁLFALVY, L. Mangrovepáfránny a Solymári-mendencéböl. – *Évi Jelent. magy. K. földt. Intéz.,* pp. 349–353. Budapest, 1966.

PALMER, A.R. Biomere – a New Kind of Biostratigraphic Unit. – *J. Paleont.,* Vol. 431, No. 3, 1965.

PANT, D.D. The Gametophyte of the Psilophytales. – *Proc. Summer School Bot. Darjeeling,* pp. 276–301, 1962.

PANT, D.D. On the Stem and Attachment of *Glossopteris* Leaves. – *Phytomorphology* **17**, Nos. 1–4 (1967), 351–359.

PANT, D.D. and D.D. NAUTIYAL. Some Seeds and Sporangia of *Glossopteris* Flora from Raniganj Coalfield, India. – *Palaeontographica,* Vol. 107 B (1960), 41–63.

PANT, D.D. and R.S. SINGH. The Structure of *Vertebraria indica* Royle. – *Palaeontology* **11**, Pt. 5 (1968), 643–653.

PARKIN, J. The Durian Theory – a Criticism. – *Phytomorphology,* Vol. 3 (1953), 80–88.

PEIVE, A.V. Tectonics and Magmatism. – *Izv. Akad. Nauk SSSR, geol. ser.,* No. 3, 1961. (Russian)

PEIVE, A.V. Aspects of Recent Tectonics. – *Izv. Akad. Nauk SSSR, geol. ser.,* No. 7 (1962), 3–9. (Russian)

PENNINGTON, W. Lake Sediments: Pollen Diagrams from the Bottom Deposits of the North Basin of Windermere. – *Phil. Trans. R. Soc.,* Vol. 233 B, 1947.

PERMYAKOV, A.I. Features of the Formation of Spore-Pollen Spectra of Recent Continental Sedimentary Deposits as Exemplified by the Yenisei River Basin. – *Trudy Inst. Geol. i Geofiz. Sibirsk. Otdel. Akad. Nauk SSSR,* No. 25, 1964. (Russian)

PERMYAKOV, A.I. Spore-Pollen Spectra of Surface Samples and the Composition of the Plant Cover in the Lower Tunguska River Area. – *Vest. Moskov. Gos. Univ., Biol. i Pochvoved.,* No. 6, 1966. (Russian)

PERVUKHINA, N.V. *Aspects of the Morphology and Biology of a Flower.* – Leningrad, "Nauka," 1970. 168 pp. (Russian)

PETTITT, J. Heterospory and the Origin of the Seed Habit. – *Biol. Rev.,* Vol. 45 (1970), 401–416.

PHILPOTT, J. Blade Tissue Organization of Foliage Leaves of Some Carolina Shrub-Bog Species as Compared with their Appalachian Mountain Affinities. – *Bot. Gaz.* **118**, No. 2 (1956), 88–105.

PIEL, K.M. Palynology of Oligocene Sediments from Central British Columbia. – *Can. J. Bot.,* Vol. 49, No. 11, 1971.

PIERART, P. Les associations de microspores et de mégaspores dans une couche (couche 70 de Beeringen) du Westphalien a supérieur de la Campine (Belgique). – *Rev. Palaeobot. Palynol.* 7, No. 4 (1968), 275–283.

PIRES, J.M., Th. DOBZHANSKY, and G.A. BLACK. – An Estimate of the Number of Species of Trees in an Amazonian Forest Community. – *Bot. Gaz.* **114**, No. 4 (1953), 467–477.

PLUMSTEAD, E.P. The Habit of Growth of Glossopteridae. – *Trans. Geol. Soc. S. Afr.,* Vol. 61, 1958.

POHL, Fr. Die Pollenerzeugung der Windblütler. – *Beih. Bot. Zbl.,* 56/A (1937a), 365–470.

POHL, Fr. Die Pollenkorngewichte einiger windblütiger Pflanzen und ihre ökologische Bedeutung. – *Beih. Bot. Zbl.,* 57/A (1937b), 112–172.

POLYAKOVA, I.D. *Lithological and Geochemical Characteristics of Cretaceous and Paleogene Deposits of the West Siberian Plain in Connection with the Reconstruction of Paleolandscapes.* – Author's Summary of Candidate Thesis. Rostov-on-Don, 1967. (Russian)

POORE, M.E.D. Studies in Malaysian Rain Forest, I. – *J. Ecol.* **56**, No. 1 (1968), 143–196.

POPLAVSKAYA, G.I. On the Distribution of Stomata in Plants. – *Dokl. Akad. Nauk SSSR, nov. ser.* **66**, No. 5 (1949), 949–952. (Russian)

POPOV, M.G. A Survey of the Vegetation and Flora of the Carpathians. – *Moskov. Obshch. Ispyt. Prirody,* 1949. (Russian)

POTONIE, H. Historisches zur Frage nach der Genesis der Steinkohle. – *Naturw. Wschr.,* No. 7 (1907a), 113–117.

POTONIE, H. Ein von der Holländisch-Indischen Sumatra Expedition entdecktes Tropen-Moor. – *Naturw. Wschr.,* No. 42 (1907b), 657–666.

POTONIE, R. Zur Mikrobotanik des eozänen Humolites des Geiseltals. – *Arb. Inst. Paläobot.,* Vol. 4 (1934), 25–125.

POTONIE, R. Moorpflanzengesellschaft des Karbons und der Rhythmus ihrer Wandlungen. – *Paläont. Z.* **24**, Nos. 3–4 (1951), 166–183.

POTONIE, R. Die Bedeutung der Sporomorphen für die Gesellschaftsgeschichte. – *C. r. Troisième congr. des études de stratigr. et de géol. du Carbon.* Heerlen, 1951, Vol. 2, 1952.

POTONIE, R. Zur Paläobiologie der karbonischen Pflanzenwelt. Ökologische Zeugnisse der Karbonflora zur Kohlenbiogonie. – *Naturwissenschaften* **40**, No. 4 (1953), 119–127.

PRESTON, F.W. The Commonness, and Rarity, of Species. – *Ecology,* Vol. 29 (1948), 254–283.

PRINADA, V.D. *The Mesozoic Flora of Eastern Siberia and Transbaikalia.* – Moscow, Gosgeoltekhizdat, 1962. (Russian)

PROCTOR, V.W. Long-Distance Dispersal of Seeds by Retention in Digestive Tract of Birds. – *Science* **160**, No. 3825 (1968), 321–322.

PROKHANOV, Ya.I. Evolution of the Leaf of Woody Dicots. – In: "*Problemy filogenii rastenii,*" pp. 71–110. "Nauka," 1965. (Russian)

PROKHOROVA, K.V. Comparison of the Composition of Recent Vegetation with Subfossil Spore-Pollen Spectra in the Northern Taiga. – *Bot. Zh.* **50**, No. 5 (1965), 626–637. (Russian)

PROSKAUER, J. Studies on Anthocerotales, 4. – *Phytomorphology*, Vol. 10 (1960), 1–19.

PURI, G. S. The Occurrence of *Woodfordia fruticosa* (L.) S. Kurz. in Karewa Deposits of Kashmir, with Remarks on Changes of Altitude and Climate during the Pleistocene. – *J. Indian Bot. Soc.,* Vol. 22 (1943), 125–131.

PURSEGLOVE, J.W. The Origin and Distribution of the Coconut. – *Trop. Sci.* **10**, No. 4 (1968), 191–199.

P'YAVCHENKO, N.I. Reflection of the Recent Composition of Forests in Recent Pollen Spectra. – *Bot. Zh.* **52**, No. 2 (1968), 174–189. (Russian)

RADCHENKO, G.P. Criteria and Methods for Paleogeographic Reconstructions of Former Conditions in an Area of Ancient Land Based on Paleontological Data. – In: *Metody paleogeograficheskikh issledovanii.* Leningrad, 1963. (Russian)

RADWANSKI, A. and P. RONIEWICZ. General Remarks on the Ichnocoenose Concept. – *Bull. Acad. pol. Sci. Cl. II. Sér. Sci. biol.* **18**, No. 1 (1970), 51.

RAHN, P.H. The Relationships between Natural Forested Slopes and Angles of Repose for Sand and Gravel. – *Bull. Geol. Soc. Am.,* Vol. 80 (1969), 2122–2128.

RAPOPORT, A. Different Approaches to the General Systems Theory. – In: "*Sistemnye issledovaniya,*" pp. 55–79. Moscow, "Nauka," 1969. (Russian)

RAUNKIAER, C. Statistik der Lebensformen als Grundlagen für die biologische Pflanzengeographie. – *Beih. Bot. Zbl.,* Vol. 27 (1910), 171–206.

RAUNKIAER, C. The Use of Leaf Size in Biological Plant Geography. – *Bot. Tidsskr.,* Vol. 34, 1916.

RAUNKIAER, C. *The Life Forms of Plants and Statistical Plant Geography.* – Oxford, 1934. 631 pp.

RAUNKIAER, C. *Plant Life Forms.* – Oxford, 1937.

RAYUSHKINA, G.S. Ecology of the Fossil Flora of the Southern Altai (Bukhtarma River). – *Synopses of Reports of the 1st All-Union Conference on Paleobiogeochemistry and Paleoecology.* Baku, 1969. (Russian)

REICHEL, W. Zyklische Sedimentation und ihre Ursachen in Unterrotliegenden des Döhlener Beckens bei Dresden. – *Geologie* **17**, No. 8 (1968), 875–884.

REID, E.M. and M.E.J. CHANDLER. The Bembridge Flora. Catalogue of Cainozoic Plants in the Department of Geology, Vol. 1. – *Bull. Br. Mus. Nat. Hist. A. Geology,* 1926.

REMPE, H. Untersuchungen über die Verbreitung des Blütenstaubs durch die Luftströmungen. – *Planta,* Vol. 27 (1937), 93–147.

REMY, W. Bemerkungen zur Stratigraphie der Ilfelder kohlenführenden Serie und Mitteilungen über einen Wierbeltierfährtenrest, *Ichnium* sp. – *Geologica, Berl.,* No. 11, 1952.

REMY, W. Die Rotliegendflora Thüringens und ihre Stellung im Gesamtbild der Perm – Karbon-Flora. – *Hallesches Jb. mitteldt. Erdgesch.* **2**, No. 1 (1953), 28–31.

REMY, W. Laubfusit, ein Beitrag zur Frage der Fusitbildung. – *Glückauf* **90**, Nos.1–2 (1954), 64–67.

REMY, W. and R. REMY. Die Verteilung der Pflanzenfossilien auf die Flöze des Plöts-Löbejüner Stefan. – *Z. angew. Geol.* **4**, No. 11 (1958), 522–524.

RENSCH, B. *Neuere Probleme der Abstammungslehre. Die transspezifische Evolution.* – Stuttgart, 1954. 436 pp.

RICHARD, P. Un spectre pollinique type de la sapinière à bouleau blanc pour la forêt Montmorency. – *Naturaliste can.* **95**, No. 2 (1968), 565–576.

RICHARDSON, J. B. Some British Lower Devonian Spore Assemblages and their Stratigraphic Significance. – *Rev. Palaeobot. Palynol.* **1**, Nos. 1–4 (1967), 111–129.

RIDLEY, H. N. *Dispersal of Plants throughout the World.* – London, 1930.

RIEK, E. F. Lower Cretaceous Fleas. – *Nature,* Vol. 227 (1970), 746–747.

RITCHIE, G. A. *Ginkgo*: Window to the Past. – *Natn. Pks. Mag.* **43**, No. 257 (1969), 12–14.

RITCHIE, J. C. and S. LICHTI-FEDOROVICH. Pollen Dispersal Phenomena in Arctic-Subarctic Canada. – *Rev. Palaeobot. Palynol.,* Vol. 3, Nos. 1–4, 1967.

ROBERTSON, T. Plant Control in Rhythmic Sedimentation. – *C. r. Troisième congr. des études de stratigr. et de géol. du Carbon.* Heerlen, 1951, Vol. 2 (1952), 515–521.

ROCK, C. Fossile Kryptogamen aus der eozänen Braunkohle des Geiseltales. – *Nova Acta Leopoldina, Neue Folge,* Vol. 3, No. 4, 1939.

ROSSIGNOL, M. Sédimentation récente dans la Mer Morte. – *Pollen et spores* **11**, No. 1 (1969), 17–38.

RUDAU, B. Über Creno-Paläontologie und ihre Bedeutung. Lebende Mikrofossilien aus paläozoischen Salzlagern. – *Geologie* **15**, No. 7 (1966), 881–885.

RUDWICH, M. J. S. The Infra-Cambrian Glaciation and the Origin of the Cambrian Fauna. – In: A. E. M. Nairn (ed.). *Problems in Palaeoclimatology.* London–New York–Sydney, Interscience Publications, 1963.

RUKHIN, L. B. *Fundamentals of General Paleogeography.* – Leningrad, Gostoptekhizdat, 1959. 259 pp. (Russian)

RUNCORN, S. K. Corals and the History of the Earth's Rotation. – *Sea Front.* **13**, No. 1 (1967), 4–12.

RUPKE, N. A. Sedimentary Evidence for the Allochthonous Origin of *Stigmaria,* Carboniferous, Nova Scotia. – *Bull. Geol. Soc. Am.,* Vol. 80 (1969), 2109–2114.

RYBNÍČEK, K. and E. RYBNÍČKOVÁ. The History of Flora and Vegetation on the Blatomire in Southeastern Bohemia, Czechoslovakia (Palaeoecological Study). – *Folia geobot. phytotax. Bohemoslov* **3**, No. 2 (1968), 117–142.

SAHNI, B. Foreign Pollen in Ovules of *Ginkgo* and of Fossil Plants. – *New Phytol.,* Vol. 14 (1915), 149–151.

SALISBURY, E. J. On the Causes and Ecological Significance of Stomatal Frequency, with Special Reference to the Woodland Flora. – *Phil. Trans. R. Soc.,* Vol. 216B (1928), 1–65.

SAMSONOV, S. K. On the Flora of the Arid Zone of the Upper Cretaceous Epoch. – In: "*Mezozoi neftegazonosnykh oblastei srednei Azii (biostratigrafiya i paleogeografiya),*" pp. 93–120. Moscow, "Nauka," 1967. (Russian)

SANGSTER, A. G. and H. M. DALE. A Preliminary Study of Differential Pollen Grain Preservation. – *Can. J. Bot.*, Vol. 39 (1961), 35–43.

SCHAFFER, L. and S. C. WILKE. The Ordination of Fossil Communities, an Approach to the Study of Species Interrelationships and Communal Structure. – *Pap. Mich. Acad. Sci., Arts and Lett.*, Vol. 50 (1965), 199–213.

SCHENKE, G. Die Einregelung von Calamitenachsen im I Flötzpaket des Döhlener Beckens. – *Geologie* **18**, No. 10 (1969), 1206–1213.

SCHIELDS, L. M. Leaf Xeromorphy. – *Bot. Rev.*, Vol. 16 (1950), 399–447.

SCHIMPER, A. F. W. *Pflanzengeographie auf physiologischer Grundlage.* – Jena, 1898.

SCHINDEWOLF, O. H. *Grundlagen der Palaeontologie.* – Stuttgart, 1950.

SCHLEGELMILCH, V. Rotfärbungen im Thüringer Schiefergebirge. – *Geologie* **17**, No. 2 (1968), 136–155.

SCHMIDT, F. H. Palynology and Meteorology. – *Rev. Palaeobot. Palynol.*, Vol. 3 (1967), 27–47.

SCHMIDT, H. *Einführung in die Paläontologie.* – Stuttgart, 1935.

SCHNEIDER, W. Cuticulae dispersae aus dem 2. Lausitzer Flöz und ihre fazielle Aussage. – *Freiberger ForschHft., Paläontol.*, C 222, 1969.

SCHOPF, J. M. Petrified Peat from a Permian Coal Bed in Antarctica. – *Science* **169**, No. 3942 (1970), 274–277.

SCHOPF, J. W. Precambrian Microorganisms and Evolutionary Events Prior to the Origin of Vascular Plants. – *Biol. Rev.*, Vol. 45 (1970), 319–352.

SCHUBERT, S. Zur Geschichte der Moore, Marschen und Wälder Nordwestdeutschlands, 2. – *Mitt. Prov. Stelle für Natur. Kampl. Hannover,* No. 4, 1933.

SCHWARZ, O. and H. WEIDE. Systematische Revision der Gattung *Sequoia* Endl. – *Feddes Repert.* **6**, No. 3 (1962), 159–192.

SCHWARZBACH, M. *Climates of the Past.* – Van Nostrand, 1963.

SCHWARZBACH, M. Das Klima des rheinischen Tertiärs. – *Z. dt. geol. Ges.*, Vol. 118 (1968a), 33–88.

SCHWARZBACH, M. Neurere Eiszeithypothesen. – In: *Eiszeitalter und Gegenwart; – Jber. dt. QuartärVerein.*, Vol. 19 (1968b), 250–261.

SCHWEGER, C. E. Pollen Analysis of Iola Bog and Paleoecology of the Two Creeks Forest Bed, Wisconsin. – *Ecology* **50**, No. 5 (1969), 859–868.

SCOTT, D. On the Occurrence of Seed-Like Fructifications in Certain Palaeozoic Lycopods. – *Ann. Bot.*, Vol. 14 (1900), 713–717.

SCOTT, D. H. *Evolution of Plants.* – Holt, 1926.

SEARCY, K. B. Variation in *Cercocarpus* in Southern California. – *New Phytol.* **68**, No. 3 (1969), 829–839.

SENYANINOVA-KORCHAGINA, M. V. *On the Victory of the Angiosperms from the Standpoint of Heliogeophysics.* – Leningrad, 1959. (Russian)

SEREBRYAKOV, I. G. Main Evolutionary Trends of Life Forms in the Angiosperms. – *Byull. Moskov. Obshch. Ispyt. Prirody, Otdel. Biol.* **60**, No 3 (1955), 71–91. (Russian)

SEREBRYAKOV, I. G. *Ecological Plant Morphology.* – Moscow, "Vysshaya Shkola," 1962. (Russian)

SEREBRYAKOV, I. G. and T. I. SEREBRYAKOVA. Life Forms of Angiosperms and their Evolution in Individual Systematic Groups. – *Bot. Zh.* **64**, No. 9 (1969), 1321–1325. (Russian)

SEREBRYAKOVA, T. I. *Shoot Morphogenesis and the Evolution of Life Forms of Grasses.* – Moscow, "Nauka," 1971. (Russian)

SERGEEV, A. N. Geological Conditions of Peat Bog Formation in the Central Parts of Western Siberia. – *Vest. Moskov. Gos. Univ., Otdel Geol.,* No. 2 (1969), 95–100. (Russian)

SEVAST'YANOV, V. F. Disproportionation of Chemical Elements in the Presence of Oxidation-Reduction Processes in Sediments of the Mediterranean. – *Litologiya i Poleznye Iskopaemye,* No. 1 (1968), 3–15. (Russian)

SEWARD, A. C. Catalogue of Mesozoic Plants in the Department of Geology, British Museum. – *The Wealden Flora, 2, Gymnospermae,* pp. 205–220. London, 1895.

SEWARD, A. C. *Fossil Plants,* Vols. 2, 3. – Cambridge, 1910, 1917.

SEWARD, A. C. *Plant Life through the Ages, 2nd ed.* – Hafner, 1933.

SHACKLETON, N. J. The Last Interglacial in the Marine and Terrestrial Records. – *Proc. R. Soc.* **174**, No. 1034 (1969), 135–154.

SHAKHOVA, O. V. and V. V. KOLPAKOV. Correlation of Spore-Pollen Spectra of Surface Samples with the Vegetation of Northern Yakutia. – *Byull. Moskov. Obshch. Ispyt. Prirody, Otdel Biol.* **71**, No. 6 (1966), 92–102. (Russian)

SHANTSER, E. V. *Alluvium of Plain Rivers of the Temperate Zone and its Importance for the Study of Patterns in the Structure and Formation of Alluvial Formations.* – Moscow, Akad. Nauk SSSR, 1951. 271 pp. (Russian)

SHARMA, B. D. Investigations on the Jurassic Flora of Rajmahal Hills, India, 5. – *Acta bot. hung.,* Vol. 14 (1968), 373–383.

SHATILOVA, I. I. Pollen and Spores from the Bottom Sediments of the Black Sea. – *Dokl. Akad. Nauk SSSR* **179**, No. 5 (1968), 1196–1199. (Russian)

SHCHEGOLEV, A. K. The Flora at the Permo-Carboniferous Boundary in the Donets Basin. – In: "*Geologiya uglenosnykh formatsii i stratigrafii karbona SSSR,*" pp. 234–243. Moscow, "Nauka," 1965. (Russian)

SHCHEGOLEV, A. K. Patterns in the Evolution of the Flora and Vegetation of the Equatorial-Subtropical Belt in the Carboniferous Period. – In: "*Tezisy dokl. k XII sessii Vses. paleont. obshchestva,*" pp. 59–61. Leningrad, 1966. (Russian)

SHEINMANN, Yu. M. *Platforms, Folded Zones, and Development of the Earth's Structures.* – Magadan, 1959. (Russian)

SHELYAG-SOSONKO, Yu. R. Classification of Phytocenological Types. – *Ukr. Bot. Zh.* **25**, No. 4 (1968), 3–10. (Ukrainian)

SHENNIKOV, A. P. *Plant Ecology.* – Moscow, 1950. (Russian)

SHILOV, M. P. On the Relationship between the Plant Cover and the Development of Segments of River Floodplains. – *Bot. Zh.* **54**, No. 2 (1969), 232–238. (Russian)

SHMITKHYUZEN, I. *General Vegetation Geography.* – Moscow, "Progress," 1966. 310 pp. (Russian)

SHNITNIKOV, A. V. On the Holocene History of Eurasian Lakes. – In: "*Ritmy i tsiklichnost' v prirode,*" pp. 51–58. Moscow, "Mysl'," 1970. (Russian)

SHOTWELL, J. A. Community Succession in Mammals of the Late Tertiary. – In: J. Imbrie and N. Newell (eds.). *Approaches to Paleoecology*, pp. 135–150. New York–London–Sydney, 1964.

SHUL'TS, S.S. Cretaceous Forests in the Lower Reaches of the Amu Darya. – In: "*Voprosy paleogeograficheskogo raionirovaniya v svete dannykh paleontologii*," pp. 30–34. Moscow, "Nedra," 1967. (Russian)

SHVETSOV, M.S. *Petrography of Sedimentary Rocks*. – Moscow, Gosgeoltekhizdat, 1958. 348 pp. (Russian)

SIEGEL, S.M. Evidence for the Presence of Lignin in Moss Gametophytes. – *Am. J. Bot.* **56**, No. 2 (1969), 175–179.

SIMPSON, G.G. The Climate during the Pleistocene Period. – *Proc. R. Soc. Edinb.*, Vol. 50, 1929–1930.

SIMPSON, G.G. Mammals and the Nature of Continents. – *Am. J. Sci.*, Vol. 241 (1943), 1–31.

SIMPSON, G.G. *Tempo and Mode in Evolution*. – Hafner, 1944.

SIMPSON, G.G. Holartic Mammalian Faunas and Continental Relationships during the Cenozoic. – *Bull. Geol. Soc. Am.*, Vol. 58 (1947), 613–688.

SIMPSON, G.G. Probabilities of Dispersal in Geologic Time. – *Bull. Am. Mus. Nat. Hist.*, Vol. 99 (1952), 163–176.

SIMPSON, G.G. The History of Life. – In: S. Tax (ed.). *Evolution after Darwin*, Vol. 1 (1960), 117–180.

SIMPSON, G.G. *The Geography of Evolution*. – Philadelphia–New York, 1965. 249 pp.

SINITSYN, V.M. *Introduction to Paleoclimatology*. Leningrad, "Nedra," 1967. 232 pp. (Russian)

SINITSYN, V.M. Geological Method of Reconstructing Past Climates. – *Vest. Leningrad Gos. Univ., geol. geogr. ser.*, Issue 1, No. 6 (1969), 55–69. (Russian)

SITÁR, V. Tertiärflora aus Umgebung von Levice. – *Acta geol. et geogr., Univ. Comen. Geol.*, No. 12 (1967), 155–162.

SITTER, L.U. de. *Structural Geology*. 2nd ed. – New York, McGraw-Hill, 1964.

SKOBLO, V.M. Fossil Lakes in Jurassic Igneous Strata of Western Transbaikalia. – In: *Mezozoiskie i kainozoiskie ozera Sibiri*, pp. 9–21. Moscow, "Nauka," 1968. (Russian)

SKOČEK, V. Fosfatisované rostlinné zbytky ve středočeském karbonu. – *Čas. Miner. Geol.* **14**, No. 2 (1969), 219–221.

SLADKOV, A.N. *Introduction to Spore-Pollen Analysis*. – Moscow, "Nauka," 1967. 270 pp. (Russian)

SMITH, A.G. Problems of Inertia and Threshold Related to Post-Glacial Habitat Changes. – *Proc. R. Soc.*, Vol. 161 B (1965), 331–342.

SMITH, A.H.V. The Sequence of Microspore Assemblages with Occurrence of Crassidurite in Coal-Seams. – *Geol. Mag.*, Vol. 94 (1957), 345–363.

SMITH, A.H.V. Palaeoecology of Carboniferous Peat Bogs. – *Nature*, Vol. 89 (1961), 744–756.

SMITH, A.H.V. and M.A. BUTTERWORTH. Miospores in the Coal Seams of the Carboniferous of Great Britain. – *Spec. Pap. in Paleontol.*, No. 1. London, 1967. 324 pp.

SMITH, D. L. The Evolution of the Ovule. – *Biol. Rev.*, Vol. 39 (1964), 137–159.

SOAD, S. I. and S. SAMI. Studies of Pollen and Spores Content of Nile Delta Deposits (Berenbal Region). – *Pollen et spores* **9**, No. 3 (1967), 467–480.

SOBOLEV, N. D. Sima. – *Geologiya i Geofizika,* No. 1, 1963. (Russian)

SOCHAVA, A. V. *Beds of the Cretaceous of Central Asia.* – Leningrad, "Nauka," 1968. (Russian)

SOCHAVA, V. B. Aspects Related to the Classification of Vegetation, Typology of Physicogeographic Facies, and Biogeocenoses. – In: "*Voprosy klassifikatsii rastitel'nosti*"; *Trudy Inst. Biol. Ural'sk Fil. Akad. Nauk SSSR,* No. 27, 1961. (Russian)

SOCHAVA, V. B. Model for a New Map of the World's Vegetation. – In: "*Geobotanicheskoe kartografirovanie,*" pp. 3–16. Moscow–Leningrad, "Nauka," 1964. (Russian)

SOCHAVA, V. B. Botanico-Geographic Relationships in the Amur River Basin. – In: "*Amurskaya taiga,*" pp. 5–15. Leningrad, "Nauka," 1969. (Russian)

SOÓ, R. de. The History of the Vegetation of Hungary. – *Collect. Bot.* **7**, No. 2 (1968), 1109–1120.

SORGENFREI, T. *Molluscan Assemblages from the Marine Middle Miocene of South Jutland and their Environments.* – Copenhagen, Peitzel, 1958. 503 pp.

SPACKMAN, W. and E. S. BARGHOORN. Coalification of Woody Tissue as Deduced from a Petrographic Study of Brandon Lignite. – In: *Coal Science Advances in Chemistry,* series 55; *Am. Chem. Soc. Wash.,* pp. 695–710, 1966.

SPECHT, R. L. A Comparison of the Sclerophyllous Vegetation Characteristics of Mediterranean Type Climates in France, California and Southern Australia, 1. Structure, Morphology and Succession. – *Aust. J. Bot.* **17**, No. 2 (1969), 269–276.

SRIVASTAVA, S. K. Palynology of Late Cretaceous Mammal Beds, Scollard, Alberta (Canada). – *Paleogeogr. Paleoclimatol. Palaeoecol.* **3**, No. 1 (1967), 133–150.

SRIVASTAVA, S. K. Pollen Biostratigraphy and Paleoecology of Edmonton Formation (Maestrichtian), Alberta, Canada. – *Palaeogeogr. Palaeoclimatol. Palaeoecol.,* Vol. 7, No. 3, 1970.

STACE, C. A. The Use of Epidermal Characters in Phylogenetic Considerations. – *New Phytol.* **65**, No. 3 (1966), 304–318.

STAHL, W. and R. JORDAN. General Considerations on Isotopic Paleotemperature Determinations and Analyses on Jurassic Ammonites. – *Earth Planet. Sci. Lett.* **6**, No. 3 (1969), 173–178.

STANLEY, E. A. Abundance of Pollen and Spores in Marine Sediments off the Eastern Coast of the United States. – *Southeastern. Geol.,* Vol. 7 (1965), 25–33.

STANLEY, E. A. The Problem of Reworked Pollen and Spores in Marine Sediments. – *Mar. Geol.,* Vol. 4 (1966), 397–408.

STEBBINS, G. L. Evidence on Rates of Evolution from the Distribution of Existing and Fossil Plant Species. – *Ecol. Monogr.,* Vol. 17 (1947), 149–158.

STEBBINS, G. L. The Probable Growth Habit of the Earliest Flowering Plants. – *Ann. Mo. Bot. Gdn.,* Vol. 52 (1965), 457–468.

STEBBINS, G. L. *Variation and Evolution in Plants.* – New York–London, 1967. 643 pp.

STEENSTRUP, J. Geognostisk-geologisk Undersögelse af Skovmoserne Vidnesdam og Lill-mose idet nordliche Sjelland. – *Vid. Selsk. naturvid. og. math. Afd. Kjöbenhavn,* Vol. 9 (1842), 17–120.

STEPANOV, S. A. Paleobotanical Description of Devonian Deposits in Southern Siberia. – In: *"Materialy po regional'noi geologii Sibiri,"* pp. 73–78. Novosibirsk, 1969. (Russian)

STEPHENS, S. G. The Potentiality for Long-Range Oceanic Dispersal of Cotton Seeds. – *Am. Nat.* **100**, No. 912 (1966), 199–210.

STEUBING, L. Grundlagen und Entwicklungstendenzen der Pflanzenökologie. – *Biol. Rundsch.* **5**, No. 5 (1967), 193–203.

STEWART, W. *An Upward Outlook in Plant Morphology,* Vol. 14, No. 1, 1964.

STIDD, B. M. and J. W. HALL. The Natural Affinity of the Carboniferous Seed, *Callospermarion.* – *Am. J. Bot.* **57**, No. 7 (1970), 827–836.

STOCKMANS, F. Végétaux Eodévoniens de la Belgique. – *Mém. Mus. r. Hist. nat. Belg.,* 93, 1940.

STOCKMANS, F. and Y. WILLIERE. Flores anciennes et climaties. – *Naturalistes belg.* **44**, No. 7 (1963), 317–340.

STONE, E. C. and R. B. VASEY. Preservation of Coast Redwood on Alluvial Flats. – *Science* **159**, No. 3811 (1968), 157–161.

STUIVER, M. Tree Ring, Varve and Carbon-14 Chronologies. – *Nature,* Vol. 228 (1970), 454–455.

SUKACHEV, V. N. Principles of Genetic Classification and Biogeocenology. – *Zhurn. Obshch. Biol.,* Vol. 5, No. 4, 1944. (Russian)

SUKACHEV, V. N. Main Concepts in Forest Biogeocenology. – In: *Osnovy lesnoi biogeotsenologii.* Moscow, "Nauka," 1964. (Russian)

SUKACHEV, V. N., V. T. SOKOLOVSKAYA, and I. A. BANNIKOVA. New Data on the Likhvin Flora in the Vicinity of Moscow. – In: *"Istoriya razvitiya rastitel'nogo pokrova tsentral'nykh oblastei evropeiskoi chasti SSSR v antropogene,"* pp. 22–44. Moscow, "Nauka," 1968. (Russian)

SUKHOV, S. V. *Seeds of Late Paleozoic Plants of Middle Siberia.* – Leningrad, "Nauka," 1969. 264 pp. (Russian)

SULLIVAN, H. J. Palynologic Evidence Concerning the Regional Differentiation of Upper Mississippian Floras. – *Pollen et spores* **7**, No. 3 (1965), 539–563.

SULLIVAN, H. J. Regional Differences in Mississippian Spore Assemblages. – *Rev. Palaeobot. Palynol.* **1**, Nos. 1–4 (1967), 185–192.

SURANGE, K. R. and H. K. MAHESHWARI. Studies in the *Glossopteris* Flora of India. 11. Some Observations on *Vertebraria* from the Lower Gondwana of India. – *Palaeobotanist,* Vol. 9 (1962), 61–67.

SWAIN, F., J. M. BRATT, and S. KIRKWOOD. Carbohydrate Components of Upper Carboniferous Plant Fossils from Radstock, England. – *J. Paleont.* **43**, No. 2 (1969), 550–553.

SZAFER, W. The Significance of Isopollen Lines for the Investigation of the Geographical Distribution of Trees in the Postglacial Period. – *Bull. Acad. pol. Sci. III. Math. Nat.,* ser. B, pp. 235–239, 1935.

SZAFER, W. Flora pliocenska z Kroscirnka nad Dunajcem. – *Rozpr. Wydz. mat.-przyr. pol. Akad. Umiejętn.*, Vol. 72, 1946.

SZAFER, W. Miocene Flora from Stare Gliwice in Upper Silesia. – *Trav. Serv. géol. Pol.*, Vol. 33 (1961), 205 pp.

SZAFRANSKY, F. Zmiany roślinności w świetle analizy pylkowej. – *Folia quatern.*, Vol. 29 (1968), 41–48.

TAGEEVA, N. V. Deltas as Indicators of Tectonic Conditions. – *Byull. Moskov. Obshch. Ispyt. Prirody, Otdel. Geol.*, Vol. 43, No. 2, pp. 16–35. (Russian)

TAKHTADZHYAN, A. L. *Aspects of the Evolutionary Morphology of the Higher Plants.* – Leningrad, Akad. Nauk SSSR, 1954. (Russian)

TAKHTADZHYAN, A. L. *Fundamentals of the Evolutionary Morphology of the Angiosperms.* – Leningrad, "Nauka," 1964. 236 pp. (Russian)

TANAI, T. and T. ONOE. A Mio-Pliocene Flora from the Ningyo-Toge Area on the Border between Tottori and Okayama Prefectures, Japan. – *Rep. Geol. Surv. Japan,* No. 187, 1961. 59 pp.

TANSLEY, A. G. The Use and Abuse of Vegetational Concepts and Terms. – *Ecology,* Vol. 16, 1935.

TAO JUN-RONG. A Late Eocene Florula from the District Weinan of Central Shensi. – *Acta bot. sin.* **13**, No. 3 (1965), 272–277.

TAUBER, H. Differential Pollen Dispersion and the Interpretation of Pollen Diagrams. – *Danm. geol. Unders.,* Vol. 2 (1965), 1–69.

TAUBER, H. Investigations of the Mode of Pollen Transfer in Forest Areas. – *Rev. Palaeobot. Palynol.,* Vol. 3 (1967a), 277–286.

TAUBER, H. Differential Pollen Dispersion and Filtration. – In: E. J. Cushing and H. E. Wright, Jr. (eds.). *Quart. Paleoecol., Proc. VII Congr. Intern. Assoc. Quart. Res.,* Vol. 7 (1967b), 131–141.

TAYLOR, J. H. Sedimentary Features of an Ancient Deltaic Complex: the Wealden Rocks of Southeastern England. – *Sedimentology,* Vol. 2 (1962), 2–28.

TAYLOR, T. N. Paleozoic Seed Structures: a Monograph of the American Species of *Pachytesta.* – *Palaeontographica,* Vol. 117 B (1965), 1–46.

TEICHMÜLLER, M. Vergleichende mikroskopische Untersuchungen versteinerter Torfe des Ruhrkarbons und der daraus entstandenen Steinkohlen. – *C. r. Troisième congr. des études de stratigr. et de géol. du Carbon.* Heerlen, 1951, Vol. 2, 1952.

TEICHMÜLLER, M. Rekonstruktionen verschiedener Moortypen des Hauptflözes der Niederrheinischen Braunkohle. – *Fortschr. Geol. Rheinld. Westf.,* Vol. 2 (1958), 599–612.

TERASMAE, J. Recent Pollen Distribution in the Northeastern District of Mackenzie; Northwest Territories (Canada). – *Palaeogeogr. Palaeoclimatol. Palaeoecol.,* Vol. 3 (1967), 17–28.

TETERYUK, V. K. Compositional Features and the Spore-Pollen Ratio in Coal Seams, Soil, and the Top of Layers. – In: *"Problemy palinologii, No. 1."* Kiev, "Naukova Dumka," 1971. (Russian)

THODAY, D. The Significance of Reduction in the Size of Leaves. – *J. Ecol.,* Vol. 19 (1931), 297–303.

THOM, B.G. Mangrove Ecology and Deltaic Geomorphology. – *J. Ecol.* **55**, No. 2 (1967), 301–343.

THOMAS, B.A. The Cuticle of the *Lepidodendron* Stem. – *New Phytol.* **65**, No. 3 (1966), 296–303.

THOMAS, H.H. and T.M. HARRIS. Cycadean Cones of Yorkshire Jurassic. – *Senckenberg. leth.*, Vol. 41 (1960), 136–161.

THOMSON, P.W. Grundsätzliches zur tertiären Pollen- und Sporenmikrostratigraphie auf Grund einer Untersuchung des Hauptflözes der rheinischen Braunkohle in Libar, Neurath, Fortuna und Brühl. – *Geol. Jber.*, Vol. 65 (1949), 113–126.

TIMOFEEV, P.P. *A Jurassic Coal-Bearing Formation of Southern Siberia and the Conditions of its Formation.* – Moscow, "Nauka," 1970. (Russian)

TIMOFEEV, P.N. and L.N. BOGOLYUBOVA. The Degree of Decomposition of Plant Material as a Factor of the Tectonic Regime in an Area of Peat Accumulation. – *Dokl. Akad. Nauk SSSR,* Vol. 144, No. 4, 1962. (Russian)

TITOV, I.A. *Interaction between Plant Communities and the Environment.* – Moscow, 1952. (Russian)

TKACHENKO, V.F. Conditions of Formation of Coal Measures in the Donets Basin. – *Geol. Zh.* **28**, No. 5 (1968), 34. (Russian)

TOKUNAGA, S. Fossil Pollen and Spores in Japanese Coals. – *J. Fuel Soc. Japan* **44**, No. 464 (1965), 792–796. (Japanese)

TOLMACHEV, A.I. Incompleteness of the Geological Record. – *Vest. Leningrad. Gos. Univ.,* No. 9, 1961. (Russian)

TOLMACHEV, A.I. *Principles of Studying Geographic Ranges.* – Leningrad. Gos. Univ., 1962. 100 pp. (Russian)

TOLMACHEV, A.I. Geographical Patterns of Evolution in the Recent and Remote Geological Past (Surveys on the Biogeography of Bygone Geological Periods). – In: *"Voprosy paleobiologii i biostratigrafii,"* pp. 25–55. Moscow, Gosgeoltekhizdat, 1969. (Russian)

TOURTELOT, H.A. and R.O. RYE. Distribution of Oxygen and Carbon Isotopes in Fossils of Late Cretaceous Age, Western Interior Region of North America. – *Bull. Geol. Soc. Am.* **80**, No. 10 (1969), 1903–1922.

TOWNROW, J. On *Lepidopteris madagascarensis* Carpentier (Peltaspermaceae). – *J. Proc. R. Soc. N.S.W.,* Vol. 98 (1966), 203–214.

TRAVERSE, A. and R.N. GINSBURG. Pollen and Associated Microfossils in the Marine Surface Sediments of the Great Bahama Banks. – *Rev. Palaeobot. Palynol.* **3**, Nos. 1–4 (1967), 243–254.

TROOP, J.E. and J.T. MICKEL. Petiolar Shoots in the Dennstaedtioid and Related Ferns. – *Am. Fern J.* **58**, No. 2 (1968), 64–70.

TRYBOM, F. Bottenprof fran svenska insiöar. – *Geol. För. Stockh. Förh.,* Vol. 10, 1888.

TSUKADA, M. Untersuchungen über der Verhältnis zwischen dem Pollengehalt der Oberflächenproben und der Vegetation des Hohlandes Shiga. – *J. Inst. Polytech. Osaka Cy. Univ.,* Vol. 9D (1958), 217–234.

TSUKADA, M. Pollen Succession, Absolute Pollen Frequency and Recurrence Surfaces in Central Japan. – *Am. J. Bot.* **54**, No. 7 (1967a), 821–831.

TSUKADA, M. Vegetation in Subtropical Formosa during the Pleistocene Glaciations and the Holocene. – *Palaeogeogr. Palaeoclimatol. Palaeoecol.* **3**, No. 1 (1967b), 49–64.

TSVELEV, N. N. Some Aspects of the Evolution of the Plant Kingdom and the "Expanding" Earth Hypothesis. – *Byull. Moskov. Obshch. Ispyt. Prirody, Otdel Biol.* **74**, No. 4 (1969), 27–36. (Russian)

TURNAU, E. Mikroflora i paleogeografia karbonu produktywnego w polskiej części Karpat. – *Biul. Inst. Geol.,* No. 235 (1970), 163–244.

TURNER, C. and R. WEST. The Subdivision and Zonation of Interglacial Periods. – In: *Eiszeitalter und Gegenwart,* Vol. 19 (1968), 93–101.

TURUTANOVA-KETOVA, A. I. Some Jurassic Seeds and Flowers from Middle Asia and South Kazakhstan. – In: *Voprosy paleontologii,* Vol. 1, 1950. (Russian)

TWENHOFEL, W. H. *Treatise on Sedimentation,* 1926.

UCHUPI, E. and G. F. JONES. Woody Debris on the Mainland Shelf of Ventura, Southern California. – *Sedimentology* **8**, No. 2 (1967), 147–151.

VAKHRAMEEV, V. A. The Role of Geological Conditions in the Development and Distribution of Angiosperm Floras in Cretaceous Time. – *Byull. Moskov. Obshch. Ispyt. Prirody, Otdel Geol.,* Vol. 22, No. 6, 1947. (Russian)

VAKHRAMEEV, V. A. Jurassic and Early Cretaceous Floras. – In: V. A. Vakhrameev et al. *Paleozoiskie i mezozoiskie flory Evrazii i fitogeografiya etogo vremeni,* pp. 213–281. Moscow, "Nauka," 1970. (Russian)

VALENTINE, J. M. Biogeographic Units as Biostratigraphic Units. – *Bull. Am. Ass. Petrol. Geol.* **47**, No. 3, Pt. 1 (1963), 457–466.

VALENTINE, J. W. Niche Diversity and Niche Size Patterns in Marine Fossils. – *J. Paleont.,* Vol. 43, No. 4, 1969.

VAL'TS, I. E. Primary and Diagenetic Changes of the Microstructure of Plant Material in the Peat and Lignite Stages. – In: "*Voprosy metamorfizma uglei i epigeneza vmeshchayushchikh porod,*" pp. 15–25. Moscow, "Nauka," 1968. (Russian)

VAN HOUTEN, F. B. Origin of Red Beds – Some Unsolved Problems. – In: *Problems in Palaeoclimatology,* ed. by A. E. M. Nairn. Wiley, Interscience, 1964.

VAN STEENIS, C. G. J. The Distribution of Mangrove Plant Genera and its Significance for Palaeogeography. – *Proc. K. ned. Akad. Wet.* **65C**, No. 2 (1962), 164–169.

VAN STRAATEN, L. M. Sedimentation in Tidal Flat Areas. – *J. Alberta Petrol. Geologists,* Vol. 9 (1960), 203–226.

VASILEVSKAYA, V. K. Anatomical Features of Some Plants of Central Kazakhstan. – *Trudy Bot. Inst. Akad. Nauk SSSR,* ser. 3, No. 3 (1965), 125–134. (Russian)

VASIL'EV, I. V. On the Flora and Vegetation of Western Siberia and Kazakhstan in the Paleogene (according to Ichnophytological Data). – *Trudy Vses. Nauchno-Issled. Geol. Inst.,* Vol. 130 (1969), 168–178. (Russian)

VASILEVICH, V. I. and V. S. IPATOV. Some Structural Features of Supraorganismic Systemic Levels. – *Zhurn. Obshch. Biol.* **30**, No. 6 (1969), 643–651. (Russian)

VASSOEVICH, N. B. Stratigraphy in the Light of Studies on Sedimentary Differentiation. – *Izv. Akad. Nauk SSSR, ser. geol.,* No. 5, 1960. (Russian)

VERNADSKII, V. I. The Biosphere. – Moscow, "Mysl'," ser. geol., No. 11, pp. 3–18. 1967. (Russian)

VERZILIN, N. N. On a New Hypothesis of Red Bed Formation, Proposed by A. V. Sochava. – *Litologiya i Poleznye Iskopaemye,* No. 3 (1970), 119–122. (Russian)

VIKHREEVA-VASIL'KOVA, V. V. Morphological and Anatomical Features of Plants of the Arctic Tundra. – In: *"Rastitel'nost' Krainego Severa i ee osvoenie,"* No. 8, pp. 126–165. Moscow–Leningrad, "Nauka," 1966. (Russian)

VINOGRADOV, A. P. Origin of the Earth's Envelopes. – *Izv. Akad. Nauk SSSR,* No. 11, 1962. (Russian)

VISHNU-MITTRE. Studies on the Fossil Flora of Nipania (Rajmahal Series), Bihar, Coniferales. – *Palaeobotanist* **6**, No. 2 (1959), 82–111.

VISHNU-MITTRE and H. P. GUPTA. A Living Fossil Plant Community in South Indian Hills. – *Curr. Sci. (India)* **37**, No. 23 (1968), 671–672.

VISSCHER, H. Permian and Triassic Palynology and the Concept of Tethys Twist. – *Palaeogeogr. Palaeoclimatol. Palaeoecol.* **3**, No. 1 (1967), 151–167.

VON POST, L. Forest Tree Pollen in South Swedish Peat Bog Deposits. – *Pollen et spores* **9**, No. 3 (1967), 335–401.

VORMAN, M. Study of a Peat Profile on the Frisian Coast of the Former "Luiderzee." – *Meded. bot. Mus. Herb. Rijsk.-Univ. Utrecht,* No. 111, 1952.

VORONOV, P. S. and S. S. NEZAMETDINOVA. Patterns in the Orientation of Rift Valleys of Mid-Oceanic Ridges. – *Dokl. Akad. Nauk SSSR* **194**, No. 4 (1970), 905–908. (Russian)

VOROSHILOVA, G. I. and G. G. BATSAMYSHINA. The Influence of Certain Factors on Stomatal Frequency in Conifers. – *Uchen. Zap. Dal'nevostoch. Gos. Univ., ser. pochv.-bot.,* Vol. 25, pp. 131–137. Vladivostok, 1969. (Russian)

VOSSMERBÄUMER, H. Paläoökologische Ausdeutung fossiler Wurzelböden. An Beispielen aus dem Lias a 1 von Halsingborg. – *Geol. För. Stockh. Förh.* **91**, No. 1 (1969), 111–126.

VRONSKII, V. A. Pollen and Spores from Bottom Sediments of the Aral Sea. – *Dokl. Akad. Nauk SSSR* **195**, No. 5 (1970a), 1163–1166. (Russian)

VRONSKII, V. A. Main Features in the Development of Vegetation from the Southwestern Caspian Lowland in the Upper Pliocene and Pleistocene. – *Bot. Zh.* **55**, No. 10 (1970b), 1432–1441. (Russian)

VUORELLA, J. The Indication of Farming in Pollen Diagrams from Southern Finland. – *Acta bot. fennica,* Vol. 87 (1970), 40 pp.

WALCH, K. M., R. ROWLEY, and N. J. NORTON. Displacement of Pollen Grains by Earth-Worms. – *Pollen et spores* **12**, No. 1 (1970), 39–44.

WALDMAN, M. and W. S. HOPKINS, Jr. Coprolites from the Upper Cretaceous of Alberta, Canada, with a Description of their Microflora. – *Can. J. Earth Sci.* **7**, No. 5 (1970), 1295–1303.

WALKER, T. R. Color of Recent Sediments in Tropical Mexico. A Contribution to the Origin of Red Beds. – *Bull. Geol. Soc. Am.* **78**, No. 7 (1957), 917–920.

WALTER, H. *Ecology of Tropical and Subtropical Vegetation.* (English translation of *Die Vegetation der Erde*). – Edinburgh, Oliver and Boyd, 1971. 539 pp.

WALTON, J. Scottish Lower Carboniferous Plants: the Fossil Hollow Trees of Arran and their Branches. – *Trans. R. Soc. Edinb.,* Vol. 58 (1935), 313–337.

WALTON, J. On the Morphology of *Zosterophyllum* and Some Other Early Devonian Plants. – *Phytomorphology* **14**, No. 1 (1964), 155–160.

WARDLE, P. Evidence for an Indigenous Pre-Quaternary Element in the Mountain Flora of New Zealand. – *N. Z. Jl. Bot.* **6**, No. 1 (1968), 120–125.

WARMING, E. *Om Platerigets Livsformer.* – Copenhagen, 1908. 86 pp.

WATTS, W. A. Late-Glacial Plant Macrofossils from Minnesota. – In: E. J. Cushing and H. E. Wright, Jr. (eds.). *Quart. Paleoecol., Proc. VII Congr. Intern. Assoc. Quart. Res.,* Vol. 7 (1967), 89–99.

WATTS, W. A. A Pollen Diagram from Mud Lake Marion County, North-Central Florida. – *Bull. Geol. Soc. Am.* **80**, No. 4 (1969), 631–643.

WATTS, W. A. and R. C. BRIGHT. Pollen, Seed and Mollusk Analysis of a Sediment Core from Pickerel Lake, Northeastern South Dakota. – *Bull. Geol. Soc. Am.* **79**, No. 7 (1968), 855–876.

WEAVER, J. E. and F. E. CLEMENTS. *Plant Ecology.* – New York, 1929.

WEBER, C. A. Über die diluviale Vegetation von Klinge in Brandenburg und ihre Herkunft. – *Englers Bot. Jahrb.,* 17, 1893.

WEBER, C. O. Die Tertiärflora der Niederrheinischen Braunkohlenformation. – *Palaeontographica* **2**, No. 4 (1852), 115–234.

WEBER, R. Die fossile Flora der Rhät-Lias Übergangsschichten von Bayreuth (Oberfranken) unter besonderer Berücksichtigung der Coenologie. – *Erlanger Geol. Abhandl.,* No. 72 (1968), 73 pp.

WEIGELT, J. Über die vermutliche Nahrung von *Protosaurus* und über einen körperlich erhaltenen Fruchtstand von *Archaeopodocarpus germanicus.* – *Leopoldina,* Vol. (1930), 269–280.

WESLEY, A. The Fossil Flora of the Grey Limestones of Veneto, Northern Italy, and its Relationships with the Other European Floras of Similar Age. – *Palaeobotanist,* Vol. 14 (1965), 124–130.

WEST, R. G. Pollen Zones in the Pleistocene of Great Britain and their Correlation. – *New Phytol.* **69**, No. 4 (1970), 1179–1184.

WESTOLL, T. S. Mountain Revolutions and Organic Evolution. – In: J. Huxley et al. (eds.). *Evolution as a Process.* London, 1954.

WHITTAKER, R. H. and G. M. WOODWELL. Surface Area Relations of Woody Plants and Forest Communities. – *Am. J. Bot.* **54**, No. 8 (1968), 931–939.

WHITTINGTON, H. B. Taxonomic Basis of Paleoecology. – In: J. Imbrie and N. Newell (eds.). *Approaches to Paleoecology,* pp. 19–27. New York–London–Sydney, 1964.

WILDE, M. H. A New Interpretation of Coniferous Cones. 1. Podocarpaceae (*Podocarpus*). – *Ann. Bot. n. sp.,* Vol. 8 (1944), 1–41.

WILLIAMS, C. B. The Logarithmic Series and the Comparison of Island Floras. – *Proc. Linn. Soc. Lond.,* Vol. 158 (1947), 104–108.

WILLIAMS, D. B. and W. A. S. SARJEANT. Organic Walled Microfossils as Depth and Shoreline Indicators. – *Mar. Geol.* **5**, Nos. 5–6 (1967), 389–412.

WILSON, J. T. A Revolution in Earth Science. – *Can. J. Min. Metal.,* Vol. 61 (1968), 190–192.

WILSON, R. L. and S. E. HAGGERTY. Reversals of the Earth's Magnetic Field. – *Endeavour,* Vol. 25, 1966.

WOLFE, J. A. Tertiary Climatic Fluctuations and Methods of Analysis of Tertiary Floras. – *Palaeogeogr. Palaeoclimatol. Palaeoecol.* **9**, No. 1 (1971), 27–57.

WOLFE, J. and E. S. BARGHOORN. Generic Changes in Tertiary Floras in Relation to Age. – *Am. J. Sci. Bradley,* Vol. 258 A (1960), 388–399.

WOLFE, J. and D. M. HOPKINS. Climatic Change Recorded by Land Floras in Northwestern North America. – In: *11th Pacif. Sci. Congr.,* pp. 67–76, 1967.

WOLFE, J., D. M. HOPKINS, and E. B. LEOPOLD. Tertiary Stratigraphy and Paleobotany of the Cook Inlet Region, Alaska. – *Prof. Pap. U.S. Geol. Surv.,* 398A, 1966.

YABE, H. The Latest Connection of the Japanese Islands to the Asiatic Continent. – *Proc. Imp. Acad. Japan,* Vol. 5 (1929a), 167–169.

YABE, H. Geological Age of the Latest Continental Stage of the Japanese Islands. – *Proc. Imp. Acad. Japan,* Vol. 5 (1929b), 430–433.

YAROSHENKO, P. D. and V. A. GRABAR'. *Successive Changes of the Plant Cover in the Transcarpathians.* – Leningrad, "Nauka," 1969. 113 pp. (Russian)

YOSHIDA, R. Nota sôbre um tufo de Glossopteridae na camada Jrapuá, Criciuma, S. c. – *Bolm. Soc. bras. Geol.* **15**, No. 4 (1966), 69–77.

YUFEREV, O. V. Paleobiogeographic Zones and Units of the Stage Scale – *Izv. Akad. Nauk SSSR, geol. ser.,* No. 5 (1969), 77–84. (Russian)

ZADDE, I. N. Comparative Anatomico-Ecological Description of Some Laurel-Leaved and Sclerophyllous Trees and Shrubs in Connection with their Geographic Distribution. – *Biologicheskie Nauki,* No. 5 (1970), 66–70. (Russian)

ZAKLINSKAYA, E. D. Late Cretaceous and Early Paleogene Floras (according to Palynological Data). – In: V. A. Vakhrameev et al. *Paleozoiskie i mezozoiskie flory Evrazii i fitogeografiya etogo vremeni,* pp. 302–331. Moscow, "Nauka," 1970. (Russian)

ZIMINA, V. G. On the Age of the Pospelovis Suite and the Time of Appearance of Gondwana Elements in the Permian Flora of the Southern Maritime Territory. – *Dokl. Akad. Nauk SSSR* **189**, No. 5 (1969), 1073–1076. (Russian)

ZIMMERMANN, W. Main Results of the "Telome Theory." – *Palaeobotanist,* Vol. 1 (1952), 456–470.

ZIMMERMANN, W. About Mesozoic Pteridophylls. – *Am. J. Bot.,* Vols. 814–819. 1969.

ZONNEVELD, J. L. S. Quaternary Climatic Changes in the Caribbean and N. South America. – In: *Eiszeitalter und Gegenwart,* Vol. 19 (1968), 203–208.

INDEX*

* Only the most important genera, families, etc. of plants appear in the index. Numbers in italics indicate that the item appears in a figure.